T0176767

NGN ARCHITECTURES, PROTOCOLS AND SERVICES

NGN ARCHITECTURES, PROTOCOLS AND SERVICES

Toni Janevski

Ss. Cyril and Methodius University, Macedonia

This edition first published 2014
© 2014 John Wiley & Sons, Ltd

Registered office
John Wiley & Sons Ltd, The Atrium, Southern Gate, Chichester, West Sussex, PO19 8SQ, United Kingdom

For details of our global editorial offices, for customer services and for information about how to apply for permission to reuse the copyright material in this book please see our website at www.wiley.com.

The right of the author to be identified as the author of this work has been asserted in accordance with the Copyright, Designs and Patents Act 1988.

All rights reserved. No part of this publication may be reproduced, stored in a retrieval system, or transmitted, in any form or by any means, electronic, mechanical, photocopying, recording or otherwise, except as permitted by the UK Copyright, Designs and Patents Act 1988, without the prior permission of the publisher.

Wiley also publishes its books in a variety of electronic formats. Some content that appears in print may not be available in electronic books.

Designations used by companies to distinguish their products are often claimed as trademarks. All brand names and product names used in this book are trade names, service marks, trademarks or registered trademarks of their respective owners. The publisher is not associated with any product or vendor mentioned in this book.

Limit of Liability/Disclaimer of Warranty: While the publisher and author have used their best efforts in preparing this book, they make no representations or warranties with respect to the accuracy or completeness of the contents of this book and specifically disclaim any implied warranties of merchantability or fitness for a particular purpose. It is sold on the understanding that the publisher is not engaged in rendering professional services and neither the publisher nor the author shall be liable for damages arising herefrom. If professional advice or other expert assistance is required, the services of a competent professional should be sought.

Library of Congress Cataloging-in-Publication Data applied for.

ISBN: 9781118607206

Typeset in 10/12pt TimesLTStd by Laserwords Private Limited, Chennai, India
Printed and bound in Malaysia by Vivar Printing Sdn Bhd

1   2014

To my great sons, Dario and Antonio, and to the most precious woman in my life, Jasmina

Contents

About the Author

Toni Janevski, PhD, is a Full Professor at the Faculty of Electrical Engineering and Information Technologies, Ss. Cyril and Methodius University, Skopje, Macedonia. He received his Dipl. Ing., MSc, and PhD degrees in electrical engineering, all from the Faculty of Electrical Engineering and Information Technologies, Ss. Cyril and Methodius University in Skopje, in 1996, 1999, and 2001, respectively.

During 1996–1999 he worked for the Macedonian mobile operator Mobimak (currently T-Mobile, Macedonia), contributing to the planning, dimensioning, and implementation of the first mobile network in Macedonia. In 1999 he joined the Faculty of Electrical Engineering and Information Technologies in Skopje.

In 2001 he conducted research in optical communications at IBM T.J. Watson Research Center, New York. During 2005–2008 he was an elected member of the Commission of the Agency for Electronic Communications (AEC) of the Republic of Macedonia, where he contributed towards the introduction of new technologies in Macedonia, such as WiMAX and 3G mobile networks, as well as new operators and services.

During the periods 2008–2012 and 2012–2016 he was an elected member of the Senate of the Ss. Cyril and Methodius University in Skopje. In 2009 he established the Macedonian ITU (International Telecommunication Union) Center of Excellence (CoE) as part of the European CoE network, and has served as its head/coordinator since than. He is the author of the book titled "Traffic Analysis and Design of Wireless IP Networks," published in 2003 by Artech House Inc., Boston, USA. Also, he is the author of the book "Switching and Routing," written in the Macedonian language, published in September 2011 by the Ss. Cyril and Methodius University in Skopje, for which in 2012 he won the "Goce Delchev" award, the highest award for science in the Republic of Macedonia.

He has published numerous research papers and led several research and applicative projects in the area of Internet technologies and mobile and wireless networks. He is a Senior Member of IEEE. His research interests include Internet Technologies, Mobile/Wireless and Multimedia Networks and Services, Traffic Engineering, Quality of Service, Design and Modeling of Telecommunication Networks, as well as Next Generation Networks.

1

Introduction

1.1 Introduction

The development of telecommunications and communication technologies in the twenty-first century, at least in its first half, has an unambiguous direction toward a single goal, and that is the Internet as a single platform for all services through a global network. However, the initial concept of telecommunications was based on real-time services such as voice communication between users over a telephone network (i.e., telephony), or diffusion of video and/or audio (i.e., television and radio). If we go even further in the past, one may mention the telegraphy in the nineteenth century as the first telecommunications technology for data transmission based on the usage of electrical signals.

However, the world of telecommunications or ICT (Information and Communication Technologies) is continuously evolving and changing, including the technologies, regulation and business aspects. Going from the telegraphy as main telecommunication service in the nineteenth century, then the telephony and television (including the radio diffusion) as fundamental telecommunication services in the twentieth century (and they continue to be nowadays), and Internet phenomenon by the end of the twentieth and beginning of the twenty-first century, telecommunications have changed and the technologies have changed. But, in such process was kept the backward compatibility for the flagship services, such as telephony and television, and their integration with the new services, such as Internet native services [e.g., World Wide Web (WWW), electronic mail (e-mail), etc.].

Regarding the development of the telecommunications so far, one may distinguish among four key phases:

1. The automation of the telephone exchanges and networks at the end of the nineteenth and the beginning of the twentieth century;
2. The transition from analog to digital telecommunication systems from the 1970s to 1990s;
3. The integration of the circuit-switched telephone networks, such as Public Switched Telephone Networks (PSTN) and Public Land Mobile Networks (PLMN), with the packet-based Internet in the 1990s and 2000s;
4. The convergence of all telecommunication services, including telecom-native services (such as telephony and TV/radio) as well as Internet-native services (such as WWW, e-mail, peer-to-peer services, etc.), over the broadband Internet as a unified global networking platform (regardless of the access network type, either wired or wireless), toward the Next Generation Networks (NGN), in 2010s and 2020s.

NGN Architectures, Protocols and Services, First Edition. Toni Janevski.
© 2014 John Wiley & Sons, Ltd. Published 2014 by John Wiley & Sons, Ltd.

The four phases of telecommunications development have resulted in exponential increase of number of telecommunication networks and number of users. The nineteenth century can be denoted as a century dedicated to telegraphy. At the end of the nineteenth century the telephony was invented and telephone networks started to be implemented around the world. The twentieth century was dedicated mainly to telephony as primary service in telecommunications worldwide. At the end of the twentieth century appeared the Internet for public usage. Nowadays, in the second decade of the twenty-first century, all telecommunications services are being transferred to Internet. Hence, from this point of view, one may say (or predict) that twenty-first century will be dedicated to Internet and will be information centric. The framework of such development is set by the ITU (International Telecommunication Union) in the NGN concept. The main requirement for accomplishment of such task is broadband access to Internet, including fixed broadband, as well as mobile broadband. The broadband is a term used to describe the Internet access data rates which can provide access to all existing telecommunication services at given time including the currently most demanding ones such as video or multimedia streaming services (e.g., TV over the Internet). The birth and rise of the Internet, as well as broadband access to the global Internet network, has influenced the "look" of the telecommunications (i.e., the ICT).

So, today we have several important segments in the ICT globally. Telephony is still one of the primary services, where one can distinguish between fixed telephony and mobile telephony. Further, Internet is usually identified by certain types of services such as WWW, e-mail, peer-to-peer services, and many more, provided in so-called best-effort manner. Best-effort principle is based on connection control by the end point of the communication (called hosts, such as computers, servers, mobile terminals, etc.), where network nodes perform basically routing of all packets from all services without differentiation among them. Finally, Internet requires broadband access, including fixed broadband and mobile broadband, with aim to provide capabilities for different types of services including the most demanding ones regarding the available data rates (i.e., the bandwidth). These five segments form the outlook of today's telecommunications. The number of users for fixed telephony, mobile telephony, individual Internet users, and users with fixed and mobile broadband access to Internet are shown in Figure 1.1 (for more details a reader may refer to [1]). It is obvious that mobile telephony has overtaken the number of telephone users from the fixed telephony a decade ago. Hence, the number of mobile users increases exponentially and it is targeting the total population on Earth. Mobile telephony is personal, while fixed telephony is related to a certain location (e.g., a home or an office). Hence, the market capacity for the fixed telephony is several times smaller than the market capacity for the mobile telephony. However, it is likely that mobile telephony will be soon saturated by reaching 100% of the world population. On the other side, the mentioned trend of integration between traditional telecommunications and the Internet, which have been developed separately at the beginning of each of them, is finally resulting in transition of telecommunication world into the Internet world, and vice versa. The number of Internet users is increasing exponentially in the past two decades, almost in parallel with the rise of number of mobile users, as it can be seen in Figure 1.1. The broadband is crucial for the Internet. However the exponential rise of the broadband access started 10 years ago, including fixed broadband access and mobile broadband, and currently it is in a similar position to mobile telephony a decade ago (Figure 1.1). Hence, the highest market potential (in the ICT world) currently is the broadband Internet. Similar to the mobile-to-fixed telephony comparison, the mobile broadband is personal and hence will have faster exponential

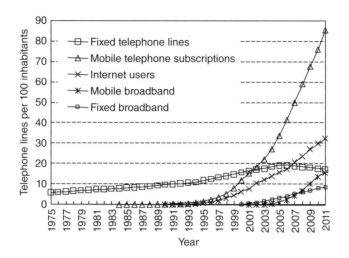

Figure 1.1   Global development of ICT

growth and higher penetration compared to the fixed broadband. On the other side, the fixed broadband will always have a higher capacity due to scarce radio spectrum resources over a given geographical area.

According to the discussions above, in the following part we will cover the main aspects of the traditional telecom world, and then the traditional Internet world, which converge nowadays to NGN as defined by the ITU (International Telecommunication Union). However, the ITU was established in 1865, when there was only the telegraphy present as a telecommunication service, and hence it was originally founded as the International Telegraph Union. Later the word "Telegraph" in the name of the ITU was replaced with the word "Telecommunication", with the aim to cover the broader range of services after the invention of telephony and later the invention of radio broadcast and television. Today the ITU is part of the United Nations (UN), as a specialized UN agency for ICT. It is main international organization for telecommunications, which provides harmonization regarding the radio frequency spectrum worldwide through the sector International Telecommunication Union-Radiocommunication (ITU-R). Also, ITU develops technical standards as well as provides harmonization for usage of ICT technologies globally via its sector International Telecommunication Union-Telecommunications (ITU-T). Finally, ITU strive to improve the access to ICTs to developing countries and underserved communities worldwide through its sector International Telecommunication Union-Development (ITU-D), because everyone in the world has a fundamental right to communicate.

1.2   Traditional Telecom World

Traditional telecom world is mainly based on the telephony, which is the most important service in it. Hence, on the way toward the NGN, the telephony is still one of the most influential services. The other important traditional telecommunication service is television (also, coming from the first half of the twentieth century, while main spreading of the television worldwide happened in the second part of the last century). However, from the beginning the television was not offered by telecom operators which provided the telephony. Instead, the television was provided via separate broadcast networks, either terrestrial or cable. Traditional telecommunication networks are in fact the telephone networks, and hence they are in the focus in the following sections.

1.2.1   History of Telephony

The current look of the telecommunications started in the nineteenth century with the invention of the telephone by Alexander Graham Bell in 1876. However, telephony as a service even at the beginning required large number of users to have telephones, so they could call each other. So, the telephony has never been a service that could be dedicated to a privilege group of users only, because in such case they could call only each other, not the rest of the people. And, the true meaning of the telephony is possibility to call anyone from anywhere in the world using a global interconnected telephone networks. But, when a large number of users already had telephones, then the problem was to connect all users in organized telephone networks, something that introduced the need for switching. One may define switching as a way to connect a certain input line to a certain output line using so-called exchange (i.e., switch) in the traditional telephony networks. However, first switches were manual, based on an operator (human) in a central switching room. The human operator provided the switching between the calling and the called user lines by plugging the jacks of the connector wires to the calling user's line on one side and to the called user's line on the other side, using a switchboard for such task. After the termination of one side of the connection, by putting the telephone on-hook (signaled usually by a lamp on the operator's switchboard), the operator would release the connection by pulling off the jacks. But, the manual approach was not incremental and thus needed automation.

The first automatic exchanges appeared at the end of nineteenth century, which were based on step-by-step switches called Strowger switches. Further, they were replaced by crossbar switches in the first half of the twentieth century, which stayed in operation worldwide until the 1990s (i.e., until the transition from analog to digital telephony). The exchanges were connected via transmission systems, which were based on multiplexing of analog voice signals over coaxial copper cables, while the access toward end subscribers (called local loop) was based on twisted pairs (i.e., a pair of copper wires between the user's telephone device and the exchange). So, for almost 100 years (from the invention of the telephone in 1876) the telephone networks were analog, which means that original analog signals (generated by human on one side of the connection) were carried over transmission systems and via established connections in the switching matrices (in each exchange on the way) to the called party, and vice versa (from called party to the calling party). Hence, the first really worldwide network for communication between people was the telephone network, consisted of many national telephone networks interconnected on national and international levels, thus providing truly global network in which each user could reach every other user on Earth. However, telephony is one of the most demanding services regarding the delay end-to-end, since it is a conversational type of service which happens in real-time, similar to two (or more) people conversation face-to-face. Telephony is two-way (full duplex) end-to-end communication. For that purpose it requires signaling before the voice connection is established, with aim to find the called party based on certain address (i.e., a telephone number) and to provide channels end-to-end (between the calling and called users) in all transmission systems and all exchanges on the way.

Further, one of the major phases in traditional telecommunications was the digitalization, which meant a transition from analog signals to digital signals in the telephone networks. That was accomplished by representing the analog voice signals with digits (mainly the digits were bits, due to simplicity of handling only two logical and physical levels, representing the binary "0" and binary "1," than multiple ones). This was made possible by the development of electronics and computer systems in 1960s. As a result of this development first appeared hybrid switching systems, characterized with computer-based management, while the switching remained analog. Such systems were called quasi-electronic systems.

Later, in the 1970s/1980s and especially in the 1990s the switching was done with digital systems. That paved the ground for telecommunications networks to become digital networks. However, the local loop in digital telephone networks remained to be analog. One may say that it happened because it was a simpler solution (due to backwards compatibility with the analog telephone handsets) as well as cheaper one (to perform analog-to-digital conversion, and vice versa, on the side of the telephone exchanges rather than the user side).

The digitalization of the subscriber lines began with the spread of ISDN (Integrated Services Digital Network) in the 1990s, a technology which provided digital local loop (over twisted pairs) with basic rate interface of 144 kbit/s, consisted of two 64 kbit/s bearer channels (for telephony or data via dial-up modem) and one 16 kbit/s signaling channel (i.e., delta channel). However, ISDN has not experienced success as one might expected in the 1990s. The main reason for that was the appearance of the Internet and its exponential growth in the world since 1993 when the WWW (which is the dominant Internet service until today) started to be used globally via browsers (i.e., WWW clients). The need for higher data speeds (i.e., rates) grew continuously, because the Internet provided different services (e.g., WWW, video streaming, multimedia, voice, peer-to-peer services, gaming, e-mail, chat, etc.) over the same network. ISDN could not satisfy such requirements for higher data rates for the emerging Internet services. That resulted in the replacement of ISDN with the xDSL technologies where DSL means Digital Subscriber Line. In the 2000s and 2010s the most widespread DSL was Asymmetric Digital Subscriber Line (ADSL), providing higher data rates in downlink compared to lower data rates in the uplink, which corresponds to the traffic behavior of the Web (i.e., the WWW) as the dominant Internet-native service.

From the start of the digitalization era in telecommunications in the 1970s one may notice the fast exponential increase of fixed telephony users worldwide until 2005, and afterwards it is slight decline (Figure 1.2; for more details a reader may refer to [1]). This is mainly due to exponential growth of mobile telephony, which is more personal and hence has higher market potential compared to the fixed telephony.

1.3   Public Switched Telephone Networks

Traditionally, telecommunication services have been associated with the existence of a certain network infrastructure. The main goal in traditional telecommunication networks is to ensure accurate and timely transmission of information from the source to the end user (i.e., the destination). The information is transmitted through copper, optical cables, or through wireless networks, which are located between the transmitter and receiver. The end-to-end channel is consisted of the transmitter, receiver, and the transmission system for transfer of

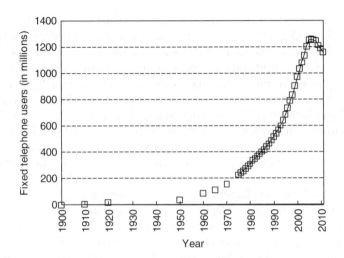

Figure 1.2   Telephone users statistics since 1900

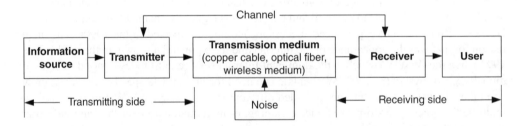

Figure 1.3   Model of communications channel

the information (shown in Figure 1.3). Transmitted signals are degraded by the noise in the communication channel. Usually, transmitters and receivers are packed into a single terminal device, such as phone, mobile terminal, personal computer, lap-top, and so on. The end terminal can be used by a human or another machine (e.g., Web server, e-mail server, automatic voice machine, etc.).

In practice, the communication channel model is not simple as shown in Figure 1.3. Namely, the path from source to destination in most cases is divided into several sections, interconnected via exchanges (i.e., switches) in traditional telephony or routers in Internet.

In classical telecommunication for the transmission of voice (i.e., telephony) telecommunication systems can be divided into two main groups:

- Switching systems;
- Transmission systems.

PSTN and PLMN, as they have existed at the end of the twentieth century, are based on switching and transmission of digital signals. Because main service was telephony, all systems were designed to fit the requirements for end-to-end delivery of voice in both directions.

1.3.1   Pulse Code Modulation

In digital telephone networks the globally accepted standard for performing A/D (analog to digital) conversion is Pulse Code Modulation (PCM), standardized by ITU-T as Recommendation G.711 [2].

Using the PCM, the standard analog voice signal in the frequency bandwidth 0–4000 Hz is transformed into digital binary signal with data rate 64 kbit/s. There are three basic steps for the PCM, and they are:

- sampling;
- quantization;
- coding.

Sampling is process of taking samples of the analog voice signal at regular time intervals. According to the Nyquist theorem for sampling of analog signals, the frequency used for sampling must be at least two times bigger than the highest frequency in the analog signal. Since the analog signals for voice are in the range 0–4000 Hz, the highest frequency is 4000 Hz = 4 kHz. Hence, the sampling frequency adopted for PCM in 2×4000 Hz = 8000 Hz, which results in taking samples every 125 μs (1/8000 Hz = 125 μs).

The range of the electric amplitude (voltage) of the analog signal is further divided into limited number of levels, called quantization levels. For voice it is set to 256 levels which can be presented using 8-bit codes ($2^8 = 256$). So, the quantization is a process of converting the amplitude of a given voice sample to one of the 256 discrete levels. However, due to the fact that probability for low-amplitude signals is higher than probability for high-amplitude signals, usually the number of levels is more concentrated at low-amplitude values, with aim to reduce their distortion (i.e., quantizing noise).

Finally, encoding is a process of assigning 8-bit code to each quantized sample. The code belongs to one of the 256 discrete signal levels. That gives as outcome the data rate for single voice channel, which is 8 bits/125 μs = 64 kbit/s. This data rate, obtained with the PCM, is further the basic rate for all transmission systems (i.e., multiplexing hierarchies) in digital telecommunication networks. Therefore, the 64 kbit/s is also denoted as Digital Signal 0 (DS-0), and it is basic granularity in circuit-switched telephone networks.

1.3.2   Architecture of the Telephone Network

The telephone network is consisted of end users who have phone handsets (i.e., telephones) connected via twisted pairs (twisted pair is a pair of copper wires) which connect the telephone in the home (or office) to the local exchange. Further, telephone exchanges are interconnected with transmission systems, following certain network hierarchy. When all switching and transmission is accomplished by using digital signals then we refer to it as a digital network.

Environment of exchanges in a telephone network is shown in Figure 1.4. It is consisted of subscribers' lines (twisted pairs), remote exchange concentrators, Public Branch Exchanges (PBXs), line concentrators, as well as circuits to and from other exchanges.

Subscriber line is the local loop. In digital telephone networks in twentieth century the subscriber line was usually analog and A/D conversion was completed at the local exchange premises (exchange to which the user was connected). We have still such networks today,

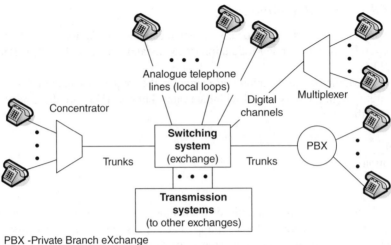

PBX -Private Branch eXchange
Trunk – group of channels shared by users

Figure 1.4   General environments for an exchange

mainly at the incumbent telecom operators, which were the national state-owned operators in the past. So, in traditional digital network, the local loop is analog. The telephone device receives its power supply from the local exchange. When a telephone is taken off-hook, the circuit between the telephone device and the exchange is closed, and electric current flows over it (which is DC – Direct Current). The two wires in the local loop are twisted to eliminate the crosstalk. Analog signals in local loop are coupled (between the telephone and the twisted pair on one side, and between twisted pair and the exchange on the other side) by using transformers. Further, analog signals are converted into digital signals at the local exchange. Then, the digital signals are transferred via the PSTN to end receiver for decoding (e.g., in local exchange on the other side which performs D/A conversion). However, due to the establishment of circuits in analog telephone networks, traditional telephony is called circuit-switching. The term continued to be used in digital telephone networks based on 64 kbit/s lines end-to-end (by using PCM). Hence, the PSTN are also referred to as circuit-switched networks.

The remote exchange concentrators and line concentrators are used to aggregate the traffic from many subscribers to smaller number of channels between the user premises and the local exchange. Such approach is possible in telecommunication networks because different users are active (have phone connections with other users) in different times. So, statistically speaking, there is always certain level of traffic in the network which is generated from a certain group of users, and it depends upon the activity of the users. Usually, the telephone networks are dimensioned by the traffic intensity (number of established telephone connections) in the so-called busy-hour (the hour with highest average traffic intensity during a day). Due to that, certain number of users can be served by a group of shared 64 kbit/s channels. For example, if we have 100 users, and average user activity in the busy hour is 20% (meaning that an average 20 out of 100 users have established connections during the busiest hour in the day), then that group of users can be served with 30 channels (with average call blocking probability of 1%, calculated by using Erlang-B formulae [3]). So, instead using 100 channels for 100 users the telecom operator may serve the users with 30 channels, thus saving the capacity and the costs for service provisioning. This simple example, in fact, explains the need and the usage of concentrators in telephone networks.

Additionally, there are also private telephone networks, called PBXs, which are used by companies with aim to manage the internal voice traffic (within the company) thus avoiding the need for switching the internal traffic via the PSTN owned by a telecom operator. However, PBX must be connected to PSTN to provide possibility for outgoing and incoming calls from other users.

Finally, each telephone network must be connected with other telephone networks via switching and transmission systems, with aim to provide global connectivity for voice communication since telephony was the main service in traditional telecommunication networks.

1.4   Signaling Network

In digital telephone networks prior to voice communications there is a need to find the called user by using destination telephone number and then to reserve 64 kbit/s channels in both directions between the local loops of the caller and called party, in all switching and transmission systems on the way. For such task there is a need of signaling in telephone networks, which existed since the analog era. Signaling can be done using analog signals (e.g., electrical signals) or digital signals (e.g., bits, bytes, or messages/packets).

Definiton 1.1: Definition of Signaling

Signaling is mediated exchange of control information signals using certain signaling alphabet (set of signals or messages).

There is line signaling between the user telephone and the local exchange, as well as signaling between telephone exchanges. In many PSTNs the line signaling is still analog, while signaling between exchanges is digital.

Another general classification of signaling in PSTN is into two groups:

- Channel Associated Signaling (CAS), where certain signaling information is associated with the voice channels over the same transmission medium;
- Common Channel Signaling (CCS), where signaling information from many users is multiplexed over a common channel and can be carried separately from the voice traffic.

1.4.1   SS7 Architecture

In digital telecommunication networks, from 1980s the most used is Signaling System No. 7 (SS7), standardized by the ITU [4], and accepted globally. It belongs to the CCS type of signaling and it is used in all PSTN and PLMN worldwide. SS7 is packet-based signaling, so it has introduced packet-switching globally in telecommunication networks for the first time in 1980s, even before the Internet growth in the 1990s. Therefore, here is described SS7 with some important details regarding the network architecture and protocols.

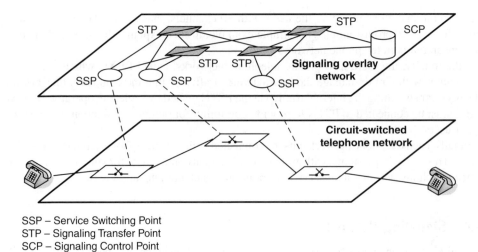

SSP – Service Switching Point
STP – Signaling Transfer Point
SCP – Signaling Control Point

Figure 1.5   SS7 overlay network

SS7 logical network architecture defines three types of nodes (Figure 1.5):

- Service Switching Point (SSP);
- Signal Transfer Point (STP);
- Signal Control Point (SCP).

Each of the three node types in SS7 has specific functionalities. So, SSP is a network element in SS7, which is integrated with local telephone exchanges (with attached subscriber lines to them). SSP converts dialed number (called B-numbers or Global Titles according to SS7 terminology) into SS7 signaling messages and establishes signaling connection with the SSP of the called user. SSP establishes, manages, and terminates voice connections. It sends signaling messages to another SSP via the STP node to which it is connected.

STP is a router and/or signaling gateway in the SS7 network. This node (STP) has main task to route signaling messages between so-called signaling points in the network. Those STPs who act as gateway nodes actually connect signaling network of one telecommunications network (e.g., belonging to a telecom operator) with signaling network of another telecommunications network (e.g., belonging to another telecom operator).

SCP provides access to certain application in SS7. In fact, SCP can be viewed as a database with an appropriate interface for database access by other signaling points in SS7. Typical usage of SCP is for special B-numbers such as the 0800 series (when the called party is charged for calls), or for the provisioning of roaming in PLMN.

Signaling nodes are logically separated from the network for transmission of voice signals. Thus, they form a signaling overlay network as shown in Figure 1.5. Separation is logical, because over the same physical transmission media (copper, optical, or wireless medium) are transmitted signaling and voice channels. Also signaling nodes are located at the same physical locations and integrated in the switching systems (i.e., telephone exchanges).

1.4.2 SS7 Protocol Model

SS7 is composed of multiple protocols which are layered according to OSI (Open System for Interconnection) model. According to OSI there are seven layers of protocols which can cover all possible functionalities in telecommunication networks. The mapping between the OSI layering model and SS7 standardized protocols is shown in Figure 1.6.

SS7 is mapped into five layers of the OSI model, as shown in Figure 1.6. Physical level in SS7 (OSI layer 1) is MTP (Message Transfer Part) layer 1 (MTP1), specified in ITU Recommendation Q.702 [5]. This layer defines physical and electrical characteristics of the signaling link.

Layer OSI-2 in SS7 is MTP layer 2 (MTP2), standardized by ITU Recommendation Q.703 [6]. This level has task to provide reliable transfer of signaling messages from the source to the destination signaling point through a signaling link that directly connects two network nodes in the SS7 network.

Network layer in the SS7 is MTP layer 3 (MTP3), standardized by ITU Recommendation Q.704 [7]. MTP3 provides functionalities for routing of signaling messages between signaling points in SS7 network. As already mentioned above, the routing in the SS7 network is performed by the STP signal points.

Above the network layer in SS7 are defined customized signaling protocols that are designed for specific types of services (e.g., telephony, ISDN, mobile networks, etc.). For example, for ISDN subscriber line is used ISUP (ISDN User Part), for traditional telephony is used TUP (Telephone User Part), for services in a mobile network is used MUP (Mobile User Part), and so on. Internally SS7 network also uses TCAP (Transaction Capabilities), which typically uses SCCP (Signaling Connection Control Part) at OSI layer 4 in SS7 protocol model.

SCCP (defined by ITU Recommendations Q.711–719) provides a connection-oriented and connectionless (connectionless means that each message is independently routed through the signaling network) network services through transfer of MTP3 signaling messages between SSP nodes in the network. Thus, MTP layers allow transfer of signaling messages from one signaling point to another (i.e., from one node to another node in the SS7 network), while

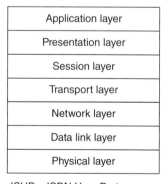
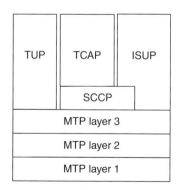

ISUP – ISDN User Part
MTP – Message Transfer Part
SCCP – Signaling Connection Control Part
TCAP – Transaction Capabilities
TUP – Telephone User Part

Figure 1.6   SS7 protocols mapped to OSI layering model

SCCP enables transmission of signaling messages end-to-end (from sender application in the sending signaling point to recipient application in the receiving signaling point).

TCAP (defined by the ITU Recommendations Q.770–Q.779) messages are used for applications in SS7. This protocol is used for communication between SCP nodes through STP nodes. Because SCP represents databases, it follows that TCAP is mainly used to access databases in the SS7 signaling network. Transmission of TCAP messages end to end is performed by using the SCCP.

At the end one may briefly summarize that SS7 is dominant and universal signaling system which is based on packet switching and transmission. The same system is used in fixed and mobile networks signaling for voice calls and supplementary services (e.g., call waiting, call forwarding, conference call, call barring, calling line identification presentation, etc.).

Due to its excellence as well as backward compatibility regarding the signaling for the telephony, SS7 is also adapted for usage in Internet environment by SIGTRAN (Signaling Transmission) family of specification created by IETF (Internet Engineering Task Force) for carrying SS7 signaling (for PSTN and PLMN) over IP networks.

However, best known equivalent of SS7 (and also its successor) in Internet environment is SIP (Session Initiation Protocol), which will be covered in the following chapters.

1.5   Transmission Systems

In telecommunications signals are carried from one node to another node in the network by using transmission systems, which are carrying signals, such as electric signals (over copper cables), optical signals (over fiber) and radio signals (over wireless links). With aim to reduce the costs of the transmission systems, as well as to manage them more efficiently, each transmission link is used to carry many signals from many users. Multiple usage of a given transmission medium is called multiplexing. Hence, systems that perform multiplexing are called multiplexers. Multiplexing is done at the sending side of a transmission link (there are always two ends – the sending and receiving ends), where many incoming signals are placed on the same transmission medium without interfering with each other. The reverse process called demultiplexing is being made always at the receiving side.

The transmission system is used to transfer N transmission signals instead of only one signal, as shown in Figure 1.7. This is done in order to more effectively use the expensive transmission links.

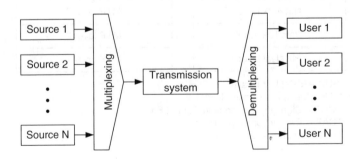

Figure 1.7   Multiplexing on a transmission system

Typically, network nodes (e.g., switches, routers) interconnected with transmission systems consist a telecommunication network. Moreover, the practice confirms that about two-thirds of the cost of a telecommunications network is related to the transmission systems and one-third of the cost is related to the switching systems. In this context, a reduction in transmission costs significantly affects the reduction of the total cost of the telecommunication network (including capital investments as well as operating costs).

Generally, telecommunication networks are designed by a trade-off between price and performance. Often it is not a question of whether something can be done or offered as a service, but it is more a question how much such service will cost and who or how many users will be able to pay for it. In that direction, to allow affordable cost of the telecommunication services to majority of the population globally, the main tendency is to minimize the overall cost of telecommunication networks as much as possible.

Transmission systems, historically speaking, were analog before the 1970s and then digital in the past several decades. In the analog era the multiplexing was done of 4 kHz wide channels by using analog modulation techniques. The 4 kHz bandwidth is related to the dedicated spectrum to voice for the analog telephony, which is in the range 300–3400 Hz, thus resulting in 4 kHz bandwidth per voice channel in one direction. In digital systems the same frequency bandwidth is digitalized with the PCM, resulting in bit rate of 64 kbit/s per channel. The multiplexing of channels in digital transmission systems is done initially by using a PDH (Plesiochronous Digital Hierarchy), and later by using a SDH (Synchronous Digital Hierarchy) or SONET (Synchronous Optical Networking).

1.5.1   Multiplexing of Digital Channels

Multiplexing is a technique of placing many signals over one transmission medium (e.g., copper, optical fiber, or radio).

There are two basic schemes for multiplexing:

- *FDM (Frequency Division Multiplexing):* Different frequencies or frequency bands are assigned to different channels over the same transmission medium.
- *TDM (Time Division Multiplexing):* Different time intervals, called time slots, are assigned to different users over the same transmission medium and using the same frequency in the case of copper cables or radio transmission, or the same wavelength in the case of fiber as a transmission medium.

However, regarding the multiplexing schemes in telecommunications one have to mention also the CDM (Code Division Multiplexing), which is based on using different code sequences for different signals over the same frequency bands and during the same time intervals. However, CDM is not used in transmission systems so far, but it is mainly used in wireless and mobile networks (e.g., Direct Sequence Spread Spectrum – DSSS in WiFi, CDMA (Code Division Multiplexing) in IS-95 standard from second generation mobile networks in America, in 3G mobile networks as Wideband CDMA, etc.). When multiplexing schemes are used in the access networks (where end users access the network) they are called multiple access schemes, such as FDMA (Frequency Division Multiple Access), TDMA (Time Division Multiple Access) and CDMA (Code Division Multiple Access).

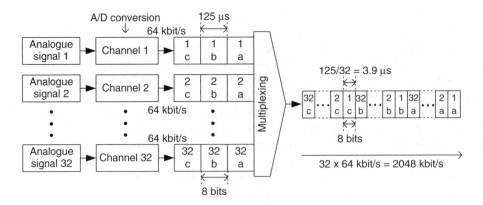

Figure 1.8   Principle of time division multiplexing

Regarding the multiplexing in transmission systems one have to mention also WDM (Wavelength Division Multiplexing). It is based on carrying user data on different wavelengths over the same fiber cable.

However, the main multiplexing techniques in transmission systems used in PSTN are based on the TDM (which is based on PCM coded voice).

1.5.2   Time Division Multiplexing in PSTN

When PCM is used for voice coding, the time duration for transmission of each 8-bits code must be less than 125 μs. However, if that time is much less than 125 μs, then there is possibility to transmit bits from other voice signals over the same transmission medium. That is called Time Division Multiplexing, illustrated in Figure 1.8, which is the basic approach in all digital transmission systems, either wired or wireless. However, it can be combined with other multiplexing techniques, such as FDM, CDM, and WDM.

ITU-T has standardized the hierarchy for bit rates in digital transmission systems [8]. In the digital hierarchy are defined four levels, where the first hierarchy level in Europe and most of the world is 2048 kbit/s, consisted of a frame with 32 time slots (32 × 64 kbit/s = 2048 kbit/s). Time slots are numbered as TS-0 to TS-31 (TS denotes Time Slot, which corresponds to single 64 kbit/s channel). From the total of 32 time slots, 30 time slots are used for voice (in telephone networks), first time slot (TS-0) in the frame is used for synchronization and alarming, and the TS-16 is mainly used for signaling related to voice connections.

In America the first hierarchy level is 1544 kbit/s and it is consisted of 24 channels, multiplexed with TDM. Although this version is developed first, the most used version on a global scale is the European primary multiplex of 2048 kbit/s.

The standardized hierarchical bit rates in digital networks are given in Table 1.1 [8].

However, to provide interconnection of digital networks based on the two different digital hierarchies, there was created hybrid hierarchy as given in Figure 1.9, which provided possibilities to map European to American digital hierarchy and vice versa.

The digital hierarchy rates in fact form the PDH, where the bit rate is controlled by a clock in the local equipment (e.g., exchange). The PDH was used in 1970s and 1980s, and later

Table 1.1 Standardized bit rates in digital transmission networks

Digital hierarchy level	Primary rate 1544 kbit/s (a.k.a. American hierarchy)		Primary rate 2048 kbit/s (a.k.a. European hierarchy)	
	Bit rate (kbit/s)	Multiplexing factor	Bit rate (kbit/s)	Multiplexing factor
0	64		64	
1	1544	24	2048	32
2	6312	4	8448	4
3	44 736	7	34 368	4
4			139 264	4

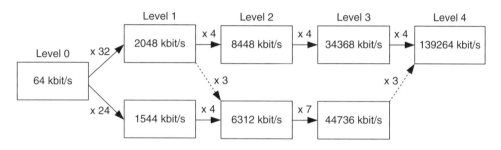

Figure 1.9 Hybrid digital hierarchies

during 1990s it was replaced by SDH [9]. In SDH there is centrally positioned primary reference clock with highest accuracy, and its reference clock is distributed to all nodes in the SDH network. The American version of SDH is called SONET. The main difference between SDH and SONET is in the basic transport levels. In SDH the basic transport module is STM-1 (Synchronous Transport Module – level 1), with data rate of 155.52 Mbit/s, while in SONET the first level is STS-1 (Synchronous Transport Signal – level 1) with data rate of 51.85 Mbit/s. However, STM-1 of SDH equals STS-3 of SONET, so they are in fact the same transmission technology and therefore often denoted as SDH/SONET. Due to the time of their standardization and implementation (end of the twentieth century), they were designed to serve mainly digital telephony based on 64 kbit/s channels. For example, STM-1 has frame duration of 125 μs (which is coming from PCM, used for digitalization of voice), with exactly 2430 time slots resulting in 2430 × 64 kbit/s = 155 520 kbit/s = 155.52 Mbit/s.

Higher SDH transport levels are obtained by using multiplexing byte by byte. In SDH transport modules there are no additional pseudo bits, so the bit rate of STM-N can be calculated as N × 155.52 Mbit/s. For example, STM-4 has bit rate of 4 × 155.52 Mbit/s = 622.08 Mbit/s. Additionally all four levels of PDH bit rates (given in Table 1.1) are mapped into the STM-1, as so-called tributary units. The standardized transport modules for SDH/SONET and associated bit rates are given in Table 1.2 [10].

Initially created digital hierarchies for voice service in telecommunication networks, such as PDH and SDH, by the end of 1990s and in the beginning of the twenty-first century started to be used for carrying Internet traffic between different access networks.

Table 1.2 Standardized SDH multiplexing levels

SDH hierarchy level	SONET hierarchy level	Bit rate (kbit/s)
	STS-1	51 840
STM-1	STS-3	155 520
STM-4	STS-12	622 080
STM-16	STS-48	2 488 320
STM-64	STS-192	9 953 280
STM-256	STS-768	39 813 120

1.6   Traditional Internet world

Development of communications and communication technologies in the first two decades of the twenty-first century has unambiguous direction toward the Internet as a single platform for all services through a global network. Moreover, the initial concept of telecommunications, which was based on telephony as well as diffusion of video and audio (i.e., television and radio), is being replaced by a completely new approach, where different types of information becomes readily available on Internet and users access them according to their own need at any time and anywhere they are. Thus, the heterogeneity of networks for different services (telephony, data, video, audio, multimedia, messaging, etc.) has vanished by convergence to a single network (the Internet) that unites all the possible heterogeneity of services, terminal devices (phones, computers, mobile devices, TVs, etc.), transmission media (copper pairs or cables, optical fiber, and wireless medium). In fact, the Internet provides approach for communications and information access and exchange, between people or machines (i.e., computers), similar to electrical distribution network for electrical appliances (toasters, washing machines, computers, etc.). For example, power plugs and sockets are the same for different electrically operated devices. In the case of the Internet, the access to the Internet is in fact a "socket" for all types of communications services that allow access to certain information (for example, to specific content on the Web) or exchange of information (e.g., telephony over the Internet, messaging).

Definiton 1.2: What is the Internet?

Internet is a network of networks that are connected to each other and that use Internet Protocol (IP) for their interconnection and exchange of information (regardless of its type) between end hosts that are attached to them.

Internet communication is created to be independent of the transport network (e.g., PDH, SDH, WDM, etc.) or access network (e.g., Ethernet, modem connection, wireless and mobile Internet access, etc.). Hence, Internet technologies include protocols at the network layer (e.g., the IP, as a fundamental network protocol) and upper protocol layers (above the network layer).

1.6.1   History of the Internet

To be able to predict future development of a certain technology it is always important to know its history. That is particularly valid statement regarding the Internet history, especially knowing its impact on the ICT of today and near future. However, one can distinguish certain periods in the development of the Internet.

First period in Internet history is early development of the project in the 1960s and 1970s. The following are few important dates from that period [11]:

- *Early 1960s*: Forerunner of the Internet was the ARPA (Advanced Research Projects Agency) project, occasionally later renamed as DARPA (Defense Advanced Research Projects Agency). The research project led by Licklider began in October 1962.
- *1960s*: Appearance of ARPANET and research on packet switching by Leonard Kleinrock (theoretical approach) and Larry Roberts (practical approach).
- *September 1969*: First Internet node was implemented at UCLA (University of California Los Angeles).
- *29 October 1969*: There were exchanged the first characters through two nodes connected by a link (established by a modem using a telephone line) between nodes placed at UCLA in the laboratory of Kleinrock, and also at SRI (Stanford Research Institute).
- *1969*: Requests For Comments (RFC) documents were invented by Steve Crocker. They were initially created as unofficial notes for ARPANET. Since than the RFCs have become official records for Internet technologies (RFCs are available on www.ietf.org, where the IETF comes from the Internet Engineering Task Force, the organization that standardizes Internet technologies).
- *1970s*: Robert Kahn began pioneering work on communication principles for operating systems that led to a change in the ARPANET toward more open network architecture (which would later result in the TCP/IP protocol model).
- *1972*: First e-mail programs created by Ray Tomlinson and Larry Roberts.
- *December 1974*: Vint Cerf and Robert Kahn created the Transmission Control Program, RFC 675 [12], where IP was the connectionless datagram oriented service within the program. Later these two programs were divided as two separate protocols: IP and TCP (Transmission Control Protocol). Therefore, IP model is often called TCP/IP.

The second period in Internet history started at the beginning of the 1980s, when most of the fundamental protocols and technologies of Internet were invented and standardized. Those protocols are used until the present day, and form the networking basis of the global Internet. Most important dates from that period are:

- *August 1980*: User Datagram Protocol (UDP) is standardized [13], which is built over the IP and provides simple transmission model for sending messages (called datagrams) via IP packets between end hosts without prior communication to set up specific data paths.
- *September 1981*: Standardization of the IP as RFC 791 [14], as well as standardization of TCP as RFC 793 [15]. The IP is the primary network protocol that enables and defines the global Internet, while TCP is the primary transport protocol that controls congestion end-to-end.
- *1982*: Invention of Domain Name System (DNS) as a distributed and scalable mechanism for mapping (resolving) host-names into IP addresses.
- *1 January 1983 (Flag Day)*: In all hosts in the Internet in a single day has been replaced NCP (Network Control Program) with TCP/IP protocol stack (standardized in September 1981). From that day onwards, TCP/IP is the basic protocol stack that defines the Internet and it is implemented in all hosts attached to the global Internet network. In 1983 the Internet had

only a few hundred hosts (mainly hosts at universities and research laboratories), so at that time it was possible to make such a change. In fact, the change was implemented in all hosts by installation of the operating system Unix BSD (Berkeley Software Distribution – from Berkeley University, USA). It was the first implementation of IP and TCP protocols, and it was first operating system containing "socket" as an abstraction of the operating system for applications that run on it.

- *1985*: NSF (National Science Foundation) in the United States began a program for the establishment of many access networks, centralized around six Supercomputing Centers (National Science Foundation network – NSFNET), which was interconnected with ARPANET by using the TCP/IP protocol stack.

The third period in the development of the Internet is marked by the invention of the WWW, which is the most important Internet-native service ever (from today's point of view). Also, in that period was standardized next version of the IP, called IP version 6. The third period covers the 1990s and its most important events are the following:

- *1989*: Tim Berners-Lee at CERN, Switzerland, sets out the basics of the most important Internet-native service – the Web (i.e., WWW):
  - Developed Web server and Web browser (called "WorldWideWeb") in 1990;
  - First web site setup on 13 November 1990.
- *30 April 1993*: CERN announces that WWW (the project started by Tim Berners-Lee) will be free to everyone, which is one of the most important events in the history of the Internet and its development since than.
- *October 1993*: Dynamic Host Configuration Protocol (DHCP) is standardized as RFC 1531 and 1541, and it was upgraded in 1997 with RFC 2131. DHCP provided possibility for automatic allocation of temporary IP addresses from dedicated server to hosts in a given network, which was later essential for the growth of the Internet.
- *October 1994*: Tim Berners-Lee founded the World Wide Web Consortium (W3C) at the Massachusetts Institute of Technology, in collaboration with CERN, having support from DARPA (the American side) and the European Commission (the European side).
- *May 1996*: standardized HTTP version 1.0 as RFC 1945 [16]. HTTP (Hypertext Transfer Protocol) is in fact application layer communication protocol for the WWW.
- *January 1997*: standardized HTTP version 1.1 as RFC 2068, and later improved by RFC 2616 in June 1999 [17].
- *December 1998*: Standardized Internet Protocol version 6 – IPv6 [18], which is intended to replace the initial version of the IP since 1981 (also known as IP version 4 – IPv4).

After the 1990s most of the Internet technologies were well standardized, going from net-working and transport layers (e.g., IP, TCP, UDP, etc.) up to the application layer [e.g., e-mail, FTP (File Transfer Protocol), WWW, etc.], as well as fundamental Internet technologies (e.g., DNS, DHCP, etc.).

The first decade of the twenty-first century was marked with development of peer-to-peer communication in Internet besides the traditional client-server architecture. There were many proprietary peer-to-peer applications, starting from Napster (for music sharing) in 1999 and eDonkey (for file sharing) in 2000, and going to applications such as BitTorrent from 2009 until today. However, there appeared proprietary real-time peer-to-peer applications, first for telephony, such as the famous Skype which started in 2003 and significantly impacted the traditional telephony world, especially in the area of international calls. In the second decade of the twenty-first century, with the spread of the broadband access to Internet worldwide there is increasing number of applications for peer-to-peer video streaming. Besides proprietary peer-to-peer applications in the past years there is standardized protocol (by the IETF) for peer-to-peer signaling (needed for real-time communication, such as Internet telephony), called Session Initiation Protocol, initially standardized in 1999 with RFC 2543 (later improved by RFC 3261 as its current version).

The Internet history after the end of the twentieth century is characterized by fast exponential growth of number of Internet users worldwide, because Internet has already won the battle against other packet-switching technologies, such as Asynchronous Transfer Mode (ATM) which has been packet-switching technology developed in Europe in the 1990s. However, ATM followed the traditional telecom approach and required setup of so-called virtual channels in so-called virtual-paths prior to communications between the end points. With such approach, ATM was less flexible and more expensive than the Internet, and hence lost the battle by the end of the 1990s. So, Internet became the only packet-switching technology, accepted globally by all standardization organizations as well as the industry.

1.6.2 Growth of the Internet

From the beginning of the Internet in the late 1960s, the Internet network grows exponentially in number of IP networks as well as number of individual users or hosts connected to Internet (Figure 1.10: for more details a reader may refer to [1]).

However, the most significant growth of the Internet is mainly due to the appearance of Web traffic (i.e., WWW), which makes it easy to search and retrieve various types of information. So, WWW started "the explosion" of the Internet in the years thereafter. However, the total traffic on the Internet network grows undoubtedly much faster than the number of hosts on the network. This is affected by the continuously increasing access bit rates to the Internet, that is,

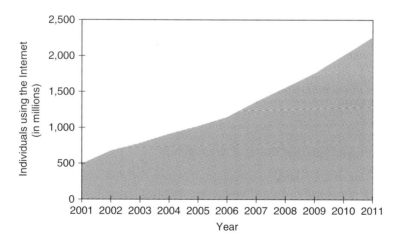

Figure 1.10 Growth of the Internet

the expansion of the broadband in fixed and mobile networks. Internet is experiencing unprecedented growth in the last decade. Therefore, development of each of today's communication technologies must take into account the growth of the Internet and its impact on consumers and telecommunication networks.

Internet is global and provides many multimedia services, which contribute for Internet network to penetrate into almost all aspects of human existence: in scientific world (where Internet was born), in social life, entertainment, information media, administration, business, governments (provisioning of public services to citizens via the Internet), and so on. Of course, the development of the Internet and its globalization were made possible with the development of personal computers and devices, and telecommunication networks to connect them. Generally, the Internet is based on the IP, which is sufficiently robust and flexible enough to allow transfer of any kind of information by using IP packets.

Traditional Internet is a networking technology based on the principle of one type of service for all, called best-effort principle. Hence, traditional Internet provides no guarantees regarding the bandwidth (i.e., the allocated bit rate to a certain connection), the loss or delay of packets. Second characteristic of today's Internet network is heterogeneity. Heterogeneity exists in the endpoints, servers, and clients, which may be personal computers, communicators, mobile terminals, powerful servers. Also there is heterogeneity in bit rates, which may be in the range from a few kilobits per second up to gigabits per second. Then, there is a difference in protocols above the IP, where can be found connectionless unreliable protocols (e.g., UDP) and connection-oriented reliable protocols (e.g., TCP). Finally, there is heterogeneity in the types of applications that use the Internet. There are applications that do not have requirements in terms of performance guarantees (e.g., WWW, file transfer, e-mail), but also they are applications that have strict requirements in terms of losses (packet losses) or delay (packet delay), such as real-time communication [e.g., voice over IP (VoIP), video streaming, etc.].

1.6.3   Internet Architecture

Telecommunications networks are classified into different categories depending on the geographic distribution, use, and implementation. Each network has an adopted architecture and protocols structure (which can be mapped on the OSI protocol layering model). Internet as a global telecommunication network follows the same principles regarding certain network architecture and certain protocol architecture. Different networks, part of the global Internet, are also called IP networks due to mandatory implementation of IP protocol in all network nodes and end hosts.

1.6.3.1   Traditional Internet Protocol Architecture

Internet technologies are related to definition of certain software modules (programs) which run on network nodes and hosts and do certain specified tasks (standardized by given RFC from the IETF). To facilitate building software for communication between processes on the same or different machines, it was necessary to separate tasks into specific processes and to define interfaces between them. Hence, Internet has also layered protocol model, which can be mapped on the OSI protocol layering model.

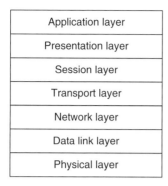
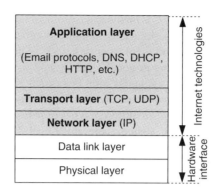

Figure 1.11   Comparison of traditional Internet protocol model with the OSI layering model

Protocol architecture that uses TCP/IP and UDP/IP protocols consists of four parts, as shown in Figure 1.11. In IP model OSI layers 5–7 are united in single level, which is the application level in the Internet. Network layer in the Internet is the IP, while transport layer is consisted mainly from TCP and UDP protocols. Bellow the network layer is a hardware interface that contains the OSI layers 1 and 2, that is, physical layer and data link layer, respectively.

1.6.3.2   Traditional Internet Network Architecture

Internet is composed of many so-called IP networks. Each IP network is separated from the other IP networks through routers. Network elements that can exist in an IP network are the following:

- *Host*: A computer-based device which contains application-layer communication protocols.
- *Repeater*: A network device used to regenerate the digital signals. Its functionalities are located only at the physical layer (OSI-1).
- *Hub*: A network device for connecting multiple devices to act as a single network segment (e.g., Ethernet segment). This device also works at the physical layer (OSI-1).
- *Bridge*: A network device used to aggregate networks or network segments in Internet to create an aggregate network. It works at OSI-2 layer, and does forwarding of data frames by using the Medium Access Control (MAC) addresses.
- *Switch*: A network device used for bridging multiple autonomous network segments (usually Ethernet network segments) and hosts, by using data forwarding based on MAC addresses. There are also network switches in Internet which process data at the network layer (i.e., the IP layer), hence such devices are called layer-3 switches.
- *Router*: A network node (i.e., a computer with multiple network interfaces) containing active applications for routing packets in the network depending on their source and destination IP addresses. Unlike traditional telephone networks, which use telephone numbers associated to the end users, the routers in Internet use IP addresses which are associated with network interfaces of routers and hosts. A router works on network layer (OSI-3).

So, a global network such as Internet is consisted of geographically distributed hardware and software elements. In such architecture, the logical end-to-end communication is performed by using same processes on the same protocol layer on different machines (e.g., hosts, routers, etc.). In other words, logical communication between nodes is implemented so that the processes of a certain protocol layer of a node or host always communicate only with processes (entities) of the same protocol level located in another node or host. For example, a given application that runs at application layer on a given machine (e.g., computer) communicates with the same application on another machine on the Internet.

In the Internet there are different types of networks depending upon the size of the geographical area they cover. Hence, there are: Local Area Networks (LANs; e.g., Ethernet, WiFi), Metropolitan Area Network (MAN; e.g., Metro Ethernet, WiMAX), and Wide Area Network (WAN; routed networks on a regional or national scale). Generally, the Internet is a collection of different types of network architectures that are interconnected (Figure 1.12). Each IP network is unique and independent, which means that it can function normally without having to be connected to another network. Interconnection of different networks in the Internet is performed by routers. Communication on the Internet takes place on best-effort principle, but some routers can have additional functionalities in terms of ensuring a certain Quality of Service (QoS). Such principles provide flexibility in the Internet for different types of services, including the existing as well as future ones.

1.6.3.3   What Are the Reasons for the Success of the Internet?

Based on the previous discussion in this chapter, one reason certainly is the low cost of network equipment and services. The network elements used in the Internet (e.g., switches, routers) typically have a much lower price than the equipment in traditional telecommunications where the dominant service was the telephony. In the balance of quality and price, the quality part

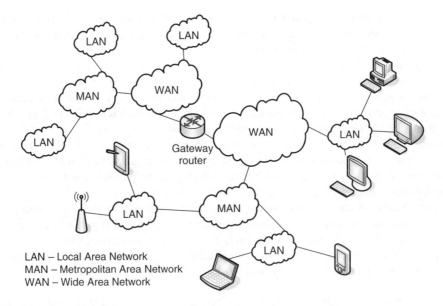

LAN – Local Area Network
MAN – Metropolitan Area Network
WAN – Wide Area Network

Figure 1.12   Basic network architecture of the Internet

during the initial design of the Internet was put in the background, so it provided much cheaper equipment and simpler way for network design and implementation of different types of services. So, today we have the so-called best-effort type of service in the Internet, which means that the available capacity of a given link in the Internet is equally shared among all connections that use this link at any given moment, using control mechanisms in transport protocols of the hosts.

Second reason for the success is the nature of the Internet as a single network for provisioning different types of services (Web, telephony, video streaming, messaging, etc.), which was an old idea in telecommunications – instead of the implementation of separate networks for different services (e.g., telephony networks for voice, distribution network for TV, separate data networks for data transfer, etc.), all services to be carried over one network by using a kind of communication socket to such network. Such telecommunication idea becomes reality with the appearance of the Internet, where the services are in fact applications running on the end hosts (e.g., computers, mobile devices, servers, etc.) while packet routing end-to-end is accomplished by the IP which is implemented in all Internet nodes.

1.7   The Convergence of the Two Worlds: Next Generation Networks

Following the discussions in this chapter, one may say that during the 1990s the Internet has established itself as the winning packet-based technology worldwide. However, it was created having in mind best-effort services, where the network is kept simple (the routers route the packets based on IP addresses) and all intelligence is in the end hosts (clients, servers, for client–server architecture), such as congestion control in transport protocols (e.g., TCP). On the other side, main telecommunication services, such as telephony and television, were usually delivered at that time via specialized networks for such services, where PSTN and PLMN were created for voice (based on circuit-switched communication) while broadcast networks (either terrestrial or cable) were designed for TV. However, the Internet and its openness to new services and technologies below the IP (the access networks) and above the IP (up to different applications, i.e., services), has provided possibility for convergence, which leads to transition of all services to the Internet. Such process creates challenges regarding the required QoS support for real-time services, as well as provisioning of terminating calls in the Internet, something that is not given by the best-effort Internet. Additionally, signaling is required for conversational services, such as VoIP, video telephony, conferencing, presence, and so on. So, there is a need of signaling overlay networks. Another important driver of the ICT technologies and their convergence is the broadband access to the Internet, including wired as well as wireless and mobile networks, which give possibility to provide different services (including the bandwidth demanding ones, such as IPTV; Internet protocol television) via the same Internet access network. Also, convergence toward the Internet provides possibilities for value-added services, by using data sharing, combining existing real-time services (e.g., VoIP, IPTV, etc.) and non-real-time services (e.g., Web, e-mail, peer-to-peer, etc.).

So, what will happen in next five years? It is certain that there will be higher level of convergence of all ICT services over all-IP networks. The NGN, specified by ITU-T [19], provides the way how it can be established for main services, VoIP and IPTV, in releases 1 and 2 of NGN, respectively. However, in the following years the voice will make a significant step toward the VoIP transition by using the standardized IMS (IP Multimedia Subsystem), the process which has recently started in many telecom operators. This is also happening for mobile networks, for example, 4G mobile networks are all-IP because that is mandatory requirements for 4G in the IMT-Advanced (International Mobile Telecommunications-Advanced) umbrella from ITU. However, this process will be further extended from VoIP to IPTV and then to non-real-time services, such as Web-based services and Internet of things, with aim to provide possibility of more complex services based on different contexts and policies. However, best-effort Internet will continue to exist in the future as well (it may be referred to as non-NGN), but it will be integrated within the global NGN architecture by the network providers (e.g., telecom operators). However, NGN brings new business models between the service providers, the network providers and the end users, because in the converged networks the main focus is put on the services.

1.7.1 NGN Perspective of Telecom Operators

The telecom operators are main building blocks of the ICT world, which provide access networks, core networks, and transit networks (networks may be neutral to services as NGN proposes, but they must be standardized first, then "properly" designed and maintained). Without network access to the Internet as a single global networking platform by the end users (either humans or machines) services cannot be used at all.

Who will pay for services? For example, WWW can survive from marketing, but VoIP cannot (there is no possibility to put marketing messages into a VoIP connection). Further, the mobile networking is taking the lead over the fixed networking regarding the number of users. But, mobile operators get most of the revenues from voice services and the predictions say it will continue in the similar manner in the following five years. Also, significant part of the revenues for mobile operators is generated by SMS (Short Message Service), and the rest is coming from mobile Internet access (regardless of the used services – WWW, e-mail, etc.). When voice goes to Internet it becomes the major service in the Internet (regarding its overall importance, not only technically), which influences the architectures of all fixed and mobile IP-based networks, due to its requirements for limited end-to-end delay, bandwidth guarantees (i.e., QoS guarantees end-to-end). That introduces mandatory admission control (which is completely different compared to traditional best-effort Internet) and traffic control, then mandatory signaling (e.g., terminating calls are very important issue), numbering, universal service, emergency calls, and many more issues which are important not only to the ICT field, but also to the society in general. Telephony is a unique conversational service which should provide the feeling to two or more people talking over the phone connection like they are talking face-to-face. This means that end-to-end delay in each direction should be less than 150 ms for best quality of user experience, but never more than 400 ms, according to ITU-T Recommendation G.114 [20]. Then, when telephony transits completely to Internet, NGN will start happening in the ICT world. In other words, putting certain service to go over IMS platform (e.g., VoIP, then IPTV, then the other ones) and using SIP for signaling, as well as providing admission control in the IP network, means that the network is in fact NGN (regardless of that whether such network or services, for example, in the marketing, is referred to as an NGN or not). We can continue this discussion with IPTV as a second service in a row of traditional telecommunication services going to the Internet (which is still not happening massively for TV, simply because analog to digital transition is still not finished for TV, and transition of television to all-IP environment will be happening in the future). There are commercial IPTV implementations even today, but they are usually closed within the telecom operators, or they are offered using the best-effort Internet principle which is without QoS support end-to-end.

1.7.2   When Will NGN Emerge?

In fact the NGN has already started to happen in practice in the recent years. For example, common IMS is being implemented by telecom operators. Traditional circuit-switching telephony is being replaced by VoIP (including fixed and mobile ones), and traditional SS7 signaling today is being carried over IP and further is being replaced by SIP signaling. On the other side, the Internet is the only packet-based network that we have today, because simply it has no competition at all. So, NGN will have more and more impact as we go toward an all-IP ICT world in the following 5–10 years.

However, the NGN is not something that will happen (or has happened) in a given moment of time, but it is an ongoing process (it is not a single standard, but umbrella of principles and recommendations for synergy of different existing and future technologies), a "living thing" which grows and adapts to the continuously changing environment regarding the technologies as well as business approaches.

This book provides unique approach for coverage of the NGN architectures, protocols, and services, as standardized by the ITU (International Telecommunication Union), by using the existing Internet technologies such as protocols from network layer up to the application layer (standardized by the IETF), as well as existing network architectures and standardized technologies for fixed and mobile broadband Internet access and transport [from 3GPP (3G partnership project), IEEE, and ITU]. Additionally, the readers will benefit from contents regarding the regulation and business aspects of NGN networks and services, besides the technology.

1.8   The Structure of This Book

The convergence toward the NGN is based on the Internet technologies, and therefore the introductory chapters of this book cover the Internet fundamentals of today, including architectures, protocols (IPv4, IPv6, TCP, DNS, etc.), Internet services (WWW, e-mail, BitTorrent, Skype, and more), as well as Internet governance. Further, the prerequisite for convergence of all ICT services over single network architectures is broadband access to the Internet. Hence, the book includes architectures of fixed broadband Internet access networks, such as DSL networks, cable networks, FTTH (fiber-to-the-home), next generation passive and active optical networks, and metro Ethernet. Also, it covers network architectures for next generation (4G) mobile and wireless networks (LTE/LTE-Advanced (long term evolution) and Mobile WiMAX 2.0), then Fixed Mobile Convergence – FMC, next generation mobile services, as well as business and regulatory aspects for next generation mobile networks and services. Further, the book focuses on the NGN as defined by the ITU, including QoS and performance, service architectures and mechanisms, common IMS, control and signaling protocols used in NGN (SIP, Diameter), security approaches, identity management, NGN service overlay networks, as well as NGN business models. Also, it covers the most important NGN services, including QoS-enabled VoIP, IPTV over NGN, Web services in NGN, peer-to-peer services, Ubiquitous Sensor Network (USN) services, VPN (Virtual Private Network) services in NGN, Internet of things and Web of things, as well as business and regulation aspects of converged services and contents. Further, the book also includes transition toward the NGN from the PSTN and from the best-effort Internet. Finally, it provides advanced topics such as IPv6-based NGN, network virtualization, and future packet based networks, as well as business challenges and opportunities for the NGN evolved networks and services.

However, this book does not refer to different ongoing research projects around the globe. This book is providing the topics in a format more appropriate for students in academia and people in the industry and ICT organizations who want to get an end-to-end view of the evolution of networks and services toward the NGN and the NGN itself, by giving also coverage of Internet technologies and broadband Internet, and putting the focus on the standardized technologies and principles on network architectures, protocols, and services which have impact today and those which will have impact in the following years.

There is no specific prerequisite knowledge for the readers (although some basic pre-knowledge about the telecommunications and the Internet in general will be beneficial) because the book is structured in such a way that it covers all required technologies and aspect for the NGN and its network architectures, protocols, and services. Also, it includes all important technologies which are used as basis for NGN, such as Internet technologies, as well as standardized technologies for fixed and mobile broadband access to Internet. The book is targeted to managers, engineers, and employees from regulators, government organizations, telecommunication companies, ICT companies, as well as students and professors from the academia, and anyone else who is interested in understanding, implementation and regulation of NGN architectures, protocols, and services, including technology, regulation, and business aspects.

References

1. ITU ICT Data and Statistics, www.itu.int/ITU-D/ict/statistics/ (accessed 4 October 2013).
2. ITU-T (1988) Pulse Code Modulation (PCM) of Voice Frequencies. ITU-T Recommendation G.711.
3. ITU-D (2003) *Teletraffic Engineering Handbook*, International Telecommunication Union.
4. ITU-T (1993) Introduction to CCITT Signaling System No.7. ITU-T Recommendation Q.700, International Telecommunication Union.
5. ITU-T (1993) Signaling Data Link. ITU-T Recommendation Q.702, International Telecommunication Union.
6. ITU-T (1993) Signaling System No.7 – Signaling Link. ITU-T Recommendation Q.703, International Telecommunication Union, March 1993.
7. ITU-T (1996) Signaling Network Functions and Messages. ITU-T Recommendation Q.704, International Telecommunication Union, July 1996.
8. ITU-T (1988) Digital Hierarchy Bit Rates. ITU-T Recommendation G.702, International Telecommunication Union.
9. ITU-T (2000) Architecture of Transport Networks Based on the Synchronous Digital Hierarchy (SDH). ITU-T Recommendation G.803, International Telecommunication Union, March 2000.
10. ITU-T (1988) Network Node Interface for the Synchronous Digital Hierarchy (SDH). ITU-T Recommendation G.707, International Telecommunication Union, January 2007.
11. Forouzan, B. (2010) *TCP/IP Protocol Suite*, Mcgraw-Hill.
12. Cerf, V., Dalal, Y., and Sunshine, C. (1974) Specification of Internet Transmission Control Program. RFC 675, December 1974.
13. Postel, J. (1980) User Datagram Protocol. RFC 768, August 1980.
14. Postel, J. (1981) Internet Protocol. RFC 791, September 1981.
15. Postel, J. (1981) Transmission Control Protocol. RFC 793, September 1981.

16. Berners-Lee, T., Fielding, R., and Frystyk, H. (1996) Hypertext Transfer Protocol – HTTP/1.0. RFC 1945, May 1996.
17. Fielding, R., Gettys, J., Mogul, J. *et al.* (1999) Hypertext Transfer Protocol – HTTP/1.1. RFC 2616, June 1999.
18. Deering, S. and Hinden, R. (1998) Internet Protocol, Version 6 (IPv6) Specification. RFC 2460, December 1998.
19. ITU (2004) General Overview of NGN. ITU Recommendation Y.2001, International Telecommunication Union, December 2004.
20. ITU (2003) One-way Transmission Time. ITU Recommendation G.114, International Telecommunication Union.

2

Internet Fundamentals by IETF

2.1   Internet Architecture and IETF Standardization

The Internet is standardized by the Internet Engineering Task Force (IETF). However, the Internet architecture is a matter of network design. Because IP (Internet protocol) traffic in general can be transferred over many access and transmission technologies, then the network architecture can be diverse. However, main rules for the Internet architectures concepts and designs are supervised by the Internet Architecture Board (IAB), which is a committee of the IETF.

2.2   Fundamental Internet Protocols

Internet protocols and technologies are positioned in upper layers of the protocol layering model, from network layer (OSI-3 layer) up to the application layer. Main protocols in the IP stack are protocols at the network layer and transport layer (OSI-4 layer). Main protocol in Internet is the IP, which exists currently in two standardized versions, IP version 4 (IPv4) and IP version 6 (IPv6). Further, on the transport layer there are two fundamental protocols, namely User Datagram Protocol (UDP) and Transmission Control Protocol (TCP).

2.2.1   Internet Protocol Version 4

IP is the main communication protocol in the IP model, which is positioned at the network layer (layer 3) according to the OSI (open system for interconnection) protocol model. It provides functions for transmission of datagrams (i.e., blocks of data) over interconnected systems of packet-switched networks from any source to any destination. Sources and destinations are hosts (e.g., computers, various devices with network interfaces) which network interfaces are identified by fixed-length addresses called IP addresses. The native Internet is built over IPv4 [1], which has been standardized in 1981 and implemented in 1983 (refer to Chapter 1). We refer to IPv4 as IP.

IP is a connectionless protocol, because it does not require establishing a connection between the source and the destination prior to the datagrams transmission, which is contrary to the traditional approach in telephone networks where signaling is required (for establishing of the end-to-end connection) before the transfer of user data. IP allows datagrams to travel from

NGN Architectures, Protocols and Services, First Edition. Toni Janevski.
© 2014 John Wiley & Sons, Ltd. Published 2014 by John Wiley & Sons, Ltd.

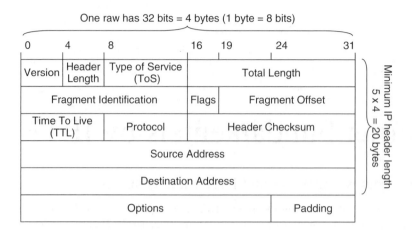

Figure 2.1   IP header format

sources to destinations using different paths (there are no logical or virtual circuits). Further, IP has no mechanisms to support end-to-end reliability, sequencing or flow control. Hence, it is unreliable protocol which provides best-effort delivery of datagrams (i.e., without any guarantees).

Internet datagrams are also called IP packets. A datagram is a variable length packet consisting of two parts: IP header and data. All functions of IP are defined by different fields in the IP header, which is shown in Figure 2.1.

IP has two basic functions: addressing and fragmentation. Addressing is provided by using source and destination IP addresses fields in IP headers. Each IP address is 4-bytes long (i.e., 32 bits). The delivery of each IP packet is performed in Internet nodes using destination IP address specified in the packet header. The selection of the path along which the packet is sent is called routing. The nodes that provide routing are called routers. The Internet, in fact, is consisted of many interconnected routers that route IP packets from their sources to their destinations by using IP addresses.

However, IP packet travels through different networks that are implemented using different technologies. For example, the most used access network for Internet is Ethernet (IEEE 802.3 standard). Each host encapsulates the IP packet into a frame on the link layer (OSI layer 2). Then, each router on the way decapsulates the received frame, then processes the IP packet (e.g., routes the packet), and then encapsulates it in another frame. The OSI-2 frames are different for different physical technologies [e.g., LAN (local area network), MAN (metropolitan area network), or WAN (wide area network) technologies]. For example, the maximum length of OSI-2 frame (known as MAC frame, where MAC is Medium Access Control) for Ethernet LAN is 1500 bytes. On the other side, the maximum length of IP packet is specified in the header field Total Length, which has 16 bits, thus giving maximum length of single IP packet of 65 535 bytes (i.e., $2^{16} - 1$). However, due to the limit of the maximum frame size in different underlying technologies (e.g., 1500 bytes for Ethernet), there is a need for fragmentation of larger IP packets according to the maximum allowed OSI-2 frame length. However, only the data are fragmented. There are three fields in IPv4 header for fragmentation purposes: Fragment Identification (16 bits), flags (3 bits), and Fragment Offset (13 bits). In IPv4 source host or any router on the path can fragment the packet (i.e., the datagram).

All fragments from a single packet carry the same unique (for the given source IP address) Fragment Identification. The Fragmentation Offset field carries the fragment offset relative to the first byte of the fragmented IP packet. It is measured in units of 8 bytes (1 byte equals to 8 bits), so it can cover with its 13 bits all possible offset values in all IP packets (which range up to size of 65 535 bytes), because $2^{13} \times 8 = 2^{16}$. Additionally, in the IPv4 header are given three flag bits from which the first bit is not used, the second bit is "not fragment bit" (if its value is 1 then the router must not fragment the packet), and the last bit (of the three flags) is a "more fragments" bit which is set if the datagram is not the last fragment. In practice fragmentation is rarely used. The common practice is to limit the IP packets below 1500 bytes since the Ethernet has become the most used local access network to the Internet.

The field Time To Live (TTL) indicates the upper bound of the time an IP packet is allowed to reside in the Internet. It is set by the sender and reduced at each network node; thus, when it reaches zero, it is destroyed. So TTL is a kind of self-destruct timer, which main purpose is to provide means to routers for elimination of "lost" packets in Internet.

Further, the Protocol field (8 bits) is defined for protocol multiplexing above the IP layer. In other words, it specifies the protocol which should process the IP payload at the receiving host.

IP does not provide Quality of Service (QoS) mechanisms, but it has the Type of Service (ToS) field defined to specify QoS requirements on precedence, delay, throughput, and reliability. The ToS field has 8 bits and it is intended to be used by gateways (edge routers in a given network) to select transmission parameters for a given network technology, or by routers to select the next hop when routing an IP packet. However, in practice all routers by default ignore or reset the ToS field if not specified otherwise in a given network. Usually, edge routers control the ToS field for all incoming packets to a given network.

Additionally, in IPv4 there is also error control coding field called Header Checksum. It is used on the IP header only. This checksum algorithm is 16 bit one's complement of the one's complement sum of all words in the header (one word has 16 bits). For the calculation of the checksum the Header Checksum value is all zeros. In IPv4 the field ToS changes in each router, hence each router must recalculate the Header Checksum value of each packet. The main value of the checksum is to protect the IP header from bit errors during transmission. Due to importance of correct information in the IP header (e.g., destination IP address) the router or host who will detect an error in the header must discard the IP packet. The mechanism to report such errors is provided by Internet Control Message Protocol (ICMP) [2].

The Options field in the IP header may appear or not. When appears it may have variable length depending upon specified options. In such case it starts with a code identifying the particular option, followed by the optional information (e.g., for security, for routing, etc.).

IP exists in each node attached to the Internet, either a host or a gateway that interconnects networks. But, it does not provide any reliability or flow control. For reliable delivery of packets, we need reliability functions to be implemented in a higher layer protocol such as TCP. The payload of each IP packet contains the data and all headers from higher layer protocols above IP (e.g., transport protocols, application protocols).

2.2.2   *Internet Protocol Version 6*

Internet Protocol Version 6 (IPv6) is the last version of the IP, and hence the newest one [3]. It is not directly compatible with its predecessor, IPv4. The address scheme of IPv6 is new and is based to serve demographic and modern networks. The address space of IPv6 is defined with 128 bits, which has enabled the existence of a huge number of addresses compared to 32 bits

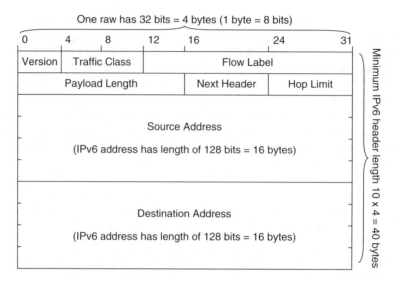

Figure 2.2   IPv6 header format

for addresses in IPv4. But IPv6 is not only developed in order to resolve the problem with the address space, because even in IPv4 it can be solved using public and private IP address space and Network Address Translation (NAT) mechanism [4]. IPv6 brought also several other important novelties, when compared with IPv4. These novelties include the novel fields Flow Label, and Next Header, as well as the exclusion of the Header Checksum. The format of IPv6 header is shown in Figure 2.2.

As a novelty compared to IPv4, IPv6 supports QoS per flow on the network layer. A flow is a sequence of related packets sent from a source to a destination. This means that the flow-based QoS (which is generally determined by losses, packet delay, and bandwidth given in bits per second) will be easier to implement in the Internet, which is especially needed for transition of traditional services, such as telephony or television, toward the all-IP environment. A packet is classified to a certain flow by the triplet consisted of Flow Label, Source Address, and Destination Address. Flow labeling with the Flow Label field enables classification of packets belonging to a specific flow. Without the flow label the classifier must use transport header and port numbers which belong to transport layer (for port numbers refer to TCP and UDP sections in this chapter). Flow state should be established in a subset or all of the IPv6 nodes on the path, which should keep track on all triplets of all flows in use. However, IPv6 does not guarantee the actual end-to-end QoS as there is no reservation of network resources (this should be provided by other mechanisms and is part of next generation networks, as discussed in the following chapters).

Other important novelty of IPv6 is the Next Header field which identifies the type of header immediately following the IPv6 header. Single IPv6 packet may carry zero, one or more next headers, placed between the IPv6 header and the upper protocol header (e.g., TCP header, UDP header). These so-called extension headers may carry routing information, authentication, authorization and accounting information, and so on, which provides better network layer functionalities in all network environments.

In order to improve the routing IPv6 header has a fixed format that allows hardware processing for faster routing. Significant changes are made regarding the fragmentation of data in IPv6, which is done at the source host contrary to IPv4 where it is performed in routers.

Header checksum in IPv6 header is omitted in order to reduce the processing of IP headers in routers (as a reminder, in IPv4 each router for each packet have to calculate a new Header Checksum, due to changes in the TTL field). The error control in IP header is redundant because it is provided in lower protocol layers (e.g., MAC layer) and upper protocol layers (e.g., TCP, UDP). Hence, checksum is omitted in IPv6.

Hop limit in IPv6 header is an eight-bit value that provides the same functions as the TTL field in IPv4.

On the other hand, IPv6 is a newer IP version that is not significantly different from the previous, IPv4. Networks still are assigned network address blocks or prefixes, IPv6 routers route packets hop by hop, providing connectionless delivery, and network interfaces still must have valid IPv6 addresses. However, IPv6 should be seen as simpler, scalable, and more efficient version of the IP.

The minimum length of the IPv6 header is 40 bytes. Hence, the IPv6 packet header is at least two times larger than IPv4 header (which has minimum header length of 20 bytes), thus introducing redundancy per packet due to longer IPv6 addresses. Namely, in real-time communications (e.g., Voice over IP) we have to use smaller packets. In such case higher header redundancy leads to inefficient utilization of the available links and network capacity than IPv4.

2.2.2.1   Discussion about IPv6 and IPv4

IPv6 will not replace the IPv4 protocol overnight. Both versions will have to coexist for a longer time although the process of transition has started to happen in the recent years. In general, IPv6 can be implemented as a software upgrade to hosts and router by replacing or updating their operating systems with embedded IPv6 protocol stack. Currently, most operators are surrounded by IPv4 routers and most of the traffic is IPv4. In order to simplify the transition to IPv6 it is important that existing IPv4 applications be able to work with IPv6 applications. There are several transition strategies from IPv4 to IPv6 that are discussed further.

In the short term, one can conclude that the introduction of IPv6 makes the Internet more complex while the benefits will depend on whether the operators will redesign their networks toward the IPv6 and its features. In long run there is no doubt. IPv6 is more complete (e.g., bigger address space, support for per flow QoS provisioning) and more efficient protocol than IPv4 (e.g., checksum is excluded), hence the Internet future belongs to IPv6 (at least regarding the network layer). However, it should be expected that the transition from IPv4 based Internet to IPv6 based Internet will last for couple of decades.

2.2.3   *User Datagram Protocol*

There are two main transport protocols (OSI layer 4) in the IP stack: TCP as a connection-oriented protocol and UDP as a connectionless protocol.

TCP is a connection-oriented protocol because before the data transmission there is required a preceding procedure for establishing a connection between the two end hosts. Unlike TCP,

UDP is a connectionless protocol because the user immediately begins sending information packed in datagrams (i.e., IP packets) without knowing whether the other side will receive them or not.

The UDP [5] was created as a simple protocol which provides a procedure for application programs to send information via datagrams with minimum of protocol mechanisms. It is a transport protocol (OSI layer 4) which assumes that IP is the underlying protocol (on OSI layer 3). Unlike the TCP, UDP does not provide flow control, reliability, and error control for data sent over IP. Because of its simplicity UDP header consists of less bytes than TCP.

UDP is useful in cases where the reliability is not needed and in cases where the higher level protocols can provide flow control and error control. It is used as a transport protocol for several well-known application protocols, such as DNS (Domain Name System), SNMP (Simple Network Management Protocol), TFTP (Trivial File Transfer Protocol), as well as real-time services over Internet, such as Voice over IP, IPTV (Internet protocol television), and so on.

UDP actually uses the IP as a network protocol, providing the same service as IP to transmit messages from one host to another, but with one important difference, and that is the ability of UDP to distinguish between different processes on the same machine by introduction of port numbers. Namely, each UDP message in its header contains two 16-bits port numbers, a source port and a destination port. Figure 2.3 shows the format of the UDP header.

The following fields are defined in the UDP header:

- Source and Destination ports are 16-bit numbers used for protocol multiplexing, meaning that a port can have a value between 0 and 65 535 (i.e., $2^{16} - 1$).
- Length is a 16-bit field that defines the length of the UDP datagram including its header and the data, so it may has a value up to 65 535. However, the UDP datagram is stored in the IP packet payload and IP packet maximum length is 65 535. Because the minimum IP header length is 20 bytes (for IPv4), the maximum size of the transport datagram is $65 535 - 20 = 65 515$ bytes.
- Checksum field has a task of error control coding, which includes both UDP header and data.

2.2.4 Transmission Control Protocol

Internet transport layer protocol for reliable transmission of information is TCP. Most popular Internet-native services today are based on TCP, such as: e-mail (electronic mail), World Wide Web (WWW), File Transfer Protocol (FTP), and so on. Unlike UDP, TCP incorporates means

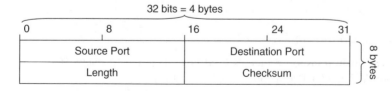

Figure 2.3 UDP header format

for congestion control in Internet, which are located in end hosts. This is also a completely different approach from traditional telecommunications where end hosts are "dumb" devices and all control resides within the network (e.g., network nodes and servers) controlled by the network operators. Due to TCP's overall importance in the best-effort Internet design, the IP stack is often referred to as TCP/IP.

TCP is a connection-oriented protocol, which provides the following main functionalities:

- *Basic data transfer*: TCP provides continuing stream of octets (i.e., bytes) from one application (which uses TCP/IP protocol stack) on one host to the same application on another host in the Internet. All octets are delivered to the receiving TCP side in exactly same order as they have been sent from the sending TCP side. So, TCP provides lossless data transmission, although losses occur all the time due to bit errors or congestion in network nodes (e.g., routers).
- *Reliability*: TCP must recover all lost data (regardless of the reason) to provide lossless transmission of data between the applications on end hosts. For that purpose TCP uses sequencing of each sent octet as well as it requires positive acknowledgment (shortly written ACK) from the receiving TCP side.
- *Flow control*: TCP provides means for flow control by using window based mechanism for sending data before receiving ACKs for previously sent data.
- *Multiplexing*: TCP provides transport level addresses called ports (as UDP also does) to allow multiple processes to use TCP on a single host. The binding of ports to processes (e.g., applications running on the host) is provided by abstractions called sockets. A socket is formed by binding of a network address (i.e., IP address) of the used network interface, and a port (e.g., TCP port). Each connection in Internet is identified by a pair of sockets, one on the sending side and the other one on the receiving side. However, sockets may be used in parallel in multiple connections from a given host.

2.2.4.1   How does TCP Work?

Before any TCP communication there should be established connection between the sending and the receiving TCP sides, which results in opening sockets for TCP on both ends. After the establishment of a TCP connection, TCP starts to accept continuing stream of octets from the application process through the application interface (i.e., socket), and then packs the data into so-called segments which are forwarded to the underlying IP (on the same host). Such TCP segments are stored into payloads of IP packets on network layer. However, the TCP decides when to forward the data to the IP layer (at the sending TCP side). So, the segments are delivered from the source to the destination via IP packets which are routed through the Internet.

At the receiving side, the destination IP module unpacks the segment from the IP packet and passes it to the receiving TCP. However, due to the nature of Internet routing (each packet may follow different path from the source to the destination), different segments may arrive out of order. Also, some segments may be lost due to congestion, and other segments may be damaged due to bit errors. Regardless of the causes for segments losses, the TCP should provide accurate in-order transmission of all octets to the receiving application. That is provided by using sequence numbers for each delivered octet (via TCP segments packed in IP packets) and applying retransmissions for each lost or damaged segment. The receiving TCP places the data received in order (in which it was sent) into the receiving TCP buffer and delivers it to the application process through the socket on the receiving host.

From the beginning TCP is specified to be a module embedded into operating systems, regardless of the host type. The interfaces are created in such manner that applications access the TCP in the similar manner as they access the file system within the host. So, interface between the application (that uses TCP as transport protocol) and the TCP include operations "open" and "close" for a TCP connection, "send" and "receive" for data (similar to "write" to a file, or "read" from a file, in a file system). But, TCP does not put message boundaries for sending the data, and it does not know which format is the transmitted data (the interpretation is left to the applications above TCP). It provides only byte stream between application and the TCP module via the sending socket (write process), and between the receiving TCP module and application via the socket on the receiving side (read process). Each endpoint may choose independently the write and read sizes. So, TCP provides byte-stream service. Hence, the sockets used by TCP are called "stream" sockets. This is completely different than UDP, which uses datagrams which are messages with defined boundaries. Hence, UDP sockets are called "datagram" sockets. Also, UDP does not need establishing a connection prior to data sending to the destination side.

TCP is a full duplex communication protocol. Between each pair of TCP end points (e.g., hosts) it allows to have two independent TCP streams in opposite directions without direct interaction. In such cases it is possible the underlying IP protocol to send control information for a given TCP stream back to the source by using IP packets carrying data in the opposite direction. However, TCP allows also half-duplex operation (i.e., one stream to be terminated and to continue with the stream in the other direction).

2.2.4.2   TCP Acknowledgment Mechanism

The reliability and flow control in native Internet communication in based on TCP, which resides on the end hosts for each TCP-based application (e.g., e-mail, WWW, FTP, etc.). In the initial IETF standard [6], TCP is standardized only with time-driven retransmission mechanism. That is, when TCP transmits a segment with data, it copies that segment in so-called retransmission queue on the sending host, and starts a timer. When the segment reaches the destination end, the receiving TCP entity immediately responds with its own segment. This segment (sent back to the TCP sender from the TCP receiver) brings confirmation in the form of so-called ACK number equal to the sequence number of the next byte that TCP receiver expects to receive. If the timer runs out, the TCP sender retransmits the segment from the retransmission queue.

2.2.4.3   TCP Header Format

Each TCP segment begins with a header which has length of minimum 20 bytes, as shown in Figure 2.4. It can also include optional fields. Then, maximum length of single TCP payload is 65 495 bytes (because IPv4 header is minimum 20 bytes long and maximum IPv4 packet size is 65 535 bytes).

Source and Destination port fields in the TCP header identify the end points for the TCP connection.

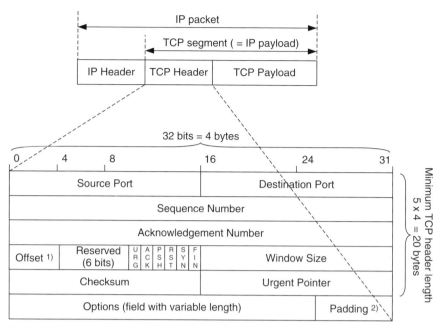

1) Offset is length of the header in number of words. A word has 32 bits (4 bytes).
2) Padding is added when Options field is shorter than than the specified Offset (i.e., the Header length in words).

Figure 2.4   TCP segment format

Sequence number is 32-bit field which is assigned by the sending side to the first byte of data in the TCP segment. However, sequence number is incremented by one for each byte of data. After the initiation of a connection, the sequence number for the first byte is random initiated.

ACK number is 32-bit field in the TCP header, which specifies (in the direction from the receiver to the sender) the sequence number of the next byte which is expected to be received. For example, if last successfully received byte (last byte of the last successfully received segment) has sequence number N, then ACK number is set to $N + 1$.

The header length is a four-bit field which contains the length of the TCP header only in 32-bit words (each word has length of 4 bytes). So, if the TCP header is 20 bytes, then the Header length field has a value of five. After the Header field there are 6 bits currently unused, reserved for future use.

Further, there are six one-bit fields called pointers. The URG bit is set when the Urgent pointer field in the TCP header is used. Similarly, the ACK bit is set to one when the ACK field is used to carry an acknowledgement. If the PSH bit is set, then the receiving TCP module should push the data from that segment directly to the application without buffering. When the RST bit is set, this indicates reset of the TCP connection due to a certain problem (i.e., TCP connection should be established again). The SYN bit is used for connection establishment together with the ACK bit. The FIN bit is used for connection termination. If one end of the TCP connection has no data to send it sets FIN = 1 to finish (i.e., terminate) the connection in that direction. The connection is terminated in both direction when the other end will also sends a segment with FIN = 1.

Window size field in the header is used by the receiving side to indicate to the sender how many bytes can be sent after the last octet that has been confirmed with the ACK number field (it is the receiver advertised TCP window). If the sender receives a segment with value 0 in the Window size field, then temporarily stops sending data until receiving a segment with Window size $\neq 0$.

The checksum field is used for error control coding with purpose of detecting corrupted segments, which are then discarded and treated the same way as lost segments due to network congestion (i.e., they are retransmitted using the TCP's ACKs mechanism).

2.2.4.4   TCP Sliding Windows for Flow Control

TCP flow control is accomplished by using a sliding window with variable length. This is accomplished by returning a "window" with every ACK from receiver to the sender, which indicated the number of octets (i.e., bytes) the receiver may send without waiting for ACKs for the already sent data. So, the TCP sender maintains three pointers associated with a given connection, as shown in Figure 2.5.

The first pointer is the left boundary of the sliding window, which marks the already acknowledged data from the TCP receiver. The second pointer marks the right boundary of the sliding window which is the octet with the highest sequence number that can be sent without awaiting ACKs for already sent data. Then, there is the middle pointer (a third one) which denotes the already sent data within the TCP sliding window. Positively acknowledged octets by the receiver indicate moving the sliding window to the right (to higher sequence numbers). There is also a window at the receiving side, which maintains the received data from the TCP connection. So, TCP maintains two windows for each direction, one at the sender and the other at the receiver [7]. Since TCP is a full duplex, that gives in total four windows associated with a given TCP connection.

2.2.4.5   TCP Mechanisms for Congestion Control

Initial TCP standard [6], is not defining a congestion control mechanism which is essential for TCP. Such mechanisms are added to the TCP afterwards. The main elements of the congestion control are [8]: slow start, congestion avoidance, fast retransmit and fast recovery. From these

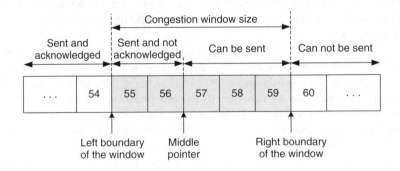

Figure 2.5   TCP sliding window

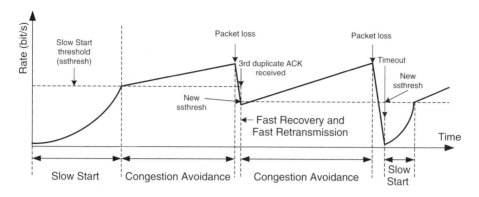

Figure 2.6 TCP congestion control

four main mechanisms the slow start and congestion avoidance must be used by TCP sending side for each connection to handle the congestion. In fact, the best-effort Internet is based on the TCP's congestion control mechanisms.

To go through the four mechanisms, one has to define initial value of congestion window (cwnd) and the slow start threshold (ssthresh). According to [8] initial value of cwnd should be not more than $2 \times SMSS$, where SMSS is Sender Maximum Segment Size (i.e., the maximum size of a segment that the sender can transmit). Because Ethernet is the most used LAN in Internet, and the maximum message transfer unit in Ethernet is limited to 1500 bytes, the SMSS is kept below 1500 bytes. However, larger segments reduce the redundancy due to headers (e.g., TCP header, IP header). Then, in slow start TCP increments the cwnd for SMSS bytes for each received ACK which acknowledges new data received. This way, the cwnd exponentially increases with time, that is, it doubles its size for each Round Trip Time (RTT). The RTT is the time needed for segments to be transmitted to the receiving end and the ACK to come back to the sending side.

The ssthresh is used as a boundary between the slow start and congestion avoidance mechanism. For example, the ssthresh may be arbitrarily high, but it may be decreased when congestion occurs. In some implementations of TCP it equals the size of the receiver advertised window.

Congestion avoidance is used when cwnd > ssthresh, as shown in Figure 2.6. In this phase the TCP increases the congestion window (i.e., cwnd) by one full-sized segment for each RTT. If RTT is constant in a given time period, then cwnd increases linearly during the congestion avoidance phase.

The available bit rate to a given TCP connection is directly related to cwnd and RTT.

Example 2.1

Let us have segment size = 1500 bytes and RTT = 200 ms, then we obtain:

- If slow start starts with cwnd = 1 segment (typical usage scenario), then TCP connection in this example will start with bit rate = 1 segment × (1500 bytes/segment × 8 bits/byte)/ 200 ms = 60 kbit/s.

- After RTT and no loss of the segment the ACK for the first segment will arrive and trigger increment of cwnd for 1 up, giving cwnd = 2 segments, which will result in two segments sent back-to-back (one after another) from sender to the receiver.
- After RTT and no loss (and no reordering) of the two sent TCP segments, two non-duplicate ACK will come back to the TCP sender, resulting in cwnd increment by 2, which will give doubled size of the cwnd (i.e., cwnd = 4 segments).
- So, each RTT the cwnd doubles, thus resulting in exponential increase of the cwnd, and hence the available bit rate to the TCP stream. For example, after 6 RTT intervals (6 × 200 ms = 1.2 s), the cwnd = 64 segments, resulting in available bit rate = 64 segments × (1500 bytes/segment × 8 bits/byte)/200 ms = 3.84 Mbit/s.

The above example shows how TCP connection adapts to the available bit rates on transmission paths from the source to the destination. However, the available bit rates are always limited regardless of the implemented technology, so increasing of the bit rates must lead to loss of a segment or segments after certain time (and it happens all the time during the connection duration). Losses occur due to two reasons: congestion (e.g., in network nodes, such as routers, IP packets carrying segments are discarded when buffer memory is full or some threshold is reached), and bit errors (due to transmission bit errors detected by the checksum control).

How does the TCP sender detect the losses? In fact, there are two mechanisms for loss detection:

- *Retransmission timeout (RTO)*: This is specified in the initial TCP standard [6], and it is dynamically derived due to variability of conditions in the networks through which the segments travel. After a time period longer than RTO value (from the time of sending a segment), the TCP retransmits the segment and goes into slow start with cwnd = 1 segment. The procedure for calculation of RTO requests measurements of the RTT and then Smoothed Round-Trip Time (SRTT) using the following equation:

$$SRTT = (\alpha \times SRTT) + [(1 - \alpha) \times RTT] \tag{2.1}$$

Further, the RTO is computed as:

$$RTO = \min [UpperBound, \max [LowerBound, (\beta \times SRTT)]] \tag{2.2}$$

where UpperBound is upper bound for the RTO (e.g., 1 second) and LowerBound is its lower bound (e.g., 0.2 second), α is a smoothing factor (e.g., $\alpha = 0.8-0.9$), and β is a delay variance factor (e.g., $\beta = 1.3-2.0$).
- *Duplicate ACK*: For each received segment TCP receiver sends ACK who carries the sequence number of next octet that is expected to be received. However, if a segment is lost and the following segments (after the lost one) continue to arrive at the receiving side that indicates that congestion has occurred. For each received segment out of sequencing order

the receiver sends ACK for the last successfully received segment in order, so such ACK will be a duplicate to a previous ACK sent back to the sender. Duplicate ACKs received by the TCP sender are treated as an indication for congestion occurrence. However, besides dropped packets due to a congestion, duplicate ACK may be caused by reordering of the segments data (which happens rarely in practice), or even by replication of a segment or the ACK itself by the network.

The duplicated ACKs are used by the fast retransmit and fast recovery mechanisms [8], usually implemented together using the following procedure:

- When third duplicate ACK is received the ssthresh is set to value:

$$\text{ssthresh} = \max (\text{FlightSize}/2; 2 \times \text{SMSS}) \tag{2.3}$$

  where FlightSize is the amount of outstanding data in the network (i.e., sent segments in the current cwnd), which is always less or equal to cwnd value.
- Sender retransmits the lost segment and sets $\text{cwnd} = \text{ssthresh} + 3 \times \text{SMSS}$, where $3 \times \text{SMSS}$ reflects the three segments which the receiver has buffered and which have resulted in duplicate ACKs. Then, for each additional duplicate ACK (which indicates another segment arrived at the receiver), the cwnd is increased by 1 segment.
- TCP sender transmits segments if allowed by the new value of cwnd and the receiver's advertised window. When the next ACK that acknowledges new data arrives the ssthresh is set to the value calculated by the Equation 2.3 in the first step above. Such ACK may also acknowledge all intermediate segments which were successfully received after the lost one.

The fast retransmit and fast recovery mechanisms provide recovery from single losses in a cwnd. The TCP version which implements all four mechanisms is TCP Reno. However, TCP Reno does not recover well from multiple lost segments in a cwnd. Hence, there are many TCP versions proposed by researchers in the past two decades. But, one has to distinguish the TCP SACK (Selective Acknowledgments) [9], among the other TCP versions. The TCP is using cumulative ACKs. However, there might be missing only certain segments in the receiver's buffer. Hence, TCP SACK is created to incorporate a mechanism which sends back to the sender so-called SACK packets informing the sender of data being received, so the TCP sender can only retransmit the missing segments. In such way, TCP SACK has higher probability to provide recovery from multiple segment losses in a cwnd. Therefore, it is used in the Stream Control Transmission Protocol (SCTP), a transport layer protocol initially created for transport of signaling information over the Internet (e.g., signaling related to voice calls).

2.2.5   Stream Control Transmission Protocol

Until the twenty-first century, Internet reliability was based on TCP as a main transport protocol, running over the IP. However, many applications and services have found TCP more limiting and have therefore used UDP as a transport protocol and have provided reliability on the application layer. The main disadvantages of TCP are the following [10]:

- Strict byte-order delivery to application (causes head of the line blocking when losses appear, resulting in received segments waiting in buffers during retransmissions of the lost segments);
- Stream orientation of TCP (no message boundaries);
- No support for multihoming, which is very important in signaling environments (e.g., for carrying SS7 signaling over the Internet).

Hence, SCTP was standardized by IETF [10] in the year 2000, designed particularly as transport protocol to be used for carrying SS7 signaling messages over IP networks. The main target was higher reliability for signaling transport, comparable with five nines reliability (i.e., 99.999% reliability) accepted as a standard in telecom sector. There are several important features provided with SCTP which result in higher transmission reliability and address the mentioned TCP disadvantages:

- *Message based protocol*: SCTP is message-based reliable protocol, instead of stream oriented as TCP, or unreliable as UDP.
- *Multihoming capability*: allows two SCTP end points to setup so-called association with multiple IP addresses on one end.
- *Multistreaming capability*: Sequenced delivery of messages over multiple streams, to reduce the impact of head-of-the-line blocking to a single stream instead of whole connection.
- *Congestion control*: This is provided by using TCP SACK mechanisms, which is mandatory in SCTP while it is only optional in TCP.
- *Security*: SCTP has enhanced security at connection establishment phase, based on so-called four-way handshake for establishing an association, instead of three-way handshake in the TCP case for connection establishment. That prevents blind Denial-of-Service (DoS) attacks, such as SYN attacks (based on three-way connection establishment of TCP with SYN and ACK fields in the header).

The SCTP multihoming and multistreaming functionalities are shown in Figure 2.7. These two capabilities of SCTP are very important on longer time scales. For example, currently SCTP is used as a transport protocol for signaling data in core network nodes in mobile networks, and its usage is standardized by 3GPP (3G partnership project). In future it is also possible to move SCTP as transport protocol toward the user's equipment (e.g., personal computers, mobile devices, etc.). Such approach is logical in cases when there is available more than one network to a given host. For example, in case of mobile users, if there are active several interfaces to different wireless or mobile networks then the SCTP could be used to provide better mobility, but also to provide aggregation of data rates available from different access networks. However, such process raises many issues to be solved (e.g., QoS support, service level agreements between the user and different networks, etc.).

SCTP certainly has bright future toward the future networks, considering all its features as well as incorporation of the best characteristics of the TCP and the UDP (which are fundamental transport protocols in Internet, implemented in all operating systems in all Internet hosts), such as best TCP congestion control approaches (the TCP SACK) and message delivery approach as found in UDP.

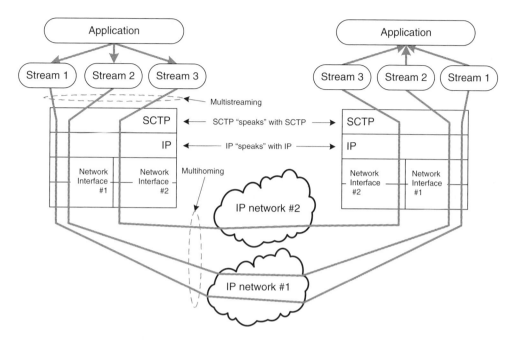

Figure 2.7   SCTP multihoming and multistreaming

2.3   Addressing and Numbering

Regarding the addressing and numbering, the traditional Internet is defined with two name spaces. One name space is the IP addressing space. The other is the domain name space (the domain name is a string of characters, dashes, and dots, used to name certain group of machines), which is implemented with the DNS.

The IP addressing name space is consisted of IP addresses, and nowadays there are two types of the IP addresses, IPv4 and IPv6 (each corresponds to one of the two standardized versions of the IP). In both cases, the IP address has two defined roles:

- *Locator role*: Looking from the bottom-up (regarding the OSI layers in a given host) IP address identifies a network interface by which a host is attached to the network, so packets can be addressed to the given interface by using its IP address as a locator in the network. Locator role of the IP address is crucial for the networking approach in Internet.
- *Identifier role*: From the application point of view, looking top-down (regarding the OSI layers) IP address is the identifier of the network interface to/from which the data is sent/received by the given application. The identifier role is implemented by the abstraction of a socket (as it is already discussed in this chapter) which binds the IP address, transport protocol (e.g., UDP, TCP), and port (used by the application in the given direction).

Network addresses (IPv4 addresses) are mapped to link-layer addresses with Address Resolution Protocol (ARP), standardized by the IETF with RFC 826 (RFC, request for comments). Regarding the OSI layering model ARP is positioned between layers 2 and 3. The functionalities of ARP (for IPv4) in IPv6 are provided by Neighbor Discovery Protocol (NDP) specified in RFC 4861.

2.3.1   IPv4 Addressing

Every host on the Internet has its own IP address, which consists of two parts: network ID (network part of the IP address) and host ID (host part of the IP address), and has a total length of 32 bits for IPv4. Network ID defines the network, and if the network should be part of the Internet it is given by a global authority, the Internet Corporation for Assigned Names and Numbers (ICANN), usually through its regional organizations. For each new network that requires access to the Internet, ICANN assigns network ID. Host ID identifies uniquely a given host in the network. IP address in a unique way identifies the specific network interface on a given Internet host (e.g., a computer, mobile device, etc.) in a given IP network. There are two types of IP addressing:

- Classful addressing;
- Classless addressing (Classless Inter-Domain Routing – CIDR).

2.3.1.1   Classful Addressing

Classful addressing is defined with five different classes of IP addresses, shown in Figure 2.8. Class-A IP address allows the existence of 126 different networks with 16 million hosts per network; class-B includes 16 382 networks with 65 534 hosts per network; class-C includes 2 million networks with 254 hosts per network.

Class-D is for multicast addresses. Besides the unicast IP address, a given host can have one or more multicast addresses of the class-D. Each datagram that contains a multicast destination

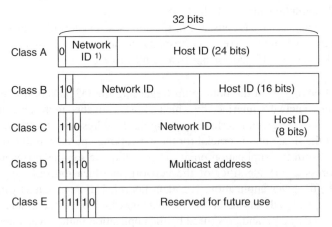

Address 255.255.255.255 (all bits "1") is the broadcast address.

1) Network ID = 127 is a loopback IP address, used only internally in a machine.

Figure 2.8   IPv4 addressing classes

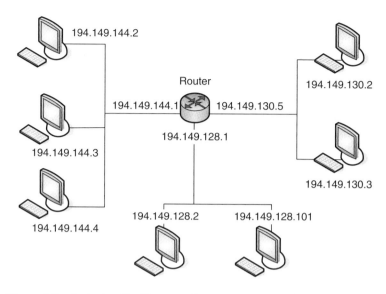

Figure 2.9   Assigning IP addresses to hosts connected to different IP networks

address simultaneously is delivered to all hosts that have been assigned to the given multicast address. Class-E addresses are reserved for future use.

IP addresses are 4-bytes in length, so the IP addresses are canonically represented in decimal-dot notation, consisted of four decimal numbers (each number ranging from 0 to 255), separated by dots (e.g., 192.168.1.1). One example of addressing in the Internet is shown in Figure 2.9, where three IP networks are interconnected through a router.

In classful addressing an organization is granted a block in one the three classes A, B, or C. Then, the network ID of the IP address is used by routers for routing a packet to its destination.

2.3.1.2   Classless Addressing

With expansion of the Internet and number of hosts in the Internet, the classful addressing could not solve the problem for larger IP address space. The solution was to change the distribution of IP addresses from classes to more flexible approach, hence classless addressing has been proposed by the IETF to replace the classful network design. The aim was to slow down the exhaustion of IP addresses as well as the growth of the routing tables. Hence, the method was called Classless Inter-Domain Routing. Again, the IP addresses are consisted of two parts (similar to classful addressing): a network prefix (in the same role as network ID) and a host suffix (in the same role as host ID). While in classless addressing the network ID can be 8, 16, or 24 bits, in classless addressing it may have any value from 1 to 32. So, if network prefix has length of n bits, then host suffix has length of $(32 - n)$ bits.

In IP classless addressing (CIDR), the IP address is represented with a 64 bit value composed of two parts:

- IP address of 32 bits;
- mask of 32 bits.

There are two commonly used approaches to denote networks addressed using classless addressing, and they are:

1. a.b.c.d/255.255.255.0 (in this case the mask is in a decimal-dot notation, the same as IP address);
2. a.b.c.d/24 (in this case the mask is a decimal number that defines the number of leftmost bits in the 32-bit mask which are set to "1"; for example, mask "/24" corresponds to a decimal-dot notation 255.255.255.0).

2.3.1.3   Allocation of IP Addresses

How host can obtain IP address? Each host must have an IP address allocated to each interface through which it connects to IP networks. There are two ways for the allocation of IP address:

- Static IP address (hard-coded) set by the system administrator;
- Dynamically assigned IP address by Dynamic Host Configuration Protocol (DHCP), which is more flexible for the network and more convenient for ordinary Internet users.

How does the network obtain the network part of the IP address? It is assigned from IP address space of the ISP (Internet Service Provider) for the given network.

How an ISP obtains a block of IP addresses? It is done through ICANN, www.icann.org. However, not all ISPs contact ICANN directly. Namely, smaller ISPs receive block of IP addresses from the major ISPs. Only the largest ISPs (which are often international organizations) communicate directly with ICANN (for assigning IP addresses) through its five Regional Internet Registries (RIRs).

2.3.2   Network Address Translation

The allocation of IP addresses by the ISPs to business and residential users created a new problem. Smaller blocks of IP addresses were insufficient to support increasing demands for IP addresses as number of Internet users increased exponentially. Hence, for purposes of scalability of IP address space and consistency of the routing principles in Internet, IP address space is grouped into two main groups: public IP addresses and private IP addresses. Private IP addresses can be reused in different private networks. Public IP addresses must be globally unique (the same public IP address cannot be allocated to two different network interfaces), because world-wide Internet routing is done using public IP addresses (in IP packet headers). The IANA (Internet Assigned Numbers Authority) has reserved the following three blocks of IP addresses for private networks [11]: 10.0.0.0/8, 172.16.0.0/12, and 192.168.0.0/16.

A standardized mechanism, called Network Address Translation [7], provides the mapping between the public and private IP addresses, and vice versa. It can be performed at any given router which is located as an edge node (i.e., gateway) between the private network and the public Internet.

NAT is used only in cases when private IP addresses are used. Unlike the public, private IP addresses can be repeated as much as we want (though in different private networks). Each private IP address must be mapped into a public IP address by using NAT in the edge router of the network that uses private IP addresses. The mapping can be of private to public IP address

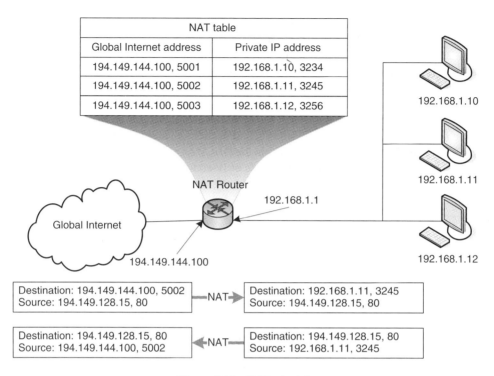

Figure 2.10   NAT principle

can be n:m (i.e., *n* private IP addresses mapped to *m* public IP addresses, where $n > m$). So, many private IP addresses can be mapped into one or a few public IP addresses (depending upon the number of the available public IP addresses).

NAT is illustrated in Figure 2.10. In the given example, three hosts with private IP addresses (192.168.1.10, 192.168.1.11, and 192.168.1.12) are mapped into a single public IP address (allocated to the interface of the router attached to the public Internet), that is the IP address 194.149.144.100 (in this example). For performing NAT the edge router (between the private network and public Internet) maintains a translation table, which translates the IP addresses as well as port numbers in IP packets (in IP headers and transport headers, respectively) from the private to the public addresses, and vice versa. Separation of connections from different private hosts (using different private IP addresses) to a single public IP address (in this example) is provided by usage of different port numbers in the NAT tables. To summarize, NAT changes the Source IP address (from private to public) for all outgoing packets from the private network to the public Internet. Also, NAT changes the Destination IP address (from public to private) for all incoming IP packets from public Internet to hosts in the private network.

2.3.3   Dynamic Host Configuration Protocol

DHCP provides mechanisms for distribution of configuration parameters to Internet hosts [12]. It is a main tool for dynamic allocation of IP addresses from a DHCP server to DHCP clients attached to the network. The DHCP consists of two components: a protocol that delivers specific user configuration parameters from a DHCP server to the user, and a mechanism for allocating network (i.e., IP) addresses to users.

DHCP supports three mechanisms for the allocation of parameters: automatic allocation, manual allocation, and dynamic allocation. With automatic allocation DHCP server assigns permanent IP address to a client. In manual allocation IP address is assigned manually by the network administrator and DHCP server is used only to distribute the chosen address to the client. In dynamic allocation, the DHCP server automatically assigns an IP address to a client from a defined pool of IP addresses, and such allocation is valid for a limited period of time (according to the configuration parameters in the DHCP server). The dynamic allocation is the only mechanism that provides automatic reuse of IP addresses.

The DHCP is based on the client–server model (Figure 2.11). There are four types of DHCP messages:

- *DHCP discover (DHCPDISCOVER)*: This message is sent from a client entering a network in which it needs to discover a DHCP server. Since the client does not have an IP address assigned to him, this message is sent from IP address 0.0.0.0 (port 68) to limited broadcast IP address 255.255.255.255 (port 67). Each client request has a unique transaction identifier (i.e., transaction ID).
- *DHCP offer (DHCPOFFER)*: The DHCP server replies to DHCPDISCOVER message with available network address and additional configuration parameters such as lease time (i.e., time duration for the assignment of the offered IP address to the client). This message is sent from DHCP server IP address (port 67) to the IP local broadcast address 255.255.255.255 (port 68), because at this phase the client has no IP address assigned to its network interface.
- *DHCP request (DHCPREQUEST)*: After receiving DHCPOFFER messages from one or in some cases more than one DHCP server, the DHCP client sends DHCPREQUEST message to the chosen DHCP server. This message is again sent to local broadcast address 255.255.255.255 (port 67), but only the referenced DHCP server will react with binding of IP address to the client.
- *DHCP acknowledgment (DHCPACK)*: After receiving DHCPREQUEST message, the DHCP server replies with DHCPACK message which indicates to the client that IP address is assigned and can be used. However, the client can accept or decline the ACK message, depending upon the probe of the IP address in the network regarding its uniqueness.

However, DHCP works initially in a single IP network, because local IP broadcast does not travel over a router in the default case. But, in many scenarios for IP network design there is a need to provide possibility for DHCP client–server communication over a router (or routers) on the way of packets (e.g., this is a scenario of a centralized DHCP server for clients found in larger routed network). In such network scenarios, there is a feature called DHCP-relay, which should be enabled specifically in all intermediate routers on the transmission path between the DHCP server and the clients.

Together with NAT the DHCP is one of the reasons for the long survivability of IPv4, because DHCP allows a smaller number of IP addresses (a pool of IP addresses) to be shared for IP addressing of larger number of users that are not continuously attached to the Internet (e.g., mobile users). Hence, DHCP is considered as one of the fundamental technologies in the Internet.

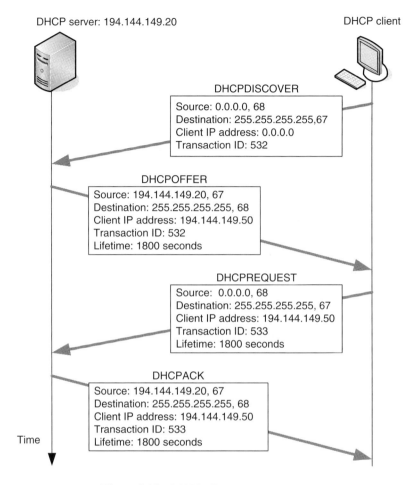

DHCP server: 194.144.149.20   DHCP client

DHCPDISCOVER
Source: 0.0.0.0, 68
Destination: 255.255.255.255,67
Client IP address: 0.0.0.0
Transaction ID: 532

DHCPOFFER
Source: 194.144.149.20, 67
Destination: 255.255.255.255, 68
Client IP address: 194.144.149.50
Transaction ID: 532
Lifetime: 1800 seconds

DHCPREQUEST
Source: 0.0.0.0, 68
Destination: 255.255.255.255, 67
Client IP address: 194.144.149.50
Transaction ID: 533
Lifetime: 1800 seconds

DHCPACK
Source: 194.144.149.20, 67
Destination: 255.255.255.255, 68
Client IP address: 194.144.149.50
Transaction ID: 533
Time   Lifetime: 1800 seconds

Figure 2.11   DHCP client–server messages

2.3.4   Domain Name System

DNS is created to provide means for usage of names (domains) instead of IP addresses to
access certain machines on the network. People more easily use names than long numbers, so
the domain name space was created.

The DNS is a distributed hierarchically-based system, which provides a Domain Name to
IP address (associated with that domain) translation, and vice versa. It is characterized by
two conceptually independent aspects. The first aspect is the definition of names and rules
for delegation of authority for the names. The second aspect is specification of a distributed
computer system that efficiently maps names to IP addresses, and vice versa.

Domain name space is defined as a tree structure, with the root on the top, as given in
Figure 2.12. Each domain name consists of a sequence of so-called labels, separated with
dots. Each label corresponds to a separate node in the tree (Figure 2.12). The domain name is
written (or read) starting from the label of a node in the tree and going up to the root (which is

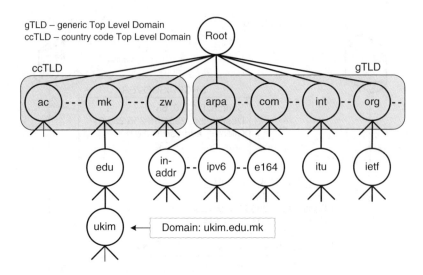

Figure 2.12 Domain name space

a null label and therefore it is not written in the domain name). For example, in domain name www.example.com one may observe that it consists of three labels divided by dots. The label on the right side is always higher in the name hierarchy. So, in the given example the top-level domain (TLD) is the domain "com."

One can refer to DNS as a protocol and as a database. DNS functionalities are provided by three major DNS components [13]:

- *Domain name space and resource records (RRs)*: This is a tree structured name space and information associated with domain names in so-called RRs.
- *Name servers*: They store information about the domain tree's structure. In general, a name server stores (or caches) only a subset of the domain space called a zone (for which it is called an "authority"). Also, a name server has pointers to other name servers that can be used to provide information about any other part of the domain name tree.
- *Resolvers*: These are programs which extract information from name servers as a response to client requests. These are typically routines which are called by the application programs on a given host in the network. For that purpose, each resolver must know at least one name server [14]. For reliability purposes, there should be two domain name servers (primary and secondary), specified in the hosts either manually or automatically (via DHCP).

2.3.5 ENUM

The numbering and naming in traditional telephone networks and the Internet is completely different. However, because the two worlds are converging over the Internet, there is also trend for convergence regarding the addressing and naming.

In traditional telephone networks addressing is provided via telephone numbers allocated to end users according to the ITU's E.164 recommendation. According to E.164 telephone numbers are based on geographical areas and they are hierarchically allocated. The telephone number structure consists of maximum 15 digits grouped in three layers: Country Code (1–3 digits), Network Destination Code, and Subscriber Number. The telephone numbers address users, not devices. The telephone users are accommodated to such numbering approach (regardless of the underlying technology for voice transmission, such as analog, digital, or Internet-based), so the standard telephone number (i.e., E.164) will not go away.

On the other side, addressing in Internet is provided with IP addresses (e.g., via DHCP), while naming is provided with the domain names. The mapping between Internet addresses and domain names is done by DNS. Hence, the DNS is fundamental Internet technology (invented by IETF) and intrinsic part of the Internet, so it will not go away either.

How the two approaches can converge? How can IETF and ITU (International Telecommunication Union) make this easier? One possible answer is ENUM [15], which provides mechanism to use DNS for E.164 numbering. It is possible to implement, because E.164 and DNS are based on hierarchical approach. In E.164 leftmost digit has highest hierarchy, while in DNS leftmost label has lowest hierarchy. So, ENUM is created to translate digits from E.164 telephone number to labels (in terms of DNS), written in the reverse order, with root "e164."

For example, let E.164 telephone number is "+389-2-12345678," then full domain name for such telephone number will be "8.7.6.5.4.3.2.1.2.9.8.3.e164.arpa."

One should note that ENUM is introducing completely different approach in domain name space, than the one used in traditional Internet for servers. With ENUM domain name space will be moved also to the client side.

Overall, ENUM is targeted toward the convergence of traditional telecommunications and traditional Internet, via the convergence of E.164 numbering scheme found in PSTN (public switched telephone network) and DNS found in Internet.

2.3.6 IPv6 Addressing Architecture

IPv6 addressing differs from the IPv4 addressing. Each IPv6 address has length of 128 bits (i.e., 16 bytes), and is divided in three parts, which is different than IPv4 addresses which have only two parts (network ID and host ID). Due to larger address, IPv6 addresses are written in colon hexadecimal notation in which 128 bits are divided into eight sections, each section with 16 bits (which equals to 4 hexadecimal digits). The preferred form is x:x:x:x:x:x:x:x, where an "x" can be 1–4 hexadecimal digits. It is less than 4 in cases when there are consecutive series of zeros in the address as shown in the example below.

Example 2.2: Example of IPv6 address

2001:0000:0000:0000:0008:0800:200C:417A, which may be written also as:
2001:0:0:0:8:800:200C:417A, and in compressed mode it will be:
2001::8:800:200C:417A (the use of "::" replaces one or more groups of consecutive zeros in the IPv6 address, and can be used only once).

There are three types of IPv6 addresses [16]:

- *Unicast*: This is identifier to a single interface in the network.
- *Anycast*: This type of address in used when identifier is given for a set of network interfaces, which may belong to different nodes. When a packet is sent to a destination anycast address

it should be delivered (by means of routing) to the nearest nodes in the set (according to the routing metrics). This type of addressing appears with IPv6 (in IPv4 are defined unicast and multicast addresses, but also local broadcast addresses which are not present in IPv6).

• *Multicast*: This is an identifier to a set of network interfaces. Packet addressed to a multicast address will be delivered to all addresses in the set.

IPv6 also allows CIDR notation (as it exists for IPv4), which is performed by using IPv6 address and a binary prefix mask, as given in IPv6 notation in Table 2.1 [16].

IPv6 addressing architecture has one hierarchy layer more than IPv4. The general format for global unicast IPv6 addresses has three parts: global routing prefix, subnet ID, and interface ID (Figure 2.13). All global unicast IPv6 addresses (other than those with leading zeros, which in fact have embedded IPv4 addresses in the lowest 32 bits) have 64-bit interface ID.

Link-local IPv6 addresses (Table 2.1) are used for so-called stateless address autoconfiguration, where 64-bits of the interface ID are obtained from the interface's link address (e.g., using 16 zeros concatenated with 48-bit Ethernet address of the given interface). IPv6 stateful address autoconfiguration is provided with DHCP in the same manner as IPv4.

2.4   Internet Routing

In Internet IP packets from a given source (host, router) should reach a given destination (host, router). Source and destination addresses are specified in the header of every IP packet. On its way from the source to the destination the IP packet may travel through multiple network nodes called routers. Each router is usually connected to several IP networks. Two interfaces of two directly connected routers are also considered as separate IP network. Each packet is routed from one router to another until it reaches the destination router which has the destination IP network attached to one of its interfaces.

Table 2.1   Types of IPv6 addresses

Address type	Binary prefix	IPv6 notation
Unspecified	00 ... 00 (128 zeros)	::/128
Loopback	0.0 ... 01 (128 bits)	::1/128
Multicast	11111111 (8 ones)	FF00::/8
Link-local unicast	1111111010 (10 bits)	FE80::/10
Global unicast (includes all anycast)	All other IPv6 addresses	–

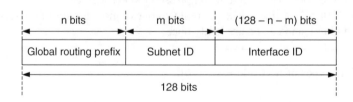

Figure 2.13   IPv6 global unicast address format

On the other side, when the network parts of the source and destination IP addresses of a given IP packet are the same, the datagram is sent directly to the destination interface without having to pass any router (because in such case both hosts are located in the same IP network).

Definiton 2.1: Definition of Routing

Routing in a packet-switched network (e.g., Internet) is the process of selecting the route (path) through which the packet will be sent. The router is a network node that makes the choice of path for the packet.

Routing is performed using routing tables (table-driven routing), which have stored information of the routes to different destinations in the Internet. However, the traditional method of routing in Internet is hop-by-hop routing (i.e., next-hop routing), in which each router routes the packet to the next router (it requires only partial routing information and reduces the size of the routing tables) and so on until reaching the destination IP network. The routing tables may be static (with manual entries) or dynamic (automatically updates when there is a change of a route, by means of routing protocols).

Each routing table usually has:

- Default routes;
- Routes specific to a particular network or host.

A set of rules that specify how the information in routing tables is used and updated via information exchange among network nodes (routers) is called a routing algorithm. The general model of routing datagrams is given in Figure 2.14.

Route-Datagram (Datagram, Routing Table)

Extract destination IP address D from the datagram and calculate network prefix N;

if {N equals network prefix of an IP network directly connected to an interface of the router}

        then {send the datagram to destination D through that network interface}

else if {routing table contains host-specific route for D}

        then {send the datagram to next hop specified for D in the routing table}

else if {routing table contains a route for network N}

        then {send the datagram to the next hop specified for N in the routing table}

else if {routing table contains a default route/gateway}

        then {send the datagram through the default route given in the routing table}

else {send a "routing error" message};

Figure 2.14   General routing algorithm

Routing tables usually contain information only for the destination network addresses, but not the destination host addresses, as the number of networks is many times smaller than the number of hosts on the Internet. Otherwise, routing tables would be too large. However, it is possible to set up routes for IP address of a host in some cases (e.g., network administration, security reasons, etc.).

If given host has multiple network interfaces (e.g., WiFi interface, Ethernet interface, etc.), then generally it should have different IP address for each network interface.

2.4.1   Routing Algorithms

The purpose of propagation mechanisms for distribution of routing information in a given network is not only to find a set of routes, but continually to update the routing information. People cannot respond to such changes fast enough, so such task in larger IP networks is accomplished with routing algorithms and protocols. To summarize, changes in routing tables may be performed in two ways:

- *Manual*: With certain commands through the user interface of the router the network administrator manually defines the routes.
- *Automatic*: Routers exchange information with each other at certain time intervals or after certain changes with the help of so-called routing protocols that are based on certain routing algorithms.

To be able to understand the essence of the routing algorithms a graph abstraction of a network (which performs routing) is shown in Figure 2.15. Namely, each of the outgoing links from a given router to other routers (with which it is connected) has a cost (or metric) associated

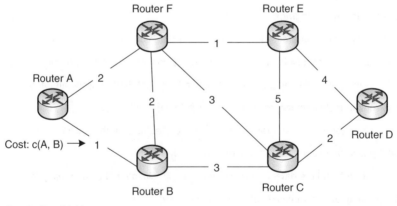

Graph: G = (N, L)

N = set of router = {A, B, C, D, E, F}

L = set of links = {(A, B), (A, F), (B, C), (B, F), (C, D), (C, E), (C, F), (D, E), (E, F)}

Figure 2.15   Graph abstraction

to it. Generally, different links (i.e., hops) may have different costs. The default cost of a link is 1. Of course, the cost may be influenced by other parameters such as traffic load on the link (link with higher traffic load will have a higher cost and vice versa) or link delay (for example, if there is a satellite link between two routers, then the cost of that link may have a higher cost than terrestrial links, since satellite links have bigger delay budget due to the longer distance traveled by the packets).

The path from a given source to a destination is composed of a sequence of links (with the exception of the case when both hosts belong to the same network, because in that case routing is not needed). The cost of the path from a given source to a given destination is the sum of the costs of all links (in the path), that is:

$$\text{Path cost}(x_1, x_2, x_3, \dots, x_p) = c(x_1, x_2) + c(x_2, x_3) + \dots + c(x_{p-1}, x_p) \qquad (2.4)$$

where $c\ (x_i,\ x_j)$ is the cost of the link between routers i and j. The purpose of the routing algorithm is to find the path with the lowest cost for a given pair of source address–destination address.

Definiton 2.2: Routing Algorithm

Routing algorithm is an algorithm that finds the path with the lowest cost.

In general, routing algorithms can be divided into the following groups:

- Distance vector;
- Link state;
- Hierarchical routing.

2.4.1.1   Distance Vector

In distance vector routing algorithms each router informs all its neighboring routers for topology changes in the network. A given router initializes its routing table that contains one entry per destination network. Each entry in the table identifies the direction to the destination network (to which next router the packet should be forwarded) and gives the distance to that network, usually measured in hops (where a hop is a link between two logically adjacent routers). Periodically each router sends a copy of its routing table (or part of it) to every neighboring router (connected on the other end of one of its outgoing links). The term distance vector comes from information sent in periodic updates to other routers. Each update message contains a list of pairs (V, D), where V denotes the destination (called a vector) and D is the distance to the destination. A typical representative of the distance-vector routing algorithms is the Bellman-Ford algorithm. According to this algorithm packets are routed by the route with the lowest cost, where the lowest cost path between two routers x and y is calculated as follows:

$$D_x(y) = \min\{c(x, v) + D_v(y)\} \qquad (2.5)$$

where $D_x(y)$ is the least-cost path from router x to y, $c(x, v)$ is the path cost from router x to router v, and $D_v(y)$ is the least-cost path from the router v to router y.

The distance vector algorithm has some disadvantages. One of them is the slower convergence of the routing information in all routers in the network, due to the propagation of routes changes on a hop-by-hop basis (which may lead to inaccurate routing information in some routers for a certain time).

2.4.1.2   Link-State

The link-state routing algorithm requires each router to have complete topological information of the network. The easiest way to understand the topological information is to imagine that each router has a map that shows all other routers and links that connect them.

A link-state algorithm performs two tasks. First, each router actively tests the status of all neighboring routers. In terms of graph abstraction of the network, two routers are adjacent if they share a link. Also, each router periodically propagates information about the status of its links to all other routers. Second, each router independently calculates the shortest path from itself to all other nodes in the network.

The most known example of a link-status routing algorithms is Dijkstra algorithm. Calculations with the Dijkstra algorithm for change of the status of the router's links result in corresponding changes in the path costs using the following formula:

$$D(v) = \min\{D(v), D(w) + c(w, v)\} \qquad (2.6)$$

where $D(v)$ is the current cost of the path from the router to the destination v, $D(w)$ is the current cost of the path from the router to the destination w, and $c(w, v)$ is the current cost of the path from w to v.

The link-status routing algorithm has its drawbacks. Namely, it may give the wrong cost for a given link. Also, rooting loops may appear. For example, routing loop appears when each of two neighboring routers thinks that the other is the best direction (i.e., best path) to a given destination.

2.4.1.3   Hierarchical Routing

Internet network is built on the principle of a network of many networks that are interconnected through routers. Certainly, there must be routers between different IP networks. However, several smaller or larger networks could be administered by a company, such as ISP. In such case we actually have independently operated networks. For example, we have a network of an ISP (which is composed of multiple IP networks), we have a network of mobile operator providing Internet service (e.g., 2.5G, 3G, or 4G mobile networks), and so on. Each of these network of networks (where a network is identified by a unique routing prefix or network ID), actually forms Autonomous System (AS). So, AS is a collection of routers under the control of one authority (e.g., a company, an operator, etc.). These ASs are connected with each other through the so-called gateway routers to Internet (i.e., to other ASs).

Each AS is uniquely identified by AS number, which are delegated by the IANA as a global authority. AS numbers were initially defined as 16-bits numbers (until 2007), which allowed existence of maximum 65 535 autonomous systems. However, due to the exponential increase of the Internet in number of hosts and networks, IANA introduced 32-bits AS numbers [17], which are being allocated nowadays.

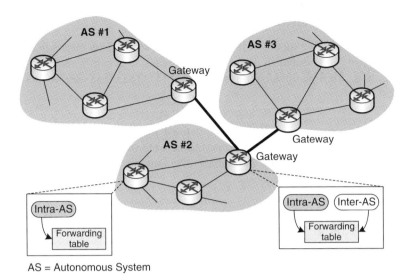

AS = Autonomous System

Figure 2.16 Hierarchical routing within and between autonomous systems (AS)

This way, with ASs as basic architectural approach in the present Internet we actually have two-tiers routing, which is referred here as hierarchical routing. Namely, it is a concept, not a specific algorithm. Thus, it is possible different algorithms to be used in each of the two tiers in the hierarchy (Figure 2.16):

- *Intra-AS routing*: This is routing between routers belonging to a single AS. Currently, the most known intra-AS routing protocols are Routing Information Protocol (RIP) [18], which is a distance-vector routing protocol, and Open Shortest Path First (OSPF) [19], which is a link-state routing protocol.
- *Inter-AS routing*: This is routing between routers belonging to different ASs. Currently, globally accepted protocol for inter-AS routing is BGP-4 (Border Gateway Protocol version 4) [20], which is neither distance-vector nor link-state, but it is referred to as a path-vector routing protocol because its routing decisions are based on several parameters, such as path, routing policies and rules.

Hierarchical routing is today practiced in the Internet inevitably. In addition, owners of ASs are left to themselves independently regulate the routing within its AS. ASs are usually owned by ISPs, companies, or institutions.

Gateway routers (which are mainly edge routers in a given AS) usually have a higher processing power and have links with higher bandwidth (in bits per second) to other ASs, because all incoming and outgoing traffic for a given AS must pass through the gateway routers. Therefore gateway routers (or shortly, gateways) must implement routing protocols that are used for intra-AS routing within the AS (to exchange routing information with other routers in the AS), as well as routing protocols for inter-AS routing (to exchange routing information with other gateways).

2.5   Client–Server Networking

Commonly used networking model in Internet is client–server. This means that some of the machines (i.e., hosts) connected to the network are ready to accept a request from another machine (i.e., host) for a particular service (Figure 2.17). The client–server model is asymmetric, as follows:

- The server provides services through well-defined interfaces (listens for requests from clients with open sockets on predefined or well-known ports).
- The client requests certain service through the given interface on the server.
- The server responds to client requests.

Generally, the communication between the client and server can use TCP, UDP, or other protocol, but both sides need to use the same type of protocol and appropriate socket interfaces ("stream" socket for the TCP and "datagram" socket for the UDP).

In the case of a connectionless-oriented communication (based on UDP/IP) there is not required explicit identification of who the server is. When sending datagram via UDP/IP the sending application needs to specify its own IP address and port number on the local machine (i.e., to open a "datagram" socket) through which the datagram will be sent. When a machine expects incoming datagrams (via UDP/IP), the receiving application must declare IP address and port on the local machine (i.e., to open a "datagram" socket) through which it expects to receive datagrams from other machines (i.e., hosts).

In the case of connection-oriented communication (TCP) the approach is different:

- The client must connect to the server before receiving or sending data from/to it.
- The server listens on a specific port and IP address (on a given interface), and must accept the communication with the client before sending or receiving data.
- The server can accept client when it receives a request to connect by the client.

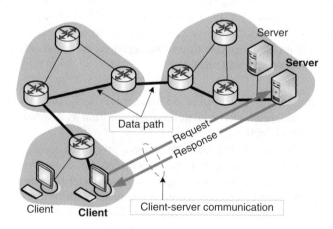

Figure 2.17   Client–server model

Well-known Internet services (i.e., applications) which use client–server communications are e-mail, FTP, Hypertext Transfer Protocol (HTTP), and so on. Also, fundamental Internet technologies, such as DHCP and DNS, are based on the client–server model.

2.6 Peer-to-Peer Networking

Peer-to-peer (shortly written as P2P) networks are self-organizing networks, generally without a centralized server. Due to such approach, P2P networks have high scalability and robustness because they do not relay on a single network host such as server in client–server network architectures. As shown in Figure 2.18, in P2P networks each participant who is connected to the network is a node with equal access to network resources and to all other users.

The owner of each node (e.g., computer) on a P2P network is supposed to set up certain resources (e.g., processing power, access data rate to Internet, memory on the hard disk, etc.) which are shared with other nodes in the P2P network. In such way P2P network is a distributed application architecture that partitions certain tasks among several peers. Hence, P2P networking is based on establishing a temporarily logical architecture of peers (nodes in the P2P networks), as an overlay network in the Internet, where peers act as clients or servers to other nodes in the P2P network allowing shared access to different resources such as files, streams (e.g., video streams), devices (e.g., sensors), and so on.

Regarding the network architectures in general, one may distinguish among two types of P2P networks:

- *Structured P2P network*: In such networks there are defined algorithms, policies, and specific network topologies. Such P2P networks are used for certain large-scale services implementation due to their higher scalability.

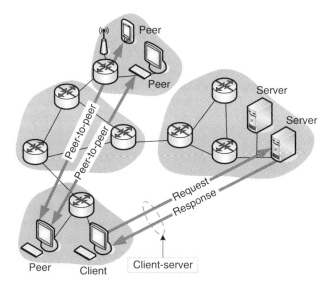

Figure 2.18 Comparison of peer-to-peer and client–server network architectures

- *Unstructured P2P networks*: This P2P networks do not define specific structure of the over-lay P2P network. However, in most of the cases there is a need for existence of centralized node for indexing or bootstrap functions of the peers. Hence, one may distinguish among several types of unstructured P2P networks:
  - *Pure P2P*: In this case all peers are equal and there are no centralized nodes. In this case connection of the peers can be done only by using IP addresses as network locators of the peers. Therefore, each peer should have knowledge of IP addresses of other peers prior to communication.
  - *Centralized P2P*: This is the case when there are centralized nodes in the network for indexing (of peers) and bootstrap functions (whom to contact when P2P application is started on a given peer host).
  - *Hybrid P2P*: In this case there are dedicated infrastructure nodes in the P2P network with defined functionalities (besides standard peers which use the P2P network). Such nodes are usually called supernodes.

There are advantages and disadvantages of P2P networking. From the side of advantages P2P networks provide a higher level of sharing resources, distributing the traffic, and processing the load among many peers. On the other side, one disadvantage of the P2P networks is a lower possibility for controlling the type of content that is shared among peers as well as its availability, which is contrary to centralized systems with a server where administrator can filter and control the files that are offered through the server. Also, due to the lack of a central authority in most P2P networks it is hard to stop file sharing regardless of the contents.

2.7   Best-Effort Internet Services

Internet is created as a best-effort network, which was initially designed for client–server communication. There are several services which are born in Internet and hence shaped its development afterwards and even success of the Internet as a packet-switched network. The most important client–server services are e-mail, FTP, and WWW. Many refer to the WWW as the main driver of Internet to its global success as a single networking platform for all telecommunication services in the first decades of the twenty-first century.

In the 2000s P2P services appeared which nowadays account for approximately half of the total Internet best-effort traffic. The other half of the best-effort Internet traffic mainly is coming from WWW as the most important best-effort service in Internet globally.

2.7.1   Electronic Mail

E-mail is one of the most known Internet services as well as the most used one. In 1982 ARPANET published proposals for e-mail (with newer versions in 2001), which defined the Simple Mail Transfer Protocol (SMTP) [21] and the Internet message format [22].

The e-mail architecture consists of two main standardized components: the User Agent (UA) and Message Transfer Agent (MTA). The standardized MTA is SMTP, which is a client–server application protocol. The network architecture for e-mail is shown in Figure 2.19. SMTP is used for sending e-mail messages from UA (on the user side) to a mail server, and sending e-mail messages from one mail server to another mail server.

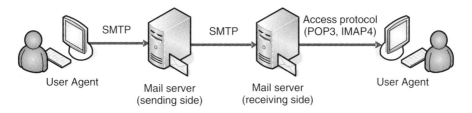

Figure 2.19 Network architecture for e-mail

Each e-mail message consists of two main parts: a header and a body. The header is separated from the message body by null line called CRLF (Carriage-Return/Line-Feed). SMTP uses TCP/IP protocol stack and well-known port 25.

The default e-mail format defined in SMTP is 7-bit ASCII text [21]. To be able to add attachments to an e-mail message the Multipurpose Internet Mail Extensions (MIME) were defined [23]. The MIME does not replace SMTP, but allows arbitrary data (text, image, video, audio, application files, message, or multipart combination) to be encoded in ASCII form and be transmitted as a standard email message.

For access to e-mail messages from UA there are two defined mail access protocols: Post Office Protocol-version 3 (POP3) [24] and Internet Message Access Protocol-version 4 (IMAP4) [25]. There is also web-based access to e-mail which is popularized via free e-mail accounts for the users (Figure 2.20). In such cases the protocol which is used for access to e-mail is HTTP.

2.7.2 File Transfer Protocol

FTP is an application for copying files from a one machine (i.e., host) to another. It is based on TCP/IP protocol stack and it using client–server communication. FTP needs two connections [26]: a data connection and a control connection, as shown in Figure 2.21.

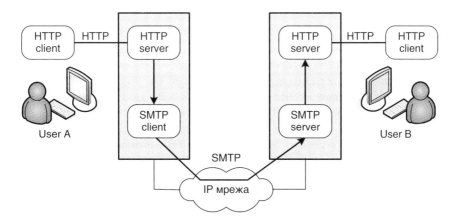

Figure 2.20 Web-based e-mail

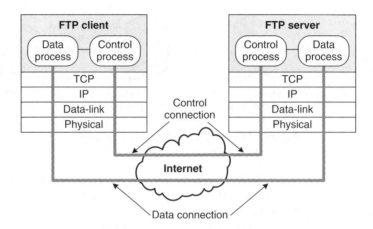

Figure 2.21   FTP control and data connections

   The control connection is always established first by using defined commands. By default, the FTP server listens on port 21 for incoming control connections from FTP clients. After establishing a control connection with the server, which includes authentication as well as definition of file type, data structure, and transmission mode, the FTP client establishes a new connection for each data transfer to be accomplished. Most implementations create a new pair of processes for new data transfer as well as establish a new TCP connection when the server needs to send new data to the client (i.e., a new file). If control connection is terminated then it means that the FTP session is terminated, which results in termination of all processes connected with FTP data transfer.

   FTP was crucial protocol for file transfer in 1980s and to some extent in 1990s. However, after the invention of the WWW, FTP usage in Internet is significantly decreased because client–server file transfer is performed by HTTP.

2.7.3   World Wide Web

WWW is a global system whose development started in 1989 by Tim Berners Lee in CERN, Switzerland. He created a protocol for the distribution of documents, where a text document could be linked to other documents or objects (e.g., images, videos, audio objects, etc.), located on the same or on another server connected to the Internet. The Web (i.e., WWW) is made up of a large number of documents called web pages. Each web page is a hypermedia document (hyper – because it contains links to other web sites; media – because it can contain other objects besides text).

   The main protocol for access to web pages is HTTP, which is based on client–server TCP/IP model. Web servers (i.e., HTTP servers) listen on well-known port 80. The web application on the user side is called browser (e.g., Internet Explorer, Chrome, Firefox, Opera, etc.), which has a HTTP client. Web servers (i.e., HTTP servers) offer web pages to web clients. The main markup language for creating web pages is Hypertext Markup Language (HTML).

HTTP is used in two different modes:

- *Non-persistent mode*: New HTTP connection is created for each object transfer (HTTP 1.0 [27]).
- *Persistent mode*: One HTTP connection is used for transfer of multiple objects between a client and a server (HTTP 1.1 [28]).

HTTP works on request-response principle. A client sends a request, and obtains a response from a server. Each response consists of three digit response code followed by blank space and readable description of the response code (e.g., "200 OK," "404 Not Found," etc.).

HTTP is a stateless protocol, does not remember a state of a connection. To be able to track transactions HTTP uses cookies. A cookie is a piece of data sent from a web server to a web client, and stored in user's client (i.e., browser). Later the server can retrieve the cookie to track the user previous activity.

HTTP also uses caching for higher efficiency (e.g., for reducing web traffic intensity on certain links or networks). Caching is implemented either locally in user's machine or on the network side via proxy servers.

2.7.4   Peer-to-Peer Services

The first famous P2P service was Napster in 1999. Then it was followed by other file-sharing systems Gnutella, Kazaa, eDonkey, and BitTorrent as the most successful one. Furthermore, Skype is also very important representative of P2P systems and networks, providing best-effort P2P voice communications with options for added video communication, chat and files exchange. Hence, in plenty of P2P networks one may distinguish Napster, Skype, and BitTorrent as the most important regarding their impact worldwide.

2.7.4.1   Napster

Napster is based on a centralized P2P network architecture, where a centralized directory server stores information about peers that have certain music files. Then, upon a request-query from a peer for a certain song (from certain artist), the centralized directory server would send a response with information about hosts that have music files matching the query. Due to copyright infringements in such approach Napster was shutdown later.

Overall, Napster has started the P2P "boom" in the Internet, which results today in significant participation of P2P service in the global Internet traffic volume (measured in bytes, packets, and connections).

2.7.4.2   Skype

Skype is another successful example of P2P networks and services. However, it is also a unique P2P service, because it targeted markets which are traditionally held by the telecom operators, such as market of voice communication (i.e., the telephony). From its first release in 2003 it has spread globally and decade after its invention it is extensively used for voice especially for international calls and for roaming replacement in mobile environments (e.g., in WiFi hotspots). However, Skype is completely different than traditional telephony. There are two main differences:

- Skype is a proprietary application (as most of the P2P applications nowadays), it is not an open standard, which is completely different than most of the technologies used in the ICT (Information and Communication Technology) world, including the Internet technologies. All technologies are standardized (e.g., by ITU, IETF, IEEE, 3GPP, etc.) and exist as open standards, so nobody get royalties for implementation of certain standard by certain vendors. On the other side, standardization is very important to provide equipment from different vendors to work in different networks and scenarios.
- Skype is offered only as best-effort service (as all other P2P services which are used nowadays), which means that there is no QoS guarantees for voice calls, which is completely different than the telephony provided by telecom operators which provide strict QoS guarantees (especially regarding the end-to-end delay and bandwidth guarantees).

Skype belongs to hybrid P2P and the client–server system. Its architecture is very similar to SIP (Session Initiation Protocol) architecture [29], which contains registrar servers and presence servers, which bind the IP address of the user with its Skype username and provide availability of the user to other users (according to certain filter criteria), and vice versa. Such nodes are called supernodes, which communicate with other supernodes as well as with ordinary peers (i.e., end users with installed Skype on their devices), as shown in Figure 2.22.

Skype nowadays takes significant part of the market of international voice calls (where choice between the price and the quality becomes more important), thus it is having global impact on the telecommunication world in which the telephony is one of the most important services.

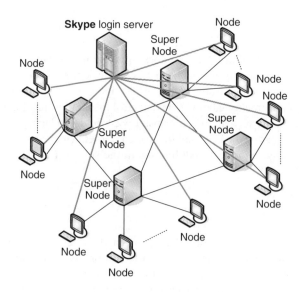

Figure 2.22   Skype architecture

2.7.4.3 BitTorrent

From many P2P services for file-sharing, regarding the traffic volume and the popularity world-wide, the most important one is BitTorrent. One of the reasons for its success lies in the network topology and in the design which provides robustness of the service [30]. With BitTorrent many users which download a given file also upload pieces (called chunks) of that file to other users downloading the same file. Each downloader reports the files it has to all of its download peers. The integrity of the chunks (i.e., file pieces) is verified by so-called torrent files. A torrent file is a so-called hash of all file chunks. A hash is an algorithm which generates a fixed length string (called "message digest" or only "digest") from a given input data. The hash function is not reversible, so a given message digest can be compared only with other message digest for verification. BitTorrent includes trackers (working on the top of HTTP servers) which are designed to help downloading peers to find each other. In summary, BitTorrent network architecture consists of three types of nodes: peers, torrent trackers, and web servers for hosting torrent files.

BitTorrent results in cost and performance redistribution for download of a given file. However, for download of a given file with BitTorrent the total upload bit rate of all peers in the network (the Internet) must be equal to total download bit rate of all peers. So, each peer must provide certain bandwidth in the uplink when it is downloading certain file in the downlink (regardless of the access network to Internet).

2.8 Internet Governance

The Internet was developed from a DARPA (Defense Advanced Research Projects Agency) project in the United States in the 1960s and 1970s, but since the 1990s it has left its initial boundaries. Officially, there is no single person or single organization or government that runs the Internet. However, during the past decades most influential in governance of the Internet were several organizations with locations in the United States, something that one may expect looking to the history of the network.

However, there are several key technical aspects that must be governed for the Internet to run. Such technical aspects are the core architecture of the Internet (organized as flat architecture of interconnected ASs), IP address space, and name space (used by DNS). In 1998 ICANN was created as a non-profit United States-based organization. ICANN has taken over most of the technical governance of the Internet, with focus of the governance of the two name spaces: IP addresses, including IPv4 and IPv6, and TLD names such as .com, .edu, .org, and so on. Assignment of IP addresses and port numbers ICANN realizes though its organization IANA. Also, IANA distributes the AS numbers, which are 16-bit or 32-bit numbers used to uniquely identify the ASs.

There are several other organizations that have important roles in developing Internet standards and policies. They are:

- *Internet Society (ISOC)*: This is international non-profit organization founded in 1992 which formally includes IETF and Internet Research Task Force (IRTF). The ISOC is overseen by the IAB, a committee initially created by the United States DARPA project. The IETF is the major voluntary organization which creates Internet standards via its working groups.

- *World Wide Web Consortium (W3C)*: Due to the importance of the WWW to the Internet, W3C was created for web development to its full potential.
- *ITU (International Telecommunication Union)*: The global harmonization and standardization in telecommunications (i.e., ICT) is done by ITU as a largest telecommunication agency in the world, part of the United Nations. Its role becomes even more significant since the whole telecommunication world, including networks and services, is converging toward the Internet.

Regarding the regulation of the contents offered through the Internet, there are different opinions and approaches. Currently, Internet is environment with an open access to all types of contents, which is further regulated by national (local) laws regarding the content publishing (e.g., on web sites) and usage (by the end users). Hence, some contents may not be accessible in some countries, due to national legislations or due to business approaches of the content providers.

However, one may also expect Internet governance to evolve further toward the future networks which are based on the Internet technologies.

References

1. Postel, J. (1981) Internet Protocol. RFC 791, September 1981.
2. Postel, J. (1981) Internet Control Message Protocol. RFC 792, September 1981.
3. Deering, S. and Hinden, R. (1998) Internet Protocol, Version 6 (IPv6) Specification. RFC 2460, December 1998.
4. Srisuresh, P. and Egevang, K. (2001) Traditional IP Network Address Translator (Traditional NAT). RFC 3022, January 2001.
5. Postel, J. (1980) User Datagram Protocol. RFC 768, August 1980.
6. Postel, J. (1981) Transmission Control Protocol. RFC 793, September 1981.
7. D.E. Comer, (2000) *Internetworking with TCP/IP – Principles, Protocols, and Architectures*, 4th edn, Prentice Hall.
8. Allman, A., Paxson, V., and Stevens, W. (1999) TCP Congestion Control. RFC 2581, April 1999.
9. Mathis, M., Mahdavi, J., Floyd, S., and Romanow, A. (1996) TCP Selective Acknowledgements Options. RFC 2018, October 1996.
10. Stewart, R. (2007) Stream Control Transmission Protocol. RFC 4960, September 2007.
11. Rekhter, Y., Moskowitz, B., Karrenberg, D. *et al.* (1996) Address Allocation for Private Internets. RFC 1918, February 1996.
12. Droms, R. (1997) Dynamic Host Configuration Protocol. RFC 2131, March 1997.
13. Mockapetris, P. (1987) Domain Names – Concepts and Facilities. RFC 1034, November 1987.
14. Mockapetris, P. (1987) Domain Names – Implementation and Specification. RFC 1035, November 1987.
15. Bradner, S., Conroy, L., and Fujiwara, K. (2011) The E.164 to Uniform Resource Identifiers (URI) Dynamic Delegation Discovery System (DDDS) Application (ENUM). RFC 6116, March 2011.
16. Hinden, R. and Deering, S. (2006) IPv6 Addressing Architecture. RFC 4291, February 2006.
17. Vohra, Q. and Chen, E.(2012) BGP Support for Four-Octet Autonomous System (AS) Number Space. RFC 6793, December 2012.
18. Malkin, G. (1998) RIP Version 2. RFC 2453, November 1998.
19. Moy, J. (1998) OSPF Version 2. RFC 2328, April 1998.
20. Rekhter, Y., Li, T., and Hares, S. (2006) A Border Gateway Protocol 4 (BGP-4). RFC 4271, January 2006.

21. Postel, J. (1982) Simple Mail Transfer Protocol. RFC 821, August 1982.
22. Crocker, D. (1982) Standard for the Format of ARPA Internet Text Messages. RFC 822, August 1982.
23. Freed, N. and Borenstein, N. (1996) Multipurpose Internet Mail Extensions (MIME) Part One: Format of Internet Message Bodies. RFC 2045, November 1996.
24. Myers, J. and Rose, M. (1996) Post Office Protocol – Version 3. RFC 1939, May 1996.
25. Crispin, M. (1994) Internet Message Access Protocol – Version 4. RFC 1730, December 1994.
26. Postel, J. and Reynolds, J. (1985) File Transfer Protocol (FTP). RFC 959, October 1985.
27. Berners-Lee, T., Fielding, R., and Frystyk, H. (1996) Hypertext Transfer Protocol – HTTP/1.0. RFC 1945, May 1996.
28. Fielding, R., Gettys, J., Mogul, J. *et al.* (1999) Hypertext Transfer Protocol – HTTP/1.1. RFC 2616, June 1999.
29. Singh, K. and Schulzrinn, H. (2004) *Peer-to-Peer Internet Telephony Using SIP*, Columbia University, New York.
30. Cohen, B. (2003) Incentives build robustness in BitTorrent. Workshop on Economics of Peer-to-Peer Systems, June 2003.

3

NGN Standards and Architectures

3.1 Main Drivers to Next Generation Networks

Traditional telecommunications, such as PSTN (public switched telephone network) and PLMN (public land mobile network), and Internet technologies, were developing in parallel during several last decades of the twentieth century. However, the choice of the Internet as a single packet-switching technology for the ICT (Information and Communication Technology) world has raised the need for standardized integration of different architectures, concepts, approaches, and services found in both types of telecommunications, traditional circuit-switched telephony and TV broadcast on one side, and best-effort Internet on the other side. Such efforts have led to start of standardization process of so-called Next Generation Networks (NGN) in the first decade of twenty-first century, standardized within the ITU (International Telecommunication Union) as the largest ICT organization in the world which provides world-wide harmonization in telecommunications.

According to the ITU's definition of NGN [1], an NGN is a packet-based network able to provide telecommunication services to users and able to make use of multiple broadband, QoS-enabled (quality of service) transport technologies and in which service-related functions are independent of the underlying transport-related technologies. It enables unfettered access for users to networks and to competing service providers and services of their choice. It supports generalized mobility which allows consistent and ubiquitous provision of services to users. Given as such, the definition of NGN is limitless and can span different networks, current and future ones, as well as different services.

Which are main drivers which led toward the NGN definition and standardization? If we go back to the beginning, the initial ITU-T (International Telecommunication Union-Telecommunications) recommendation on NGN [1], specified key factor that influenced NGN development such as competition among operators due to telecommunication markets deregulation, exponential increase in digital traffic due to the usage of Internet which increased demand and possibilities for multimedia services, increasing demand for mobility and penetration of mobile networks worldwide (in parallel to Internet growth), convergence of networks and services, and so on. Overall, the development of telecommunications started to go in information-centric networking direction. That inspired ideas for General Information Infrastructure (GII), which provided foundation of the NGN.

NGN Architectures, Protocols and Services, First Edition. Toni Janevski.
© 2014 John Wiley & Sons, Ltd. Published 2014 by John Wiley & Sons, Ltd.

The telecommunications (i.e., ICT) markets are changing rapidly from the beginning of the twenty-first century. There are two main drivers and many more or less significant ones. The most important drivers are:

- Development of mobile networks and exponential increase of the number of mobile users globally, so one may say that nowadays almost every human on the Earth has a mobile device (at least one);
- Development of the Internet as the only "survived" packet-switching technology worldwide and exponential increase of the number of Internet users, so one may expect that every human on Earth has or will have an access to the global Internet.

Traditional telecom operators, which were based on fixed services such as telephony via PSTN, have found their business under pressure with the development of the mobile networks and the Internet. Two different commercial models were confronted on the global scale: the telecom commercial model on one side, and the Internet commercial model on the other side.

The traditional telecom commercial model is based on traditional subscriptions of users to services (e.g., telephony, television, and data) and innovation based on international agreements. The main drivers of telecom model were telecom companies and telecom vendors, so individuals or groups could not have a direct impact on innovation of certain new technologies. On the other side, the Internet commercial model is more diverse from the beginning and is not based only to user subscriptions to certain service (e.g., a Web portal). Instead, Internet model uses different business models and generates revenues from different sources (e.g., marketing on the Web). But, that Internet model is not so different than TV commercial model in which TV programs are offered free to users with commercials included. So, the only solution of merging the two worlds, the traditional telecom world and the Internet world, is the convergence to a single networking platform for all heterogeneous services (existing and future ones) which is globally accepted to be IP-based (Internet Protocol) platform. Its standardized form is in fact the NGN.

As a summary, particular drivers for NGN (Figure 3.1) can be divided into several groups (but not limited only to them), such as: development of broadband Internet access including fixed broadband and mobile broadband, convergence of ICT markets to Internet, technological convergence to IP-based networks and services, requirements for end-to-end QoS provisioning in the Internet, as well as transition of PSTN, PLMN, and TV broadcast networks to the Internet environment.

3.1.1   Fixed Broadband Internet Access

Residential Internet access networks in 1990s were based on dial-up modem connections via PSTN for residential users and Ethernet based access networks connected by using leased lines [based on dedicated bit rates from the SDH (Synchronous Digital Hierarchy) and PDH (Plesiochronous Digital Hierarchy) transport networks] for enterprises. However, in the beginning of the twenty-first century started the deployment of broadband Internet access which is based on three types of transmission media: copper, fiber, and wireless. Here fixed broadband access refers to copper and fiber. Copper systems were based on twisted-pairs and copper cables. Twisted pairs were used for decades in the access networks of the PSTN, and they evolved to Digital Subscriber Lines (DSLs) by adding additional equipment at the user premises and

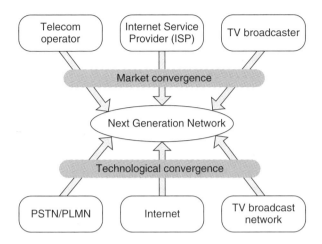

Figure 3.1 NGN drivers

on the network's side. Cable networks were initially designed for cable TV distribution, but later they were upgraded with Internet access technologies (e.g., Packet Cable). The fixed broadband access is covered in Chapter 4 in more details.

Why is broadband Internet access a driver to NGN? It is due to two main reasons: (i) the Internet itself as an adopted networking platform for services worldwide and (ii) broadband Internet access (with higher bit rates, e.g., several tens of megabits per second) provides the possibility for all existing services to be provided via the Internet access, including television and other video services (e.g., video on demand, video streaming, etc.) as most demanding ones regarding the required bandwidths.

3.1.2 Mobile Broadband Internet Access

The expansion of mobile technologies is a very important driver because mobile communication is personal communication. Mobile broadband has been established with the 3G technologies [e.g., UMTS/HSPA (Universal Mobile Telecommunication System/High Speed Packet Access), Mobile WiMAX 1.0], and evolves further with rollout of mobile networks with already standardized 4G technologies (e.g., LTE-Advanced, Mobile WiMAX 2.0). The impact of mobile broadband is different in some aspects from the fixed broadband. Users carry mobile devices all the time, which is not the case with the devices used for fixed broadband Internet access (e.g., desktop computers). Hence, certain services can be offered to the users in different contexts using their current location (e.g., so-called location-based services). Further, users have more time for usage of the mobile devices. With broadband access to the Internet, mobile users with computationally capable mobile devices (e.g., smartphones) may use all services available in the Internet regardless of their location and the time. This provides commodity and puts strong emphasis on mobile broadband as a driver to NGN. Fixed-Mobile Convergence (FMC) is directly influenced by the broadband access in fixed and mobile networks. Due to its importance for the development of the NGN, mobile broadband is covered in Chapter 5.

Why is mobile broadband important for the NGN? The first two reasons are the same as for the fixed broadband. Additionally, mobile broadband adds generalized mobility and personalized services as important drivers toward NGN.

3.1.3   Convergence to IP-based Networks and Services

Convergence is a key concept in the development of broadband infrastructure. Broadband networks are becoming economically more acceptable if the same network can be used for a greater range of applications and thus reduce the required number of separate networks. It is not only challenging from a technical point but includes technical, market, and regulatory issues. However, the convergence involves several different dimensions:

- *Service convergence*: Same content is offered on different platforms.
- *Network convergence*: Different services are offered through a unique network (e.g., telephony, television, and data services).
- *Terminal convergence*: Multifunctional terminals support a variety of services (which may be provided by different networks), which is obtained by merging devices such as fixed and mobile phones, television, and personal computer.
- *Market convergence*: A company (e.g., an operator) offers its services in various sectors. For example, telecom operators offer TV broadcast (e.g., over the IP-based infrastructure), and cable networks offer Internet access and telephony (e.g., Voice over IP – VoIP).
- *Regulatory convergence*: A regulator applies equivalent set of rules in different sectors (e.g., for PSTN and for VoIP telephony networks).

These five dimensions include technical, economic, and regulatory aspects. Technological development is often seen as the driving force that enables convergence, but on the other hand technological convergence does not always lead to a convergence of markets and regulation.

The convergence is driven by the needs of network operators and service providers as well as end users needs. However, the prerequisite for the convergence is broadband access (including fixed and mobile broadband technologies) and unified networking platform (which is adopted to be the Internet on a global scale). The technological convergence in telecommunications (i.e., ICT) with related market and regulatory convergence is one of the key drivers for the NGN. But, it is correlated with other drivers such as broadband access. On the other side, the convergence of networks and services triggers the combination of the best approaches and solutions found in traditional telecom and Internet worlds, as well as development of new ones, toward unified approach which will fit all networks and all services.

3.1.4   End-User Drivers toward NGN

End users are traditionally grouped into two main groups: business (or enterprise) users and individual (or residential) users.

Business users are seeking simplified access to the Internet with integrated voice, video, and data services with high speeds (statically higher than bit rates provided to individual users) and flexibility to adapt certain services to the business needs (e.g., cloud computing). While in traditional telecom world the access for business users was provided (and it is still being provided in many countries) via leased lines (using digital hierarchies from PDH and SDH), in Internet it is provided via Virtual Private Networks (VPN) as a standardized solution. However, the available bit rates (either via SDH/PDH or via VPN, or combined) are only a "pipe" to the Internet. Business users have certain demands for the communication services regarding the QoS, security, availability, and so on, which are also targets in the NGN framework.

Individual users are also accommodated to use telephony, TV, and data services. As all these types of services converge via the broadband Internet access, users also look for attractive pricing bundles for combined services, such as triple play (voice, TV and Internet access) or quadruple play (voice, TV, Internet access, and mobility). However, such bundles can be provided via separated networks (e.g., telephony via PSTN, Internet access via DSL over the same twisted-pair line at user's premises). With broadband access to Internet the operators can provide such bundles for integrated services over the same IP-based network, which is the ultimate goal today. The convergence of different services offered to the individual users should be done in transparent manner regarding the end users. For example, in some cases a consumer will not be aware whether a voice call is established via the PSTN or by using QoS-enabled VoIP. One may give similar example regarding the TV and IPTV (Internet protocol television; i.e., TV distribution over IP networks).

For both general types of users (business and individual) there is increasing cross-border communication for business or personal purposes, including conversational services (e.g., telephony) as well as data services (e.g., cloud computing, etc.) which demand high availability, high performance (i.e., QoS support), and security. All such requirements by the end users are addressed in NGN.

3.1.5   Operator Drivers toward NGN

The main driver toward the NGN from the operator's point of the view is market convergence as well as growing competition in the liberated ICT markets from new services and innovations. For many decades in the twentieth century the telephony (via PSTN and PLMN) was the main source of revenues for telecom operators, either fixed or mobile. Cable operators for distribution of TV channels were different type of operators than telecoms. With the rapid development of the Internet infrastructure (consisted of interconnected routers organized in autonomous systems) in the 1990s and 2000s, the need for Internet Service Providers (ISPs) increased exponentially. Although ISPs in the beginning were treated as different types of operators than telecom operators and cable TV broadcasters, they were using the existing telecommunication infrastructure in transport networks (e.g., SDH and PDH) and in access networks (e.g., via dial-up modems for individual users and via leased-lines for business users). At the bottom end, for any Internet service (in the past and today) is needed a telecommunication transport infrastructure, however with different networks nodes (i.e., routers) and different user terminals (i.e., hosts, including client machines and server machines). Naturally, the telecom operators have become also ISPs, because they already had the infrastructure which could be reused for Internet access besides the telephony. The Internet development has speed up the telecommunication market liberalization worldwide, which provided basis for new operators to appear besides the so-called incumbent operator in each country. So, the telecommunications (i.e., ICT) markets have become more competitive. On the other side, also market convergence toward the Internet is very important driver for operators to make progress toward the NGN. All operators, including fixed and mobile ones, are providing the same voice, video, and data services over IP networks.

Another important driver for the operators, especially for the incumbent operators, are decreasing revenues from telephony and increasing revenues from broadband access to best-effort Internet services (e.g., e-mail, Web-based services, etc.). Such trend drives the operators to rethink their business and redesign their networks to a completely IP-based infrastructure. Such architecture is targeted to provide different types of services, including VoIP, video (e.g., IPTV, video on demand, etc.), and best-effort services, over a single network, thus reducing the costs for network deployment and operation. Hence, an operator does not need to run different networks for different services, but it has to maintain only a common IP-based core infrastructure. Implementation of an IP network costs less than PSTN, because the network is consisted mainly of simpler network nodes (routers), and more intelligent hosts (e.g., computers, mobile devices, etc.). Additionally, with unified all-IP network infrastructure the operators can increase the service offering and subscriber base over time. However, design and implementation of all-IP based infrastructure for fixed and mobile access networks and for all types of services with different performance and security requirements drives toward the NGN. One should highlight that NGN is mandatory for operators to provide real-time services over IP-based networks, such as telephony (i.e., VoIP) and TV (i.e., IPTV), with required signaling, QoS, and security support. However, since the same IP infrastructure is used for all Internet traffic, the common subscriber databases shall be used also for handling non-real-time services such as best-effort Internet services (e.g., AAA – Authentication, Authorization, and Accounting for such services).

Another driver to NGN for operators is strategic orientation of vendors for telecommunication equipment, which are focusing on IP-based network equipment and service platforms.

Finally, different operators as well as different vendors may choose to implement NGN solutions partly. In most of the scenarios NGN implementation in fact starts with transition of telephony to all-IP network with QoS support end-to-end and SIP (Session Initiation Protocol) used for signaling via standardized IP Multimedia Subsystem – IMS (covered in the following chapters).

The leading idea and main driver for NGN is separation between transport on one side, and services and applications on the other side (Figure 3.2), via means of interoperability of IP-based networks to support services by using the standardized NGN framework.

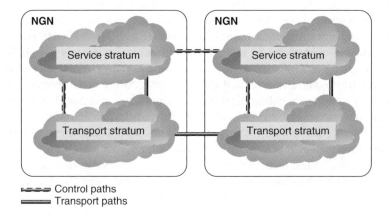

Figure 3.2   Service and transport stratums in next generation networks (NGN)

3.2 ITU-T NGN Standards

The realization of the idea for NGN has started in 2003 (Figure 3.3). In each standards body has been organized a standard group dedicated to NGN.

The ITU-T has started its work on NGN in 2003 by creating the so-called Joint Rapporteur Group on Next Generation Network (JRG-NGN), which targeted to study several main subjects such as: NGN requirements, the general reference model, NGN functional requirements and functional architecture, and evolution to NGN (from existing telecommunication networks at that time). This group delivered two fundamental NGN recommendations:

- General overview of NGN (Y.2001 [2]);
- General principles and general reference model for NGN (Y.2011 [3]).

The JRG-NGN was formed from experts of ITU-T Study Group 13 (SG13), which is targeted to future networks, including cloud computing, mobile, and NGN.

In 2004 the ITU-T Focus Group on Next Generation Network (FG-NGN) was established, which lasted for 1.5 years (until November 2005). Its main target was to accelerate the activities

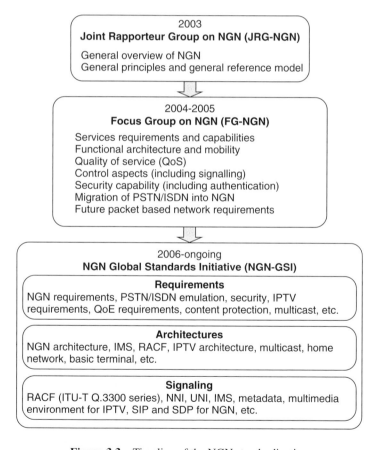

Figure 3.3 Timeline of the NGN standardization

Table 3.1   ITU-T NGN recommendations

ITU-T Y-series	Main NGN recommendations groups
Y.1900–Y.1999	IPTV over NGN
Y.2000–Y.2999	Next generation networks
Y.2000–Y.2099	Frameworks and functional architecture models
Y.2100–Y.2199	Quality of service and performance
Y.2200–Y.2249	Service aspects: service capabilities and service architecture
Y.2300–Y.2399	Numbering, naming, and addressing
Y.2400–Y.2499	Network management
Y.2500–Y.2599	Network control architectures and protocols
Y.2600–Y.2699	Packet-based networks
Y.2700–Y.2799	Security
Y.2800–Y.2899	Generalized mobility
Y.2900–Y.2999	Carrier grade open environment
Y.3000–Y.3499	Future networks
Y.3500–Y.3999	Cloud computing
Y supplements	Supplements to the Y-series recommendations

started by the JRG-NGN. In the period 2004–2005 several key aspects regarding the NGN were addressed by this group, such as: functional architecture and mobility, QoS, control and signaling aspects, service and security capabilities, the migration path from PSTN/ISDN networks to NGN, as well as future packet-based networks and their requirements. The work of FG-NGN was published in proceedings in 2005 [3].

From 2006 the work on NGN standardization within the ITU-T was transferred to the Next Generation Network Global Standards Initiative (NGN-GSI), which is working on detailed standards for NGN deployment by collaboration with different ITU-T Study Groups. The established objectives and goals of NGN-GSI were to further strengthen the leading role of ITU-T among other standardization bodies involved with NGN, to coordinate work between different ITU-T Study Groups working on various NGN recommendations, and to produce global standards based on the telecommunications market needs (which are continuously changing and often cannot be precisely predicted).

Since the start of the NGN standardization process in ITU many operators have started to migrate to all-IP networks, including the telephony migration to VoIP with similar QoS, security, and signaling support as in PSTN. The PSTN transition to all-IP environment was the most important driver for NGN from the start, thus NGN release 1 was mainly focused to VoIP. However, since the realization of NGN is an evolutionary process and different operators and administrations will start it at different times, there is a need to have different approaches for building a NGN or transition from PSTN and PLMN (which both have high investments, either is equipment or other resources such as frequency bands for mobile networks) toward the NGN.

The main recommendations for NGN belong to Y-series of ITU-T, as given in Table 3.1. However, there are also ITU-T recommendations in other series such as Q-series for signaling in NGN, which are necessary for real-time services VoIP (in NGN Release 1) and IPTV (in NGN Release 2).

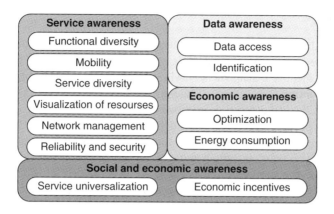

Figure 3.4 Future networks topics

There are several study topics of NGN within the ITU-T, which resulted into several recommendations groups under the NGN umbrella, including NGN architectures, end-to-end QoS, security, generalized mobility, network control architectures and protocols, service platforms, service capabilities, and service architectures, as well as interoperability of services and networks in NGN.

Toward the development of Future Networks, or in other words, continuing work on NGN, additional topics appear on the panels such as (Figure 3.4): service awareness including virtualization of resources, data awareness including identification, environmental awareness including energy consumption, as well as social and economic awareness including service universalization.

3.2.1 NGN Architectures

There are functional requirements for given types of networks (e.g., fixed, mobile) that should be implemented with aim to have NGN. All functional entities in NGN framework are built by using the Internet technologies. However, different NGN networks can have different physical network topologies and different access networks, including non-NGN architectures as well (since they will continue to exist in parallel to NGN architectures, at least during a certain time period).

3.2.2 End-to-End Quality of Service

The end-to-end QoS support in all-IP networks is essential for real-time services, such as VoIP and IPTV, which have strict requirements regarding the QoS (e.g., guaranteed bitrates, losses, delay, and delay variation – jitter). NGN provides standardized implementation of QoS in all-IP networks in the NGN framework (instead of proprietary case-by-case implementations), which is mandatory for the transition from PSTN and PLMN to all-IP networks. Additionally, NGN standardizes performance measurements and monitoring, because it is not only a goal to provide QoS solutions, but also to monitor their operation.

3.2.3   Security

Security is one of the crucial topics related to usage of the network infrastructure, services, and applications. However, security issues are related to several different NGN topics, such as QoS support, mobility, identity management (IdM), network management, authentication, authorization, and accounting (i.e., billing). There are security challenges of network and service providers on one side, and enterprises and users (i.e., consumers) on the other side. Efficient implementation of security in NGN requires defined security architecture. But, NGN is developing on the concept of open networks and service platforms, so there are no more strictly standardized interfaces between certain nodes in the network (as it was the case in traditional telecom networks). The services in NGN (as all-IP network) are provided via applications on both sides (client and server, i.e., customer and service platform) which use the Application Programming Interfaces (APIs) as common approach in the Internet. In such scenarios, the security architectures in NGN shall be built by using "components of the shelf" such as NGN components with standardized security protocols from the IETF (Internet Engineering Task Force).

3.2.4   Generalized Mobility

In the NGN term "generalized mobility" means possibility to use different access technologies at different locations while the user and/or the terminal is moving and at the same is using certain services (e.g., voice, video, data, multimedia, etc.). However, different access networks have different capabilities (e.g., WiFi and 3G/4G mobile networks significantly differ regarding the capabilities for QoS support). Additionally, there might be different Service Level Agreement (SLA) between a given user and different visited networks. However, generalized mobility in NGN provides possibility to access services regardless of the terminal equipment, while in transit or even while changing the access means (i.e., devices). So, NGN is targeted to provide capabilities for support of personal and/or terminal mobility.

   Overall, generalized mobility requires evolution of the network architectures by separation of transport stratum and service stratum and providing the same services over different wireless and mobile access networks (including horizontal mobility within a given access network, and vertical mobility between different access technologies) and fixed networks. This includes convergence of fixed and mobile telecommunications (i.e., FMC). The evolution of core networks toward the generalized mobility as well as FMC is something that is happening with the Fourth Generation of mobile networks (4G).

3.2.5   Network Control Architectures and Protocols

In the converged NGN world all services use the same network infrastructure, including access, core, and transit networks. In such case network control architectures are overlay architectures (over the established physical network infrastructure) with functions located in specified network nodes. Control architectures are targeted to resource allocation and QoS support in access and core networks, media processing, and transcoding (e.g., between different types of codecs used by the two ends of a given connection, or between different types of networks such as NGN and non-NGN), call/session control (e.g., for VoIP), and service control. Again, regarding the network control protocols the NGN framework is focused on using available "components of the shelf" (already standardized protocols), such as H.248 (from ITU-T) for media gateway control, or SIP (from IETF) for call/session control.

To have standardized approach for network control architectures, NGN standardizes different types of interfaces including user to network interface, network to network interface, as well as interfaces between network operators and service providers, which are possible to specify in NGN due to the decoupling of the transport and service stratums.

3.2.6   Service Capabilities and Service Architectures

NGN also addresses definition of service capabilities for different types of services, including real-time (e.g., VoIP, IPTV, etc.) and non-real-time (e.g., WWW, etc.) between human and machine, machine and human, and between machines (i.e., machine-to-machine), to maintain separation between the services and the underlying networks. To be able to support different business models between customers (i.e., users), network operators and service providers, as well as seamless communications for different service types, NGN includes standardization of service architectures with defined interfaces. However, besides the new approaches in NGN, it also gives attention to backward compatibility with existing services and systems or platforms. Also, service continuity over different networks is important to meet the requirements of the service providers and end users.

3.2.7   Interoperability of Services and Networks in NGN

Interoperability is very important because NGN assumes heterogeneous networks (access and core) and heterogeneous services (including heterogeneous user equipment). In IP-based networks (including NGN) interoperability is done on the networking layer, which is IP. Although initial assumptions in NGN were targeted to open approach regarding the networking protocols, currently there is convergence toward all-IP networks and NGN is the framework for such developments in telecommunications worldwide. But, there are two standardized versions of the IP, namely version 4 and version 6, which will have impact on the networks in NGN due to necessity of transition from IPv4 (i.e., IP) to IPv6. Such process has already started because IPv4 address space is exhausted by IANA (Internet Assigned Numbers Authority) and there are only unallocated parts are left in couple of the RIRs (Regional Internet Registries). However, interoperability between non-NGN networks and services (e.g., PSTN, PLMN, digital TV broadcast, etc.) and NGN networks and services, as well as service continuity [e.g., voice call transfer from PSTN to VoIP and vice versa, or from GSM (Global System for Mobile communication) to LTE (Long Term Evolution) and vice versa] is essential. However, such process involves plenty of protocols (standardized either by ITU or IETF), which are put into a certain context and architecture in NGN.

3.2.8   Future Networks

The evolution of NGN within the ITU-T is called Future Networks. The background for the need of Future Networks developments is the pace of development of new services, continuously increasing access bitrates and huge amount of network resources already built. Hence, fundamental change of telecommunications networks in short period is less likely to happen due to enormous amount of resources needed to build, operate, and maintain them. Also, architectures of networks are flexible enough to provide flexibility required by different types of services nowadays. It is provided by using the IP on the network layer in the telecommunications infrastructure, which hides all different underlying protocols (below the OSI-3 layer), and at the same time has high scalability (regarding the Internet as interconnected Autonomous Systems consisted of interconnected routers) with possibility to add QoS support and security where it is needed. However, it is not known how the networks can continue to fulfill service requirements in the future. Of course, it will depend upon the markets of such demanding services and applications and possibility to cover implementation and operation costs of eventually changed network infrastructure. In that direction are ongoing research efforts of different research groups toward network virtualization, content-based networking, energy-efficient networks, and so on. Hence, there are expectations that some future network architectures can be put into trial deployments in the period 2015–2020, with the possibility to become functional after 2020. ITU-T has denoted networks based on such recommendations as "Future Networks."

Four objectives are defined for the Future Networks [4]:

- *Service awareness*: The services are expected to increase exponentially in the future and such a trend has already started with the spreading of broadband Internet access. Future networks should provide services without significant increase in network deployment and operational costs.
- *Data awareness*: Future Networks are expected to carry huge amounts of data in access, core, and transit parts. However, users should be able to access different data (it includes all information transferred over the network, such as audio, video, Web, etc.) easily, quickly, accurately, safely, with the desired quality, regardless of the access network and their location.
- *Environmental awareness*: Future Networks are recommended to have lowest as possible consumption of materials and energy, which means to be environmental friendly.
- *Social and economic awareness*: Future Networks should be developed with awareness regarding the costs and competition, so their services should be accessible to all players in the Future Networks ecosystem, including end users, vendors, network operators, and service providers.

3.3   Standardization Synergy of ITU, IETF, 3GPP, and IEEE

Several other standardization organizations had important roles for NGN, either directly or indirectly. Some of them are regional organizations such as ETSI (European Telecommunication Standardization Institute) in Europe, CJK (China Japan Korea) in Asia, while others are global standardization organizations such as ITU, IETF, 3GPP (3G Partnership Project), and IEEE, as well as alliances such as ATIS (Alliance for Telecommunications Industry Solutions) and GSMA (Global system for mobile communication Association). The global leader in NGN standardization and its harmonization is the ITU (Figure 3.5). Besides the ITU, the most significant role in NGN development is ETSI.

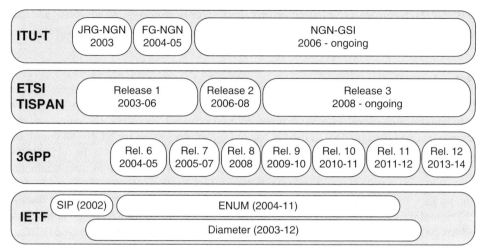

ITU-T	JRG-NGN 2003	FG-NGN 2004-05	NGN-GSI 2006 - ongoing

ETSI TISPAN	Release 1 2003-06	Release 2 2006-08	Release 3 2008 - ongoing

3GPP — Rel. 6 2004-05 — Rel. 7 2005-07 — Rel. 8 2008 — Rel. 9 2009-10 — Rel. 10 2010-11 — Rel. 11 2011-12 — Rel. 12 2013-14

IETF — SIP (2002) — ENUM (2004-11) — Diameter (2003-12)

3GPP – 3G Partnership Project
ETSI – European Telecommunications Standardization Institute
IETF – Internet Engineering Task Force
ITU-T – International Telecommunication Union – Telecommunication

Figure 3.5 Activities toward NGN standardization between different standards bodies

On the other side, ATIS and CJK made contributions to NGN in its early years. For example, ATIS filled some gaps in the standardization by promoting IP interoperability, as well as triple-play implementations. CJK contributed toward building a consensus in Asia over the NGN framework by work on the rich experience in VoIP and broadband.

3.3.1   IETF Role

Prior to NGN standardization the IETF has standardized SIP [5], which is later adopted in the NGN standards for signaling purposes. Nowadays SIP is an intrinsic part of the NGN, and main replacement for SS7 (Signaling System No. 7) signaling that is used in circuit-switched networks such as PSTN and PLMN. So, looking into the past one may conclude that SIP is indirect contribution to NGN by IETF. Similar example is Diameter protocol used for AAA in all-IP access networks including mobile networks. Diameter is also standardized by IETF in 2003 and later updated by a newer version of the protocol in 2012 [2]. For example, the Diameter is also a specified protocol in NGN standards for communication over several standardized interfaces such as interfaces with subscriber's databases.

However, one might consider a big picture related to the Internet technologies. Since NGN is based on all-IP networks, it is using Internet technologies which are standardized by IETF. The most important is IP, in both versions (version 4 and version 6), then transport-layer protocols (OSI-4 layer) such as TCP (Transmission Control Protocol), UDP (User Datagram Protocol), and SCTP (Stream Control Transmission Protocol), then routing protocols and principles, VPN, applications [e.g., WWW, e-mail, FTP (File Transfer Protocol), peer-to-peer applications, and many more], fundamental Internet technologies such as DNS (Domain Name System) and DHCP (Dynamic Host Configuration Protocol), security protocols and solutions, and many more which are incorporated into the NGN recommendations in standardized contexts.

3.3.2   ETSI Role

ETSI contributed to NGN standardization via its technical committee called TISPAN (Telecommunications and Internet converged Services and Protocols for Advanced Networking), created also in 2003. TISPAN has led global discussion on NGN and contributed to NGN Releases 1 and 2, and it is continuing work on NGN Release 3. In fact, TISPAN developed the initial IMS specification, aimed to be used for fixed VoIP as a replacement of PSTN. But, later, TISPAN transferred the IMS standardization to 3GPP with aim to have one common IMS standard for all networks and services. TISPAN NGN Release 1 was finalized in 2005, and it focused primarily to VoIP based on IMS standard for SIP-based applications. TISPAN NGN Release 2 was targeted to IPTV including non-IMS and IMS-based IPTV, and aimed to answer emerging market needs in the converged ICT world, such as triple-play and quadruple-play service offers. Release 2 was finalized by TISPAN in 2008.

Regarding the NGN standardization ETSI has closely collaborated with 3GPP (3GPP standardizes mobile technologies such as GSM in 2G, UMTS in 3G, and further evolution with LTE/LTE-Advanced in 4G). In that manner, regarding the IMS standardization, TISPAN NGN Release 1 specifications are mapped on corresponding 3GPP Release 7 specifications, and TISPAN NGN Release 2 specifications are mapped on corresponding 3GPP Release 8 specifications.

3.3.3   3GPP Role

3GPP has played important role in the NGN standardization by generating specifications for so-called common IMS in 3GPP Release 8, finished at the beginning of 2009. It is the same 3GPP release which standardized the LTE in the radio access network (Evolved UMTS Terrestrial Radio Access Network – E-UTRAN) and System Architecture Evolutions (SAE) based on all-IP principle (in access and in the core networks) which is also compatible with the NGN specifications. Also, 3GPP closely collaborated with the TISPAN (from ETSI) regarding the TISPAN NGN Releases 1 and 2 and relevant ETSI specifications for them, which were incorporated in 3GPP Releases 7 and 8, respectively.

So, one may note that 3GPP in fact created the LTE (Releases 8 and 9) and later the LTE-Advanced (in Releases 10–12 and onwards) according to the NGN framework. So, 4G mobile networking (including LTE, although it is referred to also 3.9G since it is not fulfilling all requirements set by ITU for a given technology to belong to 4G) is an NGN network. It is an all-IP based network, with mobile broadband access, with NGN functionalities located in dedicated standardized network nodes, with full separation of transport and service stratum, as well as usage of SIP for signaling within the IMS for real-time voice and multimedia calls and flows.

3.3.4   IEEE Role

IEEE is not directly involved in NGN, but more indirectly, something similar to the IETF role. The importance for the IEEE for NGN is due to their standards on access networks which are IP-native from the beginning, such as Ethernet (IEEE 802.3 standards group) and wireless networks including WiFi (IEEE 802.11 standards group) as Wireless Local Area Network (WLAN) and WiMAX (IEEE 802.16 standards group) as Wireless Metropolitan Area Network (WMAN). Ethernet is a major wired access network (to Internet) globally, while WiFi is the most spread wireless access networks including business, home, and public environments. The success of Ethernet and WiFi (which is a kind of wireless "equivalent" for the Ethernet) is due to their specification only on the lowest two OSI protocol layers (OSI-1 and OSI-2) and positioning of the IP as single networking layer (OSI-3). However, they also have lower costs (compared to the costs of the equipment in traditional telecom networks) due to asynchronous access (absence of Time Division Multiplexing-TDM), mainly based on contention schemes between the attached subscriber stations. Ethernet-based access is standardized in several NGN specifications by ITU-T, targeting QoS provisioning over such access networks.

3.4 All-IP Network Concept for NGN

NGN are all-IP networks. However, they are established as an evolution of PSTN and PLMN on one side and best-effort Internet on the other side to a single network which will be a networking platform for all types of services, existing ones and future ones. However, traditional telecommunication networks have influenced the NGN network concept, particularly for QoS-enabled end-to-end VoIP and IPTV, since it is an evolution, not a revolution in telecommunication infrastructure and services.

As shown in Figure 3.6 the evolution from PSTN and PLMN toward the NGN is performed by gradual replacement of TDM-based transport systems and interfaces (e.g., SDH), circuit-switching (e.g., phone exchanges), and SS7 signaling (packet-based, separated from the user traffic, but it is not IP-based), with IP-based transport, for example, over DWDM (dense wavelength division multiplexing), IP/MPLS (Multi-Protocol Label Switching), and IP-based signaling (e.g., SIP), respectively.

TV broadcasting "the old way" is to distribute TV over separate broadcast networks, such as terrestrial broadcast networks, satellite broadcast networks, or cable broadcast networks. In all such cases there are separate networks for TV broadcast and separate networks for voice. Due to higher bandwidth demands per channel, TV broadcast was the last technology to perform analog to digital transition. However, regarding the future, one may consider TV as completely digital technology, so the telecommunications world has become a fully digital one.

The data part in the "service trilogy" (voice, video, and data) in telecommunications is Internet, which was initially considered and deployed as a network for data transfer (e.g., file downloading or sharing, messaging, etc.) that requires IP-based network infrastructure, different than telephony networks and TV broadcast networks (Figure 3.7). Internet is a digital packet-switching network from the beginning. However, in the access networks and in the transport networks Internet has used resources from the PSTN and PLMN via modem dial-up in the access circuit-switched networks, as well as IP traffic transport over TDM-based transport networks (e.g., dedicated digital hierarchies such as 2 Mbit/s, 34 Mbit/s, etc.). On the other side, the next step in the development of the PSTN and PLMN was going from circuit-switching toward packet-switching technology as a more efficient approach regarding the utilization of available resources due to advantages of so-called statistical multiplexing (a source sends packets only when it has data to transmit, so multiple sources can efficiently share given digital resources, i.e., the available bandwidth). Hence, the convergence over a single

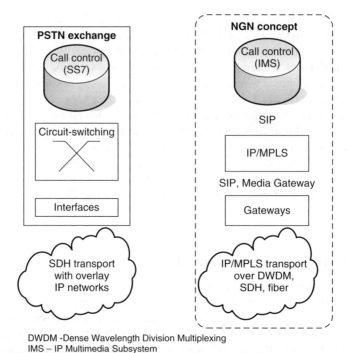

DWDM -Dense Wavelength Division Multiplexing
IMS – IP Multimedia Subsystem
MPLS – Multi-Protocol Label Switching
SDH – Synchronous Digital Hierarchy
SIP – Session Initiation Protocol
SS7 – Signaling System No.7

Figure 3.6   NGN concept

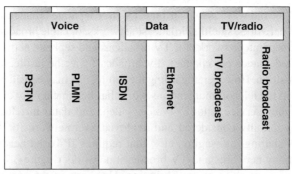

ISDN – Integrated Services Digital Network
PLMN – Public Land Mobile Network
PSTN – Public Switched Telephone Network

Figure 3.7   Separation of networks for voice, TV, and data (the old way)

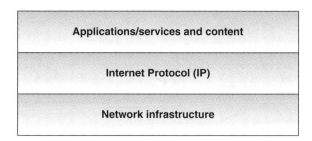

Figure 3.8   Horizontal separation of services/applications and networks (broadband IP networks – the new way)

networking layer, the IP, in all telecommunication networks globally was a "natural" evolution for both, PSTN/PLMN and Internet. The broadband access to Internet provided possibility to add video services (as most demanding regarding the required bandwidths), including TV transmission over the Internet (i.e., IPTV). So, the new telecommunication networks are horizontally split now, by separation of the applications, services and devices on one side, and network infrastructure on the other side, as shown in Figure 3.8.

Since all networks have IP protocol on the networking payer (OSI-3 layer) it is referred to as the all-IP principle. Currently, there is no alternative option on the networking layer and that is good for the convergence. And the convergence is going under the NGN framework. However, NGN is not a type of system, such as GSM, ADSL (asymmetric digital subscriber line), or WiFi, but it is a framework where different architectures can be built and different services can be offered by using the same rules for their implementation, given in the NGN recommendations.

Considering the service stratum in NGN one may notice the existence of services and applications. However, applications do exist in best-effort Internet. For example, WWW is the most important best-effort application in Internet, which is based on client-server principles (clients are browser applications, installed in user's devices such as desktop computers, mobile devices, etc., while servers are usually hosts located on the network's side). WWW communication does not require any signaling between the client and the server prior to their communication (of course, it is assumed that both, client and server, have already access to the Internet). Generally, there are many client-server and also peer-to-peer applications in Internet, which are provided in best-effort manner. On the other side, for certain services (e.g., QoS-enabled VoIP) there are requirements for existence of the call/session control applications (e.g., signaling protocols such as SIP), which are located in the network in the service stratum (consisted of set of servers and databases with standardized interfaces). The control part is standardized by 3GPP in Release 8 as common IMS and adopted by all other standardization organizations, including the ITU which included the common IMS as mandatory part of the NGN. So, one may refer to QoS-enabled VoIP in the NGN as a service, because it includes a service platform for call/session control. On the other side, IMS is not needed for WWW when it is provided via standard best-effort principle. In such case WWW is an application, while QoS-enabled VoIP is a service.

NGN specifies IP on network layer end-to-end, and such requirement defines the all-IP network concept. That means that all access, core (or backhaul), and transit networks in NGN must be IP based. IP hides the lower protocol layers (OSI-1 and OSI-2) from the upper layers,

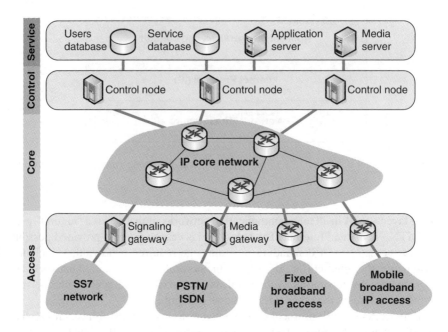

Figure 3.9 All-IP network concept for NGN

so an NGN can have single network infrastructure and single service platforms for different fixed and wireless access networks. In all-IP transport networks (used partially for core and transit networks in NGN) usually MPLS is used between OSI-2 and OSI-3 layers (therefore, MPLS is noted as an OSI layer 2.5). In all-IP principle the interconnection between different networks is also done via IP links established between pairs of gateway routers or controllers. The all-IP network concept in NGN is illustrated in Figure 3.9.

3.5   NGN Functional Architecture

The basic reference model for NGN is defined in 2004 [6], at the start of the NGN standardization process. The defined reference model is the most important novelty introduced with NGN in the telecommunications (i.e., ICT) world – the separation of services and transport in separate so-called stratums:

- *NGN service stratum*: This part of the NGN network provides user functions which transfer service-related data to network-based service functions which manage service resource and network services with a goal of enabling user applications and services. Supported services in NGN service stratum may include voice, video, data, and multimedia (as a combination of the first three), but in all cases it is related to services between peer entities. That means that service stratum provides originating and terminating calls/session between end peers (which is different than client-server model in best-effort Internet, such as WWW, where clients always initiate connections while servers always receive connection requests).
- *NGN transport stratum*: This is stratum that provides user functions that transfer data on the user's side. On the network's side transport stratum provides functions that control and

manage transport resources for carrying the data. The related data may be user, management and/or control information. To carry the data (i.e., information) between end peer entities the NGN transport stratum may provide different static or dynamic associations.

Figure 3.10 show the separation of service and transport stratums, as well as architectural concept regarding the user plane (which refers to data generated or received by the user, such as voice, video, Web pages, e-mails, etc.), control plane (which refers to call/session control, such as signaling), and management plane (which refers to communication between different entities in the service and/or transport stratum regarding the fault management, configuration, accounting, performance, and security management). However, NGN control plane is defined as a union of service stratum control plane and transport stratum control plane. The same approach (for control plane) is applied for NGN management plane in both stratums.

NGN functional architecture includes several functionalities or principles, which are further specified in several different NGN recommendations. The principles of the functional architecture are the following:

- *Support for multiple access technologies*: NGN functional architecture supports different access technologies, including wired and wireless/mobile access.
- *Open service control*: NGN includes control environment (e.g., IMS) which is opened to different types of service and different service providers including third parties (e.g., network provider and service provider can be different companies).
- *Independent service provisioning*: The transport and service stratums are separated in NGN by means of open service control. Such an approach creates ground for service competition in NGN environment and shorter time from service creation to service delivery to end users.
- *Support for services in a converged network*: NGN provides efficient convergence of different access networks to generalized services, which are offered equally in fixed and mobile converged networks.
- *Enhanced security and protection*: The consequences of the open control of services in NGN are the security treats. Therefore it is necessary to provide enhanced security solutions.
- *Functional entity characteristics*: The NGN is composed of many functional entities implemented in certain nodes in the network (e.g., routers, servers, etc.). However, the term "entity" refers to a functionality that cannot be distributed over several physical units (e.g., computers). On the other side each functional entity can have multiple instances.

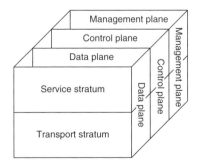

Figure 3.10   NGN basic reference model

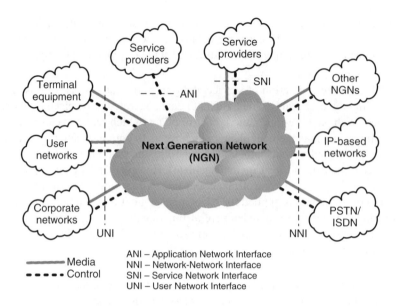

Figure 3.11   Control and media connectivity to NGN

NGN has standardized functionalities. However, there are users, applications, networks, and other networks that should connect to the NGN. According to Figure 3.11 there are four different types of interfaces to/from NGN:

- *User–Network Interface (UNI)*: provides connectivity between NGN and end terminals of users, user's network (e.g., home networks), and corporate networks (e.g., a corporate Ethernet).
- *Network–Network Interface (NNI)*; provides an exchange of control information for calls/sessions, as well as interaction regarding the media level or data plane (e.g., different codecs on different networks used in a given call/session). The NNI is used for connection of NGN with other NGN, with IP-based networks, and with traditional PSTN/PLMN.
- *Service–Network Interface (SNI)*: provides control level (i.e., control plane) and media level (i.e., user plane) interactions and exchanges between the NGN and other service providers (which use open service control functionality in NGN), such as content providers (e.g., IPTV content from a third party delivered to users via the NGN).
- *Application–Network Interface (ANI)*: an interface for interaction between the network (i.e., the NGN) and the applications. However, contrary to SNI, ANI supports interactions only in control plane (not in user plane). So, in other words, ANI provides connectivity of user in NGN to other service providers (in this case referred to as application providers). An NGN operator itself can be an application provider (e.g., IPTV stream generated "in house" by the NGN operator and delivered to its users via the NGN).

Figure 3.12 shows NGN functional architecture. According to the basic reference model, all functions are divided into service stratum functions and transport stratum functions. Main targets of the functional architecture are real-time services, including conversational multimedia (e.g., VoIP with video or multimedia telephony), and content delivery services (e.g., IPTV).

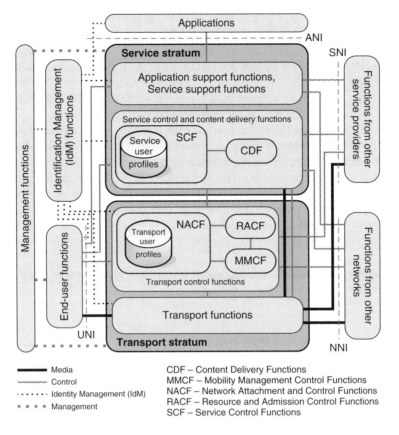

Figure 3.12   NGN functional architecture

3.5.1   Transport Stratum Functions

Transport stratum functions are divided into two groups: transport functions and transport control functions.

3.5.1.1   Transport Functions

Transport functions are related to physical network resources. The physical architecture of the NGN (e.g., interconnected routers) is divided into two main network parts: access network and core network. The routers on the edges between access networks (which are allowed to be heterogeneous in NGN) and the core network are called edge routers. The routers on the edge of the core network in NGN are referred to as gateways. Accordingly, the transport functions are grouped into functions for each network part and interconnection routers as follows:

- *Access network functions*: These functions are performing user traffic aggregation and control from access network toward the core, and vice versa. Also, these functions provide QoS support in the access networks (e.g., scheduling, buffer management, traffic classification, packet filtering, traffic shaping, and policing, etc.) as well as mobility support in certain wireless networks. In general, the access network functions are related to four different types of access networks: xDSL (DSL over twisted-pairs), Cable (over coaxial cables), wireless access (e.g., IEEE 802.11, IEEE 802.16, UMTS, LTE, etc.), and optical access (e.g., FTTH; fiber to the home).

- *Edge functions*: These functions are used for traffic aggregation from several access networks to a core network, as well as for interconnection of core networks.
- *Core transport functions*: These functions provide transport of the information in the core networks, including QoS provisioning for user traffic (with the same set of QoS mechanisms as those outlined for the access network functions).
- *Gateway functions*: There are two type of gateways in NGN, one at the user premises, and other at the edge of the core network. Hence, such functions provide capabilities for interworking with end-user functions as well as with other NGN and non-NGN networks (e.g., PSTN, best-effort Internet, etc.). Gateway functions may be controlled either by transport control functions (in transport stratum) or by service control functions (in service stratum).
- *Media handling functions*: These functions are located in transport stratum only and provide media-related functions (e.g., transcoding, etc.).

3.5.2   Transport Control Functions

Transport control functions are covering functions related to network attachment and control, mobility management (in mobile networks), as well as resource and admission control.

Network Attachment and Control Functions (NACF) provide functions needed for user attachment to the NGN and usage of its services. Such functions include dynamic allocation of IP addresses, auto-discovery of user equipment capabilities (e.g., which voice and video codecs are supported by the user device), mutual authentication (between the user and the network) and then authorization of the user (based on user profiles stored in a database), access network configuration, and location management (at the IP layer).

Mobility Management and Control Functions (MMCF) provide support for seamless IP mobility in the transport stratum, including horizontal handovers (between cells of a same radio access technology) and vertical handovers (between different types of radio access networks). However, it does not guarantee the persistence of the QoS provisioning before the handover and after its execution (in the post-handover phase). MMCF considers mobility in NGN environments as a service, independent from the type of the access technology.

Resource and Admission Control Functions (RACF) are the link between the Service Control Functions (SCF) and transport control functions in NGN. Their main goal is to provide admission control regarding a given call/session to/from end users. So, a user connection can be admitted in the transport network or rejected (or connection request can be modified) according to several parameters such as transport subscription information, SLAs, different policies and service priorities, and so on. For example, in NGN are defined three priority levels for the RACF [7]:

- *Priority level 1*: Traffic with highest priority, targeted to emergency telecommunications over NGN.

- *Priority level 2*: Traffic with higher priority than lower priority level, which is targeted for real-time services (e.g., VoIP, IPTV, video streaming, etc.), data services with required QoS support (e.g., cloud computing), and VPN services.
- *Priority level 3*: Traffic with least guarantees regarding the QoS, and least assurance for admission control in the network. This level includes best-effort Internet traffic (e.g., Web, e-mail, etc.).

The RACF hides network topology from the service stratum, providing abstract view of transport architecture to SCF. However, usage of RACF is not mandatory when there is no need for admission control (e.g., best-effort Internet traffic, such as Web). On the other side, RACF interact with SCF for applications that require certain QoS support which results in a need for admission control. Such examples are QoS-enabled VoIP (e.g., SIP signaling is used), IPTV, and so on. Besides QoS support (which is essential in NGN), RACF also provides NAPT (Network Address and Port Translation) traversal functions which are necessary when NGN operator uses private IP addresses for addressing end user terminals.

RACF interacts with NACF for exchange of user subscription information, as well as with RACF in other NGN for delivering services over multiple network operators or service providers.

3.5.3   Service Stratum Functions

Service stratum has two main groups of functions. First group is consisted of application and service support functions which have interface with applications. Second group of functions in service stratum is consisted of SCF and Content Delivery Functions (CDF; including service user profile database), which interact with transport stratum and its functions via the control plane. Also, service stratum functions interact with management functions including IdM functions.

3.5.3.1   Service Control and Delivery Functions

SCF include specific functions for resource control (via interaction with the RACF in transport stratum), as well as authentication and authorization (AA) mechanism by using so-called functional databases.

CDF interact with application support functions and service support functions to obtain the content (e.g., IPTV content) that should be delivered to the end users. On the other side CDF also interact with end-user functions by using the transport functions and their capabilities (of course, controlled by SCF).

3.5.3.2   Application and Service Support Functions

Application support functions and service support functions interact with applications via the ANI interface on one side, and with end-user functions via UNI interface on the other side. They include functions for registration, authentication, and authorization at the application layer and they provide services to the end-users in cooperation with the SCF.

3.5.4   Management Functions

With aim to provide the necessary functions, the NGN must efficiently manage quality (i.e., QoS and performance), security (due to open control of services and Internet openness globally) and reliability (i.e., how much time the service is available, which should not less than traditional telecom networks for different services migrated to NGN). Management functions in NGN are allocated in each functional entity, with interaction with network element management, network management, and service management. Management functions are applied in both stratums (i.e., service and transport stratum). They include management of faults, configurations, accounting, performances, and security [8].

3.5.5   Identity Management Functions

NGN provides functions for IdM regarding the identities of the entities. According to the ITU [9], an entity is considered something that has a distinct existence that can be uniquely identifies (e.g., a network interface of host or router, user, etc.). Additionally, each entity can be associated with multiple identities of different types, which are grouped as [10]:

- Identifiers (e.g., telephone number, IP address, Uniform Resource Identifier – URI, e-mail address, etc.);
- Credentials (e.g., username/password, digital certificate, token, smart card, etc.);
- Attributes (e.g., location, pattern, context, etc.).

In general, IdM is a set of functions and corresponding capabilities (e.g., administration of users, authentication, binding of different identifiers, etc.) which provide assurance of the identity of an entity, as well as its storage, usage, and distribution. Additionally, IdM in NGN allows identity information to be shared by a federation of business partners (i.e., federated identity information).

3.5.6   End-User Functions

End-user functions refer to functions located in end-user equipment (i.e., user terminals), which may be fixed (e.g., a desktop computer) or mobile (e.g., a mobile phone or device). These functions interact with all functional groups in service and transport stratum, as well as with management functions and IdM functions. In user (i.e., data) plane they interact with transport functions via the NNI interface for carrying user data. Regarding the call/session control traffic (e.g., signaling) end-user functions communicate with transport control functions in transport stratum, and with SCF and CDF in service stratum. However, the prerequisite for control and data traffic from/to end-user equipment is the IdM which manages identification of the end-user and the end-user equipment, including authentication (e.g., with telephone number, username/password, digital certificate, etc.) and authorization (e.g., to which services the end-user subscribes).

3.5.7   NGN Configuration and Topology

NGN includes logical decomposition of the network into sub-networks. It is not a physical decomposition because toward the future development it should be possible for a single phys-ical network to have capabilities of an access network and a core network.

An NGN configuration (Figure 3.13) on a network level includes three types of sub-networks as follows:

- *Customer network*: This is a network deployed at home or at enterprise, which is connected via the UNI with NGN access network.
- *Access network*: This network collects the end-user traffic from customer network to the core network in both directions. It also aggregates the traffic from different end-users to the core network. Also, it can be divided into several domains, and in such case network elements within a single domain are interconnected with so-called INNI (Internal Network–Network Interface). Connections between network elements from different domains as well as between access networks and core networks are realized via the NNI.
- *Core network*: This type of network provides transport of user data as well as control and management information between different network elements (e.g., between core routers). A core network may have several domains. Within a single domain network elements are

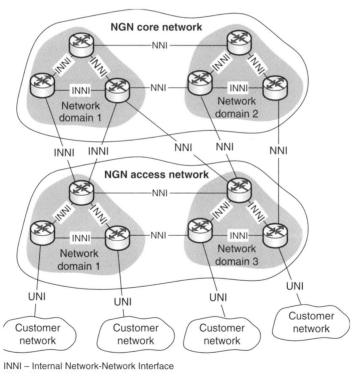

INNI – Internal Network-Network Interface
NNI – Network-Network Interface
UNI – User Network Interface

Figure 3.13   NGN network configurations

interconnected via INNI, while NNI is used between different core network domains or between a core network and other network (e.g., access network, transit network).

Regarding the NGN stratums separation, it is important to note that core networks belong to both, the service stratum and the transport stratum. Different NGN core networks are connected via so-called transit networks. On the other side, access networks belong only to the transport stratum.

3.6   NGN Control Architectures and Protocols

The central position in the NGN architecture for control functions is reserved for the NACF [11]. Figure 3.14 shows the interaction of NACF with other functional entities in transport stratum (RACF and MMCF, as well as several transport functions), with service stratum (SCF) and with customer equipment. Single NACF can be used for multiple access networks, but it is not required in the NGN architecture (it is up to network design choice). All control functional entities in NGN are shown in Figure 3.14.

3.6.1   Network Access Configuration Functional Entity

Network Access Configuration Functional Entity (NAC-FE) is responsible for allocation of IP addresses to end-user terminals (e.g., via DHCP), as well as IP addresses of DNS servers and addresses of specific service stratum components (e.g., signaling proxies in IMS).

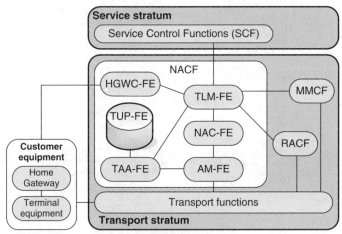

AM-FE -Access Management Functional Entity
AR-FE -Access Relay Functional Entity
HGWC-FE – Home Gateway Control Functional Entity
MMCF – Mobility Management Control Functions
NACF – Network Attachment and Control Functions
NAC-FE -Network Access Configuration Functional Entity
RACF – Resource and Admission Control Functions
TAA-FE -Transport Authentication and Authorization Functional Entity
TLM-FE -Transport Location Management Functional Entity
TUP-FE -Transport User Profile Functional Entity

Figure 3.14   NGN control architecture

NAC-FE may allocate two different types of IP addresses: persistent IP address and temporary IP address. Typical usage of the two addresses is in cases of user mobility when user is changing IP networks. In each IP network there are different IP address prefixes, so the terminal should obtain different IP address in different IP networks. However, each terminal has its own home network. There is a standardized solution from IETF which manages macro-mobility in IP networks [12], the Mobile IP (MIP). In MIP the permanent IP address (which belongs to the home network) is called Home Address, while the temporary address obtained in visited network is called Care-of-Address (CoA). Of course, MIP manages a binding between the permanent and temporary IP addresses. In general, a binding must be maintained between the persistent and temporary IP addresses in NGN to support mobility of users through different IP networks.

3.6.2 Access Management Functional Entity

This entity is located between the Customer Premises Equipment (CPE) and the NACF. It terminates OSI layer 2 connections from the user side toward the NACF and vice versa. The connection between the CPE and the NACF is logical, so there can be different access technologies and different protocols between them. For example, when PPP (Point-to-Point Protocol) is used (which was initially used for dial-up modem connections) then AM-FE (Access Management Functional Entity) terminates the PPP on the network's side [13]. Afterwards, the AAA is performed between AM-FE and NACF by using RADIUS (Remote Authentication Dial In User Service) [14], or Diameter [15], as AAA protocols (all IETF protocols). However, for IEEE 802 access network (e.g., Ethernet, WiFi, etc.) the IEEE 802.1X/PANA [16], solution can be applied for authentication. For mobile networks (e.g., UMTS, LTE/LTE-Advanced) 3GPP defines its own OSI layer 2 mechanisms [17].

3.6.3 Transport Location Management Functional Entity

This entity provides an association between the allocated IP address (via DHCP) to the end-user equipment (i.e., the CPE) and related network location information given by the NAC-FE. The TLM-FE (Transport Location Management Functional Entity) obtains identifiers of users and user equipment from the transport subscription profile stored in TAA-FE (Transport Authentication and Authorization Functional Entity).

There are two types of roles for TLM-FE, namely home and local role, or both. However, this division of roles to home and local has similarities with Home Agents and Foreign Agents in MIP [12], respectively. For a given user, the closest TLM-FE which plays the local role binds the IP address of the user with CPE and location information. The home TLM-FE for that user maintains a pointer to its current local TLM-FE. When the user moves within a given domain, then only location binding information of the local TLM-FE needs to be updated. However, when the user moves from one domain to another domain (which is under the jurisdiction of a new local TLM-FE), then the user obtains a new temporary IP address and the context information (e.g., QoS-related information, user preferences, etc.) for the given user is transferred from the old to the new local TLM-FE. To provide service functions the TLM-FE responds to queries from the service control functionalities (e.g., from Proxy – Call Session Control Function in the IMS). For mobility support between different NGN networks (i.e., roaming) the SCF in home network may contact (e.g., query) TLM-FE in a visited network.

3.6.4   Transport Authentication and Authorization Functional Entity

This entity provides user AA. The data for these AA functions TAA-FE obtains from the transport user profile stored in TUP-FE. Additionally, TAA-FE may be assigned with IP address allocation role. However, this entity may also request NAC-FE to allocate the IP address to the user equipment. There is possibility to have several TAA-FE proxies that communicate with a centralized TAA-FE server which contains the TUP-FE subscription data (for AA purposes).

3.6.5   Transport User Profile Functional Entity

Transport User Profile Functional Entity (TUP-FE) is a database in the transport stratum which contains subscription authentication data (e.g., user identifiers, supported authentication methods, etc.) and transport subscription profile (it contains network configuration information for the terminal equipment, such as IP address or IP prefix). The transport profile may be divided into several sub-profiles. The access to all data stored in the TUP-FE in the access network is provided via the TAA-FE (for AA purposes).

3.6.6   Home Gateway Configuration Functional Entity

The main purpose of Home Gateway Configuration Functional Entity (HGWC-FE) is control of the home gateway (deployed at the user premises) regarding initialization and/or updating. This entity may also provide certain configuration information considering the QoS (e.g., IP packet marking, filtering, etc.). Additionally, HGWC-FE may provide notifications regarding the availability of different terminal equipment in the home network (e.g., mobile terminal, lap-top, desktop, different sensors, TV set, etc.). On the network's side the HGWC-FE interacts with the TLM-FE to obtain information about the access network of the NGN to which it is connected to (such information may include identifiers for the subscriber, for different connections, etc.).

3.6.7   Access Relay Functional Entity

The Access Relay Functional Entity (AR-FE) has a role of relay entity between the customer equipment and the NACF. It includes different relay functions depending upon the access technology. For example, if PPP is used then AR-FE is in fact PPPoE (Point-to-Point Protocol over Ethernet) or PPPoW (Point-to-Point Protocol over Wireless) relay. In the case when DHCP is used for allocation of IP addresses to user equipment, the AR-FE acts as a DHCP relay.

3.7   Numbering, Naming, and Addressing in NGN

Currently there are different numbering, naming, and addressing schemes in traditional telecom networks (e.g., PSTN, PLMN) and traditional Internet. Traditional telecom operators (including fixed and mobile ones) use the E.164 numbering scheme [18], standardized by

ITU-T. On the other side, traditional (best-effort) Internet is using different naming and addressing schemes, such as DNS schemes (for domain names, in URI format) and IP addressing schemes (IPv4 and IPv6). Since NGN is a convergence of the two worlds, it is expected that it should include the best from both approaches for numbering, naming, and addressing in the NGN and future networks.

PSTN and PLMN (e.g., GSM-based) were primarily designed for voice communications. In those networks users are assigned unique numbers in E.164 format. Also, the routing of the calls within the PSTN is based on E.164 numbers assigned to end users. In mobile networks (e.g., GSM, UMTS, 4G) the E.164 numbers are called MSISDN (Mobile Subscriber Integrated Services Digital Network numbers), with the aim to specify the mobile character of the number as well as the ISDN (Integrated Services Digital Network) character of the number (i.e., fully digital line from the mobile terminal to the mobile network). Another implicit (to the end users) identifier in mobile networks is the IMSI (International Mobile Subscriber Identity) [19] which is an identifier used for unique registration of a subscription (for a given user) as well as for roaming purposes (when a user visits another mobile network) [20]. IMS usage in mobile networks (from 3GPP Release 8 onwards) introduces also ISIM (IM Service Identity Module), which in such a case is also stored (together with IMSI) on the UICC (Universal Integrated Circuit Card) in a mobile terminal.

On the other side, traditional Internet uses a naming addressing scheme with alphanumeric labels in the form of domain names (e.g., www.example.com) which are used to address different servers for different applications (e.g., Web, e-mail, FTP, etc.). Domain names are used in the creation of URI identifiers for different protocols [HTTP (Hypertext Transfer Protocol), SIP, etc.]. There are also other naming addressing schemes in the Internet world, such as e-mail addresses as another standardized type of identifiers. However, there are also other non-standardized types of common naming addressing used in messaging, chat, gaming, and so on (which make use of e-mail or other names chosen and registered by the users in a relevant service platform).

However, the Internet is based on routing of IP packets based on the IP addresses assigned to the network interfaces of the devices (at the end users) or servers (usually located on the network's side). Domain names are mapped to IP addresses via the DNS. This leads to the general conclusion that all naming schemes in Internet (e.g., URIs, telephone numbers, or other types of names) that are used by the end users to contact given destination/called entity must be mapped to IP addresses by using the DNS.

3.7.1   Numbering Scheme

The migration of circuit-switched networks, such as PSTN and PLMN, to all-IP based NGN, is an ongoing process which will last for some years. Because E.164 is a dominant numbering scheme for voice users in all types of networks, one may expect that it will continue in NGN, at least for some time (e.g., short or medium term). However, it will not disappear in NGN, but is expected E.164 to adapt to the IP environment with the newer numbering (E.164), naming (DNS), and addressing (IP addressing) scheme called ENUM (reader may refer to relevant section in Chapter 2), which is being standardized by IETF [21] and ETSI (for the NGN usage) [22].

3.7.1.1 Public ENUM

The advantage of the ENUM is the possibility to have a single number for a range of devices and terminals, including telephony, e-mail, Web-based services, and so on. The public ENUM provides translation of E.164 numbers into domain names tree structure of the DNS architecture, with public root "e.164.arpa." Hence, ENUM may provide possibilities for combination of various applications including voice, video, instant messaging, and so on. However, such numbering via ENUM should be coordinated, managed, and monitored by a regulatory body in a given country.

3.7.1.2 Carrier ENUM

Carrier ENUM is targeted to exchange of IP traffic (voice, data, etc.) between the operators, such as NGN operators. Using IP peering, a given operator may locate a destination operator for IP traffic exchange by using Carrier ENUM. This approach for IP interconnection between network operators is specified by the GSMA.

The carrier ENUM is based on three tiers of registries, namely Tier 0 at the root, then Tier 1 and Tier 2 in the hierarchy tree from the root (shown in Figure 3.15). Tier 0 is internationally centralized root directory for Carrier ENUM. Operators, carriers, and hubs (for traffic exchange) are supposed to query ENUM Tier 0 to find authoritative sources of routing information (for Carrier ENUM) distributed around in the world. Tier 1 is consisted of national registries for ENUM (in each country). Tier 2 DNS registries are distributed on national level, under Tier 1 registry in a given country. Overall, Carrier ENUM maps the hierarchy of numbering in E.164 from the PSTN/PLMN to NGN environment (which is all-IP).

3.7.2 Naming and Addressing Schemes

In NGN environment, which is IP-based, the addressing schemes are no longer restricted to numbers such as telephone numbers. Hence, there are also alphanumeric naming and addressing schemes such as e-mail addresses or URIs (e.g., for access to web sites). Domain names are translated to IP addresses by using DNS architecture (one should note that the same DNS

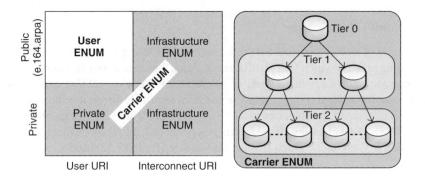

Figure 3.15 Public and carrier ENUM

architecture is used also for ENUM, but with a dedicated root of the DNS tree). Such domain name addressing schemes are used for e-mail communications in Internet, as well as for VoIP communication based on SIP or H.323 (from ITU-T). SIP and H.323 addresses can also be used to identify users (or servers) to deliver services (e.g., terminating voice calls) to the end users.

3.7.3   Numbering, Naming, and Addressing Scheme for NGN

Regarding the NGN concept of convergence from end-user perspective it is desirable to have also a converged numbering, naming, and addressing schemes toward a unified scheme for all services. However, number of services and applications within NGN is expected to increase in the future, and they all shall be supported regarding the numbering, naming, and addressing schemes, via either existing or newly developed ones.

3.7.3.1   Identifiers in NGN

Numbering, naming, and addressing schemes in NGN are directly related to identifiers. There are three types of identifiers specified in NGN [20]:

- *Home domain name*: This identifier is used to identify the operator for routing the SIP registration requests to home operator's IMS network (IMS is mandatory in the service stratum of NGN). It is specified in a DNS format (e.g., "NGNoperator123.com") and it is stored in the ISIM (if there is no ISIM application, i.e., no IMS-based module, the Home Domain Name must be derived from the data which are available to the end-user equipment).
- *Private identifiers (Private IDs)*: They are used to identify user's subscription. So, each user should have in NGN at least one private ID. The syntax of private ID is the form "username@realm", and it is a permanent ID (not related to a given session or a call) and it is stored locally in the ISIM. The "realm" in the private ID is the Home Domain Name. Private IDs are network (operator) aware.
- *Public identifiers (Public IDs)*: Each IMS user may have one or several public IDs. There is at least one public ID in the ISIM. These identifiers are used for message routing (e.g., SIP message routing to the end-user for initiation of a call/session, or notification, etc.). These IDs may take form of tel URI (e.g., E.164-based) or SIP URI (e.g., "sip:user@domain"). Public IDs are user aware.

Numbering, naming, and addressing are related to public IDs (i.e., user-aware IDs). Regarding the NGN stratums, all identifiers may be divided into two groups: IDs in the service stratum, and IDs in the transport stratum. General overview of some of the public identifiers in NGN is given in Table 3.2 [20].

3.7.3.2   ENUM Role in NGN

Naming-numbering and Address Resolution (i.e., NAR) for NGN is defined by TISPAN [23]. It refers to the usage of ENUM in NGN environment. The specified NAR process is shown in Figure 3.16.

According to Figure 3.16 the NAR starts with processing of the dialed string (e.g., a sequence of digits) by the user through a user interface. The string can also be an alias (e.g., "User123")

Table 3.2   NGN public identifiers

NGN stratums	Public ID	Format of the public ID
Service stratum (user/service identifiers)	Name(s)	SIP URI
	Number(s)	Tel URI, SIP URI (with operator's domain)
Transport stratum (network identifiers)	Address	Routing number, IP address

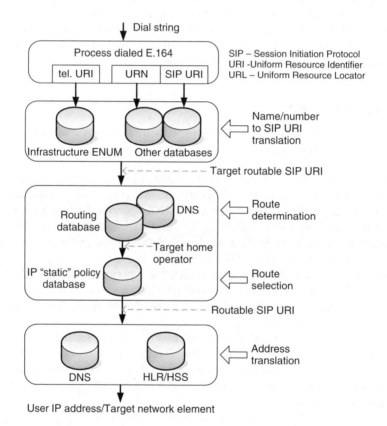

Figure 3.16   Naming-numbering and address resolution (NAR) in NGN

which is translated to a dial string by the terminal. The dial string is sent in Request URI format from the user terminal to the home network. Further, the dial string is translated to a tel URI (i.e., E.164 number in ENUM form) or a tel URI as a local number with phone context. Afterwards the target name is translated into an address, which can be a tel URI or a SIP URI. The SIP URI may be in the format:

- "sip:user@ExampleProvider.net"
- "sip:+389nnnnnn@ExampleProvider.net; user=phone"

The result of SIP URI translation is a routable SIP URI (Figure 3.16). The next step is route determination, which is based on the address (tel URI or SIP URI). It results in a route toward the IMS in the given domain, or a route toward a gateway to PSTN or PLMN, or route to another NGN. Afterwards, a route selection is performed based on the target home domain, a task completed by the nodes in the IMS (in particular, by the Serving Call Session Control Function – S-CSCF) by using DNS or a routing database. Finally, address resolution (from domain name to an IP address) is done by standard DNS. If the SIP URI belongs to the given IMS domain, then address resolution is completed within that IMS domain.

3.7.4   Discussion

Nowadays, dominant numbering scheme in all networks is still E.164, including PSTN and PLMN as well as NGN (in the form of ENUM), and it is widely adopted by the operators and the telecommunication industry. On the other side, dominant naming and addressing schemes in Internet are DNS-based schemes (e.g., URI, e-mail address) and IP addressing (IPv4 and IPv6), respectively. One may expect that these numbering, naming, and addressing schemes will continue to be dominant in near future in NGN. However, with development of new services and applications it is likely to expect that some new or modified numbering, naming, and addressing schemes will evolve.

References

1. ITU-T (2004) General Overview of NGN. ITU-T Recommendation Y.2001, December 2004.
2. Fajardo, V., ed., Arkko, J., Loughney, J., and Zorn, G., ed., (2012) Diameter Base Protocol. RFC 6733, October 2012.
3. ITU-T (2005) ITU-T NGN FG Proceedings.
4. ITU-T (2011) Future Networks: Objectives and Design Goals. ITU-T Recommendation Y.3001, May 2011.
5. Rosenberg, J., Schulzrinne, H., Camarillo, G. *et al.* (2002) SIP: Session Initiation Protocol. RFC 3261, June 2002.
6. ITU-T (2004) General Principles and General Reference Model for Next-Generation Networks. ITU-T Recommendation Y.2011, October 2004.
7. ITU-T (2006) Admission Control Priority Levels in Next Generation Networks. ITU-T Recommendation Y.2171, September 2006.
8. ITU-T (2006) Principles for the Management of Next Generation Networks. ITU-T Recommendation M.3060/Y.2401, March 2006.
9. ITU-T (2010) Functional Requirements and Architecture of Next Generation Networks. ITU-T Recommendation Y.2012, April 2010.
10. ITU-T (2009) NGN Identity Management Framework. ITU-T Recommendation Y.2720, January 2009.
11. ITU-T (2010) Network Attachment Control Functions in Next Generation Networks. ITU-T Recommendation Y.2014, March 2010.
12. Perkins, C. (2002) IP Mobility Support for IPv4. RFC 3220, January 2002.
13. Simpson, W. (1994) The Point-to-Point Protocol (PPP). RFC 1661, July 1994.
14. Rigney, C., Willens, S., Rubens, A., and Simpson, W. (2000) Remote Authentication Dial In User Service (RADIUS). RFC 2865, June 2000.
15. Calhoun, P., Loughney, J., Guttman, E. *et al.* (2003) Diameter Base Protocol. RFC 3588, September 2003.

16. Yegin, A., ed., Ohba, Y., Penno, R. *et al.* (2005) Protocol for Carrying Authentication for Network Access (PANA) Requirements. RFC 4058, May 2005.
17. 3GPP TS (2013) Evolved Universal Terrestrial Radio Access Network (E-UTRAN) Access (Release 12). 3GPP TS 23.401, V12.0.0, March 2013.
18. ITU-T (2010) The International Public Telecommunication Numbering Plan. ITU-T Recommendation E.164, November 2010.
19. ITU-T (2008) The International Identification Plan for Public Networks and Subscriptions. ITU-T Recommendation E.212, May 2008.
20. ETSI (2006) Identifiers (IDs) for NGN. ETSI Technical Specification 184 002, October 2006.
21. Faltstrom, P. and Meallingm, M.(2004) The E.164 to Uniform Resource Identifiers (URI) Dynamic Delegation Discovery System (DDDS) Application (ENUM). RFC 3761, April 2004.
22. ETSI (2011) Requirements and Usage of E.164 Numbers in NGN and NGCN. ETSI Technical Specification 184 011, February 2011.
23. ETSI (2008) Naming/Numbering Address Resolution (NAR). ETSI Technical Report 184 007, November 2008.

4

Broadband Internet: the Basis for NGN

4.1 ITU's Work on Broadband Internet

The impact of telecommunication networks and services on society and on each human globally gives an important view toward their future development. The ITU (International Telecommunication Union) as the world's largest organization for telecommunications (i.e., ICT-Information and Communication Technology) nowadays highlights the importance of broadband. The term broadband shortly denotes broadband access to the Internet as a single converged network infrastructure.

The work on broadband in ITU is covered by its three sectors, that is: International Telecommunication Union Telecommunication (ITU-T) standardization sector, International Telecommunication Union Radiocommunication (ITU-R) sector, and International Telecommunication Union Development (ITU-D) sector.

4.1.1 ITU-T Work on Broadband

In the second decade of the twenty-first century more than 90% of the total Internet traffic worldwide is going through undersea fiber-optic cables. Regarding the core and transit networks, this approaches almost 100% of Internet traffic. The access networks to the Internet are more diverse by including wireless/mobile access, fiber-optic access, and the copper access (via twisted pairs and coaxial cables). Optical technologies are providing high bandwidth needed by the bandwidth demanding applications such as video streaming, IPTV (Internet protocol television), data transfer, cloud computing, and so on. The broadband access networks nowadays can be supported only by optical transport networks, which are therefore an intrinsic part of the NGN (Next Generation Network). Providing broadband access and broadband transport belong to major targets of the ITU standardization sector (i.e., ITU-T). For example, in the world where most of the users were connected to the telecommunication networks in the past century via twisted pairs, the ITU's standard on ADSL (Asymmetric Digital Subscriber Line), which is targeted for usage over twisted pairs in the last mile/kilometer, has sped-up tremendously the introduction of global broadband access to the Internet. Since the highest cost in network deployment is the fixed access network, the ADSL [as well as other

NGN Architectures, Protocols and Services, First Edition. Toni Janevski.
© 2014 John Wiley & Sons, Ltd. Published 2014 by John Wiley & Sons, Ltd.

standardized DSL (Digital Subscriber Line) technologies] provided migration of narrowband access networks [either 64 kbit/s in PSTN (Public Switched Telephone Network) or 128 kbit/s in ISDN (Integrated Services Digital Network)] to broadband access networks by providing several megabits per second in the downlink direction (from the Internet toward the end-user equipment). Furthermore, ITU has standardized Fiber-To-The-Home (FTTH) access networks and their interfaces, which are further increasing the capacity given to the consumers. Additionally, ITU-T is performing standardization in several other areas related to the broadband Internet, such as IPTV, Internet of Things, cloud computing, IPv4 to IPv6 transition, and so on.

All of these broadband technologies and applications/services are targeted to creation of fully networked society where people can access high-speed services regardless of their current location, current time, or used device.

4.1.2   ITU-R Work on Broadband

Mobile and wireless technologies have developed to a state where almost each human on the Earth has a mobile device. The work in the mobile and wireless world is toward standardization and deployment of the mobile broadband access everywhere. Broadband mobile networks provide high data rates access to Internet on the move. On the other side fixed wireless broadband provides solutions that are easy to deploy and fast to the market, especially in rural areas or areas without a deployed transport network infrastructure (e.g., developing regions).

Due to personal character of mobile communications (each user has its own mobile terminal and subscriptions to services) and capabilities for mobile broadband added by 3G and 4G mobile technologies, the role of ITU-R is becoming more important than ever for global allocation and management of the radio-frequency spectrum. Because mobile broadband access demands more bits per second per hertz, but also more frequency bands are used to increase the available data rates toward the end-users, ITU-R has important role to synchronize globally the allocation and usage of frequency bands for the mobile broadband. On the other side, ITU-R developed the IMT-2000 (International Mobile Telecommunications-2000) umbrella of mobile standards commonly known as 3G (third generation). So, all standards that satisfied requirements set by IMT-2000 are referred to as 3G standards [e.g., UMTS/HSPA (Universal Mobile Telecommunication System/High Speed Packet Access) from 3GPP (3G Partnership Project), IEEE 802.16e, a.k.a. Mobile WiMAX 1.0, etc.]. Further, ITU-R also set the requirements for 4G mobile systems (i.e., next generation mobile networks), which are defined within the umbrella called IMT-Advanced (International Mobile Telecommunications – Advanced). All mobile standards that comply with all IMT-Advanced requirements are referred to as 4G (e.g., LTE-Advanced from 3GPP, IEEE 802.16m, a.k.a. Mobile WiMAX 2.0).

Additionally, ITU-R also manages orbital allocations for satellite systems and provides standards on mobile-satellite and fixed-satellite broadband systems and services, which are important for deployment of broadband systems in those parts of the world that lack terrestrial broadband infrastructure.

On regular World Radiocommunication Conferences (WRC) as well as on regional ones, ITU-R approves allocation of certain frequency resources for broadband access, especially for mobile broadband technologies (e.g., 4G), and at the same time provides global synchronization for usage of different frequency bands for different technologies. For example, by completion of the analog to digital (DVB-T, i.e., Digital Video Broadcasting – Terrestrial) transition for television globally, frequency bands (below 900 MHz) are becoming available for new mobile broadband technologies, such as LTE/LTE-Advanced and Mobile WiMAX 2.0. However, ITU-R also focuses on different emerging radiocommunication technologies such as cognitive radio, free-space optical communication, short-range devices, and so on.

If one summarizes, ITU-R provides ground for affordable and synchronized global broadband access to Internet worldwide by using fixed-wireless, satellite, and mobile broadband technologies.

4.1.3   ITU-D Work on Broadband

Telecommunications have become a vital part of the society since the invention of telegraphy and later telephony. However, the Internet has increased the impact of telecommunications (i.e., ICT) in other areas including administration (e.g., government, regional offices, etc.), financial area (e.g., banking, trading, shopping, etc.), social networking, collaborations, working at a distance, all types of entertainment, and many more. Such activities demand broadband access to applications and services. Hence, the world societies and economies are becoming more dependent on communication networks with high data rates with aim to provide access to all customers to different types of services over the given access network to the Internet. In that manner ITU-D works with different organization in all countries in the world, including individual users and enterprises, to examine the factors and key aspects for success of broadband deployments. At the beginning of the second decade in this century the ITU-D has pointed to broadband rollout in the world as a key target in the following years.

The role of ITU-D is also shown through its assistance to countries in the creation of broadband strategies and policies, as well as the regulation of broadband networks and services. In such process ITU-D helps via building local or regional capacity for investments and deployment of broadband technologies and applications, with aim to allow people around the world to benefit from it in personal lives and in the society in general. The final goal of ITU-D is connecting everyone everywhere via broadband, including developed and developing countries in the world.

4.2   DSL and Cable Access Networks

Broadband access to the Internet is offered to individual (i.e., residential) users and to business (i.e., enterprise) users. Fixed broadband access is provided via copper lines (e.g., twisted pairs and coaxial cables) and fiber-optic cables. Then, the main fixed broadband access technologies over copper lines are DSL technologies as well as cable access networks.

4.2.1   ADSL Success Story

ADSL is a technology which provides higher data rates over traditional telephone line (twisted pair). It belongs to the family of DSL technologies, where the signal is transmitted over the line as digital while in PSTN the telephone signal is transmitted as analog and converted to digital signal in the telephone network. ADSL provides asymmetric bit rates in downlink and uplink direction, and that is the reason for its name (Asymmetric DSL). The asymmetry fits very well the client–server Internet applications where clients at user premises request certain data from servers located in the network, and such data (as a response from a server to a given request from a client) is delivered in the downlink direction (from the network to the end-user). Also, video streaming and IPTV is delivered in the downlink direction from servers in the network to clients located at the subscribers. So, Internet traffic is asymmetrical from the beginning due to client–server nature of the native Internet services [e.g., WWW (World Wide Web), e-mail, ftp, etc.]. Another important capability of the ADSL is that it supports already implemented lengths of telephone lines (twisted pairs) in the access part (between the user and the operator's network).

ADSL uses the upper frequency band which is not used for the telephony calls, by means of splitting the frequency bandwidth of the twisted pair line using a so-called splitter (or DSL filter). So, it has been easy to be implemented without any interference with the existing PSTN, by simply adding splitters and ADSL equipment on both sides (i.e., at the end-users and at the operators). Transparent implementation of ADSL over the existing PSTN network infrastructure, especially in the access network, has provided grounds for success of ADSL from its standardization at the end of 1990s [1–3]. So, ADSL was the first global technology that provided broadband access to the Internet, with bit rates many times higher than maximum 56 kbit/s provided via modem dial-up over telephone line or maximum 144 kbit/s provided via ISDN access network. One may note that with the standardization and implementation of the ADSL has started the broadband development and deployment worldwide.

The initial standard for ADSL was created by ANSI (American National Standardization Institute) [1], but the most commonly used standard for ADSL on a global scale is ADSL standard by the ITU-T [2], also known as G.DMT [because it is using DMT (Discrete Multi-Tone) modulation]. For longer lengths of the subscriber's local loop there is also ITU-T standard [3], known as G.Lite (due to its lower data rates, up to 1.5 Mbit/s in downlink, as a tradeoff for longer lengths of the subscriber's line).

4.2.2 ADSL Access Architecture

There are several key components that participate within the ADSL architecture: ADSL transceiver unit centrally, ADSL transceiver remote unit (at the user), splitters for providing both the telephone service (i.e., POTS – Plain Old Telephony Service) and the ADSL service, and a multiplexer of DSLs called DSLAM (Digital Subscriber Line Access Multiplexer) on the side of the operator.

Figure 4.1 illustrates the architecture of DSLAM. In the direction from the network toward the terminal (downlink), the telephone (i.e., POTS) signals and ADSL signals are transferred to the users. Splitters on both sides of the subscriber line (i.e., local loop) are used for frequency multiplexing/demultiplexing of POTS and ADSL signals. This is achieved by using a low pass filter with upper frequency boundary around 4 kHz (frequency range of the POTS subscriber line is 0–4 kHz). The splitter is used to ensure the operation of POTS services even in case of failure of ADSL service.

In the uplink direction, POTS and ADSL signals are multiplexed on the side of the user (again with the splitter) and the signals are transmitted by the same twisted pair telephone line to the ADSL network elements on the operator's side (shown in Figure 4.1). On the side of the operator, multiplexed POTS and ADSL signals are transmitted to the DSLAM, which first uses a splitter to separate (i.e., to demultiplex) POTS and ADSL signals. There are two outputs of the splitter: one is for POTS signal and the other for ADSL signal that is carried

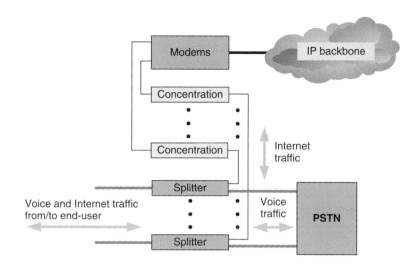

Figure 4.1   Digital subscriber line access multiplexer (DSLAM)

to an ADSL modem. On the operator's side of the local loop between the splitters and the ADSL modems are positioned so-called traffic concentrators. In the uplink direction aggregated traffic (by using concentrators) from all users connected to the DSLAM is transmitted over IP (or ATM-Asynchronous Transfer Mode) backbone network which is connected to the Internet.

In the downlink direction the aggregated traffic is separated to individual ADSL modems used for individual local loops. The DSLAM splitters in this case are used to multiplex POTS and ADSL signals on the subscriber line (i.e., local loop) in the downlink direction. At the end-user premises the received signal is demultiplexed by using splitter, so POTS signals are transferred to a telephone device connected to the splitter while ADSL signals are transferred to the user's ADSL modem.

4.2.3   ADSL Frequency Bands and Modulation

The physical layer (i.e., OSI-1) of ADSL is designed to be able to coexist with standard POTS spectrum. The two services can coexist because ADSL is using a higher frequency spectrum than the baseband spectrum used for POTS. In fact, the frequency band dedicated for voice in POTS is 0.3–3.4 kHz, but the bandwidth between 3.4 and 4.0 kHz is needed as guard band (to avoid interference) for dial-up modem communication specified in V.90 standard [4]. ADSL is full duplex communication achieved either by FDD (Frequency Division Duplex), TDD (Time Division Duplex), or ECD (Echo-Canceling Duplex). However, the implementations of ADSL equipment by vendors usually are based on FDD approach as shown in Figure 4.2. In this case, the ADSL uses spectrum between 25 and 1104 kHz. The lower band between 25 and 138 kHz is used for upstream communication (i.e., uplink), and the higher band between 138 and 1104 kHz is used for the downstream communication (i.e., downlink).

In general there are two different procedures for line encoding in ADSL: DMT modulation, and Carrierless Amplitude Phase (CAP) modulation. CAP was used for first implementations

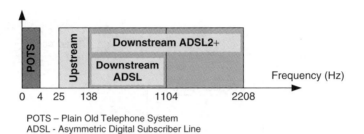

POTS – Plain Old Telephone System
ADSL - Asymmetric Digital Subscriber Line

Figure 4.2   ADSL frequency bands

Table 4.1   Main ADSL standards and bit rates

ADSL name	Downstream bit rate (max) (Mbit/s)	Upstream bit rate (max) (Mbit/s)	Standard name
ADSL	8	1	ANSI T1.413-1998
ADSL (G.DMT)	12	1.3	ITU G.992.1
Splitterless ADSL (G.Lite)	1.5	0.5	ITU G.992.2
ADSL2	12	1.3	ITU G.992.3
Splitterless ADSL2	1.5	0.5	ITU G.992.4
ADSL2+	24	3.3	ITU G.992.5

of ADSL in the 1990s. CAP divides the frequency spectrum of the telephone line in three bands. The band 0–4 kHz is used for POTS, the band 25–160 kHz is allocated for upstream data and the band 240 kHz to 1.5 MHz is used for downstream data transmission. However, the ITU-T standards for ADSL (G.991.1 [2] and G.992.1 [3]) have chosen the DMT. The DMT systems encode data on 256 frequency subcarriers which have width of 4 kHz each, from which 32 are reserved for sending data upstream. Each of these channels independently sends data using Quadrature Amplitude Modulation (QAM).

The successor of ADSL is ADSL2, defined in G.992.3 [5], and its splitterless version defined in G.992.4 [6]. With G.992.5 (i.e., ADSL2+) the used bandwidth is extended from 1.1 to 2.2 MHz, which results in two times higher data rates than ADSL2 (as shown in Table 4.1).

4.2.4   Other DSL Technologies

There are various DSL technologies defined, shown in Table 4.2. The technology that will be applied depends on the service to be provided. For example, ADSL is suitable for asymmetric services such as video on demand (VoD) and WWW. On the other side, for symmetrical services (e.g., video telephony) are more convenient DSL technologies that have symmetric flows in both directions, upstream and downstream. However, important factor for the choice of DSL technology is also the supported length of the telephone line (i.e., local loop length). Nowadays, from all standardized DSL technologies ADSL has incomparably the greatest success on a global scale.

Table 4.2   Comparison of different DSL technologies

DSL type	Maximum rate downstream (Mbit/s)	Maximum rate upstream (Mbit/s)	Maximum local loop length (m)	Number of lines	Shared access with PSTN
ADSL	8	1	5500	1	Yes
HDSL	1.54	1.54	3650	2	No
MSDSL	2	2	8800	1	No
RADSL	7	1	5500	1	Yes
SDSL	2.3	2.3	6700	1	No
VDSL	52	16	1200	1	Yes

ADSL, asymmetric DSL; VDSL, very high bit rate DSL; RADSL, rate adaptive DSL; HDSL, high bit-rate DSL; SDSL, symmetric DSL; and MSDSL, multi rate symmetric DSL.

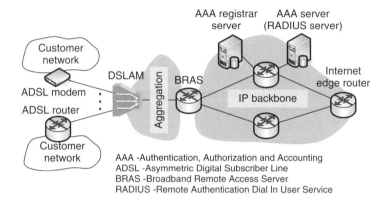

Figure 4.3   ADSL network architecture

4.2.5   ADSL Network Architecture

In Figure 4.3 is shown architecture of an ADSL network. It can be divided into access part and core part of the network. In the access network, on the network's side (on the other end of the local loop) is located the DSLAM. Several local loops end into single DSLAM, which provides aggregation of the traffic from users, and vice versa (toward the end users). Network node between the access part and the core network is ADSL Broadband Remote Access Server (BRAS). In fact, BRAS is a router which is responsible for routing traffic between the DSLAM-enabled access network and the core IP network of the operator (e.g., the Internet Service Provider). It performs aggregation of user sessions from the access network, and vice versa.

The core network for ADSL usually includes two servers, one proxy server and one AAA server (for Authentication, Authorization, and Accounting). The proxy server is so-called access registrar server, while the AAA server in ADSL networks is RADIUS server (Remote Authentication Dial In User Service). RADIUS is a client–server protocol standardized by

IETF (Internet Engineering Task Force) for AAA functionalities in dial-up access to the Internet. It is based on the client-server model [7, 8]. The RADIUS client is located in the BRAS router.

There are two possible implementations depending on whether the ADSL device works in bridged mode or routing mode.

4.2.5.1   ADSL in Bridged Mode

In the case of ADSL in bridged mode the device on the user's side is called ADSL modem and works on OSI-2 layer. ADSL modem needs a PPPoE (Point-to-Point Protocol over Ethernet) client. Since it is not possible to have PPPoE client working on the ADSL modem, it is necessary to start it on a computer (usually that is the computer in which ADSL modem is installed). Most of the operating systems have embedded PPPoE clients.

The PPPoE client encapsulates all Ethernet frames generated by the ADSL modem (from the user's side) with a PPP (Point-to-Point Protocol) header (as overhead) which includes a username and a password entered by the user (Figure 4.4).

PPPoE client sends frame toward the ADSL modem. The modem splits the PPPoE frame (which includes 1500 bytes of Ethernet payload and added PPP header with username and password) into ATM packets, each with fixed length of 53 bytes. The transport between the ADSL modem and the DSLAM is typically carried by ATM technology. Additionally, the ATM may provide a certain Quality of Service (QoS) to the user, which is needed when ADSL is used for triple play services, such as VoIP (voice over IP), IPTV, and best-effort Internet services over ADSL. ATM uses virtual paths and virtual channels, where each path consists of one or more virtual channels. Therefore, on both sides of the ATM connection (i.e., the modem on the user's side and the DSLAM on the operator's side) should be used configurable parameters VPI (Virtual Path Identifier) and VCI (Virtual Channel Identifier). Commonly used default values for these are: VPI = 1 and VCI = 32. On the Internet side the DSLAM merges ATM packets into PPPoE packets, since the transmission after the DSLAM is Ethernet. However, the transport network could be also ATM, but nowadays IP/MPLS (Multi-Protocol Label

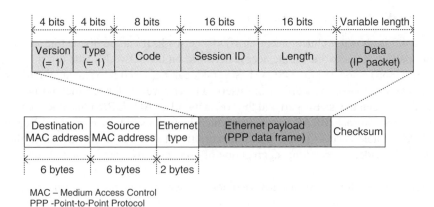

Figure 4.4   PPP over Ethernet (PPPoE) frame

Switching) solutions are almost a default in the design of transport networks while Ethernet is the default choice in the design of local area networks (LAN). Next element in the ADSL network architecture is BRAS. This device is actually a router that acts as a PPPoE server. To handle authentication of the users by username and password carried in PPPoE frame BRAS router is configured as a RADIUS client that sends authentication requests to a RADIUS server. Generally speaking, in ADSL architecture the BRAS router is usually configured to send authentication requests to a proxy RADIUS server. The proxy server separates the username and the domain (by using @ as a delimiter) and then, based on the domain after the delimiter, sends the authentication request to the appropriate RADIUS server, which may be a local RADIUS server or a remote RADIUS server. The AAA server (i.e., RADIUS server) queries its internal database to authenticate the user. Depending on the outcome of the authentication check (based on username and password) proxy server returns the appropriate RADIUS response (i.e., Access-Accept or Access-Reject) to the RADIUS client in the BRAS. In this network architecture DHCP (Dynamic Host Configuration Protocol) server is usually integrated in the BRAS node. So, after authorization and assignment of an IP address the ADSL user can access the global Internet network.

4.2.5.2 ADSL in Routing Mode

For ADSL in routing mode the end-user device works as a router (called ADSL router). In such case the PPPoE client is running in the ADSL router. Usually the authentication credentials (username and password) are set in the ADSL router via its management interface (which uses a Web-based graphical user interface and is accessible via a private IP address such as 192.168.1.1). Simultaneously with the PPPoE client starts NAT (Network Address Translation) process in the ADSL router, which translates private IP addresses allocated to computers and devices in the home network (at the user's premise) to the public IP address allocated to the ADSL router on its interface toward the Internet, and vice versa.

After configuring the PPPoE client in the ADSL router, it accomplishes the same authentication process (via BRAS node and RADIUS server) as when the ADSL device works in the bridged mode.

4.2.6 Cable Access Network

The cable networks, which use coaxial cables as their media, initially were designed to deliver broadcast TV services, first analog TV and then digital TV over the cable. However, with the development of the Internet on a global scale the focus of cable networks has shifted from networks dedicated for TV broadcast over cable to end users, toward triple-play networks which can deliver telephony (VoIP), TV, and data Internet services. The coaxial cable as a medium has better transmission characteristics (much wider frequency spectrum for signal transmission on one side) than twisted-pair, and such bandwidth was needed because TV as a service demands more bandwidth (either frequency bandwidth or bit rates) than voice. On the other side, cable networks were initially designed to carry the TV in downstream direction, which can cause problem for two-way communications (downstream and upstream) as used for all Internet services (e.g., most Internet services are based on a client–server paradigm which uses the request–response principle in opposite directions of the communication path).

There are different models of network access to the Internet through cable networks. Basically, such models differ regarding the implementation of data transmission in the downstream and upstream directions.

One approach is a hybrid network, where data in the downstream direction (i.e., toward the end user) is sent via cable TV network, and for the upstream direction is used another network such as dial-up access via a telephone line. In this approach the bit rate in the downstream can be several tens of megabits per second and cable network continues to operate with unidirectional amplifiers that normally exist in the cable networks to enhance the TV signals. In such case there is no need for additional investments in the cable network, but there is a need for alternative connection for the upstream. Additionally, another drawback of this approach is that the data is received by all users on the same cable in the downstream direction.

Approach which is most used nowadays for Internet services over cable access network is two-way data transmission over the cable network, based on DOCSIS (Data-Over-Cable Service Interface Specifications) [9].

4.2.6.1   Cable Network Architecture

Specifications and architectures for data transmission over cable networks are developed by CableLabs. In fact, CableLabs developed architecture for integrated cable network through which different services are offered to the end users, such as: analog TV and radio, digital TV, HDTV (High Definition TV), VoD, data transfer (i.e., best-effort Internet), and telephone services by using VoIP. The architecture of the integrated wired network is shown in Figure 4.5.

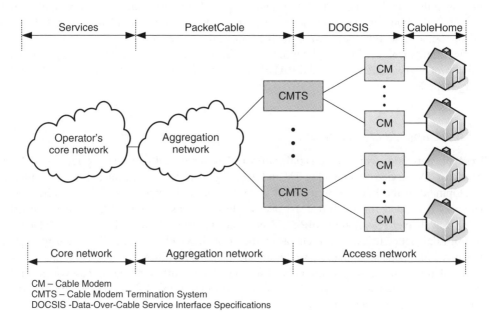

CM – Cable Modem
CMTS – Cable Modem Termination System
DOCSIS -Data-Over-Cable Service Interface Specifications

Figure 4.5   Architecture of integrated cable network

As shown in Figure 4.5 the integrated network with cable access can be divided into several segments, and they are (going from user's side toward the core network of the cable operator): home IP network, DOCSIS, PacketCable, and core IP network, which constitute a functional environment that provides transparent transfer of various types of data from different services, as well as implementation of new services in the future over the same access network.

4.2.6.2   DOCSIS Access Networks

DOCSIS specifications cover the part of the integrated cable network that is used for distribution of services from the cable provider to the end users. Generally, this section covers the Hybrid Fiber-Coaxial (HFC) network equipment that is owned by cable operators, the termination system for the cable access network called CMTS (Cable Modem Termination System) on the operator's side, and the Cable Modem (CM) on the user's side. DOCSIS has several versions of the specification so far:

- *DOCSIS 1.0*: This is the first standard for broadband Internet access via cable access network. It provides the possibility for several devices in the user's home network to be connected via the cable modem to the Internet (via the operator's core IP network), as well as the possibility for the cable operator to limit the access bit rates to the Internet for the given user.
- *DOCSIS 1.1*: This version of the DOCSIS is an enhanced version of the previous one by adding QoS support for throughput and latency control (via configurable parameters), thus providing more flexibility to cable operators and possibility to implement different service level agreements (SLAs) with end users and different business approaches for commercial delivery of the services.
- *DOCSIS 2.0*: Version 2.0 of the DOCSIS standard is targeted to upstream direction, and it enhances the bit rates in the upstream up to 27 Mbit/s. Hence, this version provides possibility for symmetric IP services over the cable network (e.g., video conferencing with higher resolution). However, higher upstream bit rates are necessary for cable access networks due to competition of FTTH access network with high bit rates in both directions (downstream and upstream).
- *DOCSIS 3.0*: The newest DOCSIS standard, notably 3.0, provides several important novelties such as: support for IPv6 (which certainly will be needed toward NGN development in near future), support for IPTV (it provides higher flexibility than traditional TV broadcast over cable networks), as well as higher data rates to and from the end users by using channel bonding on the coaxial cable (in this notation, a channel is a frequency band which is used for delivery of one analog TV channel or multiplexed digital TV channels over the cable). With the channel bonding in DOCSIS 3.0, cable networks can provide bit rates up to at least 170 Mbit/s in downstream and 120 Mbit/s in the upstream for a given end user.

The bit rates of different versions of DOCSIS standards are summarized in Table 4.3. In the downstream direction one may distinguish between DOCSIS (6 MHz bandwidth per TV channel) and Euro DOCSIS (8 MHz bandwidth per TV channel). The given bit rates are aggregate on a given coaxial cable, so they are shared between users which are using single coaxial cable in the DOCSIS access network (e.g., in a residential building).

Table 4.3 DOCSIS bit rates including overhead

Version	Downstream		Upstream
	DOCSIS (Mbit/s)	Euro DOCSIS (Mbit/s)	Both (Mbit/s)
DOCSIS 1.0 and 1.1	42.8	55.62	10.24
DOCSIS 2.0	42.8	55.62	30.72
DOCSIS 3.0 (four channels)	171.52	222.48	122.88
DOCSIS 3.0 (eight channels)	343.04	444.96	122.88

4.2.6.3 PacketCable

The PacketCable specification defines the interface used to enable interoperability of equipment for the transmission of packet-based voice, video and other broadband multimedia services over an HFC network using the DOCSIS. In fact, PacketCable is used to create a super network that unites more two-way broadband cable access networks. Main reason for the development of PacketCable was provisioning of packet-based voice communication for users connected to cable networks. Besides versions for delivery of VoIP over the cable access network, there is also specified PacketCable Multimedia as QoS architecture for real-time IP services (and not only for voice). However, CableLabs considers PacketCable more as a phone-to-phone system (similar to PSTN approach) than voice over public Internet. Its architecture includes call signaling, accounting, configuration management, security, as well as PSTN interconnection. To guarantee the QoS over the DOCSIS access network (which is also used for access to best-effort Internet services, such as Web, e-mail, etc.) the PacketCable services are delivered with guaranteed priority in the DOCSIS access part and that ensures guaranteed bit rates and controlled latency (i.e., packet delay).

There are several versions of PacketCable, as follows:

- *PacketCable 1.0 and 1.5*: The first version defines end-to-end architectures targeted to voice service (i.e., telephony) over DOCSIS-based cable access networks. The voice in this case is provided using so-called managed IP networks instead of public Internet. The exchanges (i.e., switches) and signaling nodes found in PSTN are located in several less expensive general purpose servers in the network. Overall, the architecture of PacketCable is consisted of three networks: DOCSIS HFC access network, managed IP network, and PSTN (Figure 4.6). The node which interconnects the DOCSIS access network and core network (i.e., managed IP network) is CMTS. The managed IP network has a Signaling Gateway (SG) for the control plane, and a Media Gateway (MG) for handling data traffic from/to the users (i.e., the user plane). Version 1.5, among other improvements, extends PacketCable functionalities with SIP (Session Initiation Protocol) used for signaling.
- *PacketCable multimedia*: This is so-called initiative that defines the QoS-based service platform for delivery of different multimedia services over DOCSIS 1.1 (or higher version of DOCSIS) access networks, and it is generally based on PacketCable 1.0. But, it expands certain functionalities of the PacketCable, such as authorization for QoS-based services, admission control functionalities, higher flexibility for accounting and billing, security, and so on.

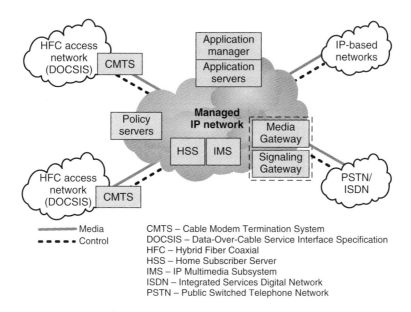

Figure 4.6   PacketCable network architecture

- *PacketCable 2.0*: This version includes an all-IP functionality to the core part of the cable networks, such as SIP functionalities as well as IMS (IP Multimedia Subsystem) [10], as shown in the PacketCable reference architecture shown in Figure 4.7. This way the transport network and services are split, so application servers can deliver services independently of the access network, thus including non-cable access (e.g., wireless access). With the introduction of IMS (which is based on SIP), CableLabs has defined several voice services (or applications) such as SIP telephony, HD (High Definition) voice, and so on.

4.3   FTTH Access Networks

The vision for the future in telecommunications (on the physical layer) is all optical networks (regarding the fixed access networks) with aim to maximize the bit rates. In the past 20 years optical networks have become standard for the transport networks that transfer aggregate traffic from many end-users, regardless of the access networks (fixed or wireless). However, as an evolution, the fiber is moving from transport networks toward the core networks and access networks in the last mile/kilometer. So, broadband in the last mile is targeted to replacement of the copper based access networks (e.g., twisted pairs, coaxial cables) with fiber. There are different architectures for fiber implementation and design in the last mile, which are generally denoted as FTTx (Fiber-To-The x), where x stands instead of cabinet or curb (FTTC), building (FTTB), home (FTTH), premises (FTTP), desk (FTTD). With FTTH optical connection reaches the home of the end user (Figure 4.8), and FTTP (Fiber-To-The Premises) and FTTD (Fiber-To-The Desk) are targeted for usage by small enterprises. However, such approach requires large capital investments, since on average the last mile implementation is two-thirds of the total cost of a given network (including core network, networks nodes, and servers). For example, FTTC will bring optical cable to a service node location near location

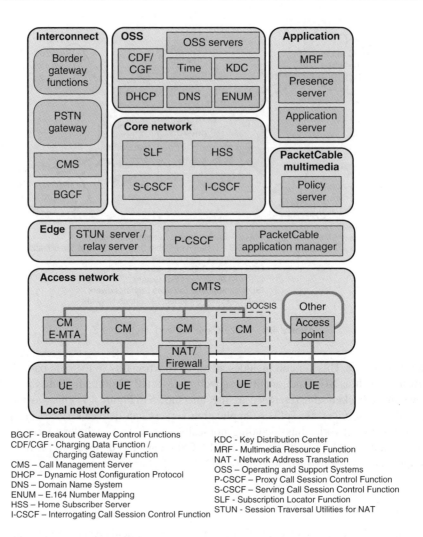

BGCF - Breakout Gateway Control Functions
CDF/CGF - Charging Data Function /
      Charging Gateway Function
CMS – Call Management Server
DHCP – Dynamic Host Configuration Protocol
DNS – Domain Name System
ENUM – E.164 Number Mapping
HSS – Home Subscriber Server
I-CSCF – Interrogating Call Session Control Function

KDC - Key Distribution Center
MRF - Multimedia Resource Function
NAT - Network Address Translation
OSS – Operating and Support Systems
P-CSCF – Proxy Call Session Control Function
S-CSCF – Serving Call Session Control Function
SLF - Subscription Locator Function
STUN - Session Traversal Utilities for NAT

Figure 4.7   PacketCable reference architecture

of the user, which is a more effective solution in terms of cost. Combinations of optical access networks with traditional twisted-pairs broadband solutions (e.g., xDSL) or coaxial access networks (HFC) are intermediate solutions. Hence, the challenges of cost-effective FTTH deployments include minimizing the number of fibers (i.e., the cost of the installed fiber) as well as optical-electronic conversions. All processing of the data (e.g., buffering, scheduling, etc.) is done in the electronic devices, while only the transmission is carried over the fiber.

The speed of deployment of FTTH is driven by the economics, especially by comparison of its cost to competitive technologies such as HFC with cable access networks, and xDSL technologies over twisted-pairs in the local loop (i.e., last mile). The cost of the FTTH network drives the chosen architecture. FTTH architectures can be characterized by: location of electronics (on the user's side, on the operator's side), location of bandwidth aggregation

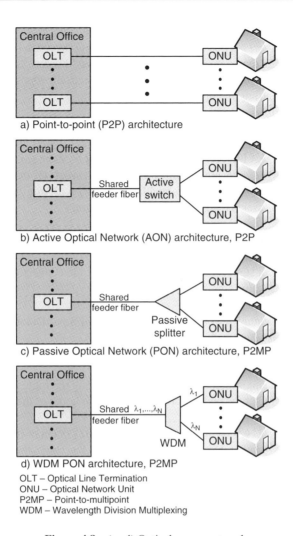

a) Point-to-point (P2P) architecture

b) Active Optical Network (AON) architecture, P2P

c) Passive Optical Network (PON) architecture, P2MP

d) WDM PON architecture, P2MP

OLT – Optical Line Termination
ONU – Optical Network Unit
P2MP – Point-to-multipoint
WDM – Wavelength Division Multiplexing

Figure 4.8   (a–d) Optical access networks

(on the operator's side of the network), bandwidth allocation to the end-user, and applied protocols. Regarding the topology, the fiber access networks can be point-to-point (P2P) or point-to-multipoint (P2MP). Based on these characteristics, there are four basic architectures for the FTTx access network, also called the Optical Distribution Network (ODN):

- *Point-to-point (P2P) architecture*: Uses a separate direct fiber link between the Central Office (CO) of the operator and the home of the user. In this architecture the number of fiber links is equal to the number of homes. This architecture can provide maximum capacity, but is least cost-effective due to dedicated fibers per user. In this architecture electronics is located only at the CO (in Optical Line Termination – OLT) and the home (in Optical Network Termination – ONT). In the case of FTTC the termination node is the Optical Network Unit (ONU).

- *Active Optical Network (AON) architecture (i.e., active star)*: Uses shared P2P fiber link (also called feeder fiber) between OLT and an active remote switch (i.e., curb switch), and P2P links between the remote terminal and ONTs at users' homes. Since active switch performs O/E (Optical/Electronic conversion) and E/O (Electronic/Optical conversion), this is P2P architecture regarding the optical paths.
- *Passive Optical Network (PON) architecture (i.e., passive star)*: Uses shared P2P fiber link between OLT and passive splitters ("passive" means that splitting is performed only by using optics, without any power supply to the unit) between the shared fiber and ONTs. This is P2MP architecture regarding the optical path between the CO and end-users.
- *Wavelength division multiplexing (WDM) PON architecture*: Uses PON architecture (listed above) with WDM (uses several wavelengths per fiber) with aim to increase further the bandwidth to the end user. This is also P2MP architecture.

Nowadays the most important P2MP configuration optical access network is the PON, based on TDM (Time Division Multiplexing). PON is composed of fiber optic cables, passive splitters that distribute optical signals through tree topology of the connectors that terminate each fiber segment. PON has some significant advantages:

- P2MP installations require fewer optical cables to cover a particular area rather than P2P fiber which is used for each individual user.
- Equipment for CO has also a lower price because an optical interface serves a network (with aggregated traffic from many users) rather than a single user.
- PON approach, with small number of active devices along the route, means that the power supply is only required at the ends of optics (CO and the home of the user).

Compared with other access technologies, PON eliminates much of the cost of installation, maintenance, and management. However TDM-PON also has some disadvantages:

- The capacity has to be shared between given number of users.
- Optical splitters divide the optical power between output ports which brings huge losses to the signal. This limits the maximum possible distance between OLT and ONU/ONT.
- PON allows multiple users simultaneously, which can be abused by malicious user that would transmit continuous light in the upstream.
- All ONU/ONT that are connected to the same optical distribution frame receive the same optical signal. This is a benefit in case of multicast traffic, but in the case of unicast traffic it affects the security because all signals are received by all transceivers (in the downstream).

Legacy TDM-PON standards define line bit rates up to 2.5 Gbit/s and a maximum length of links of 20 km. Splitter used for PON is up to 1:32 (one feeder fiber split to 32 connections toward the end users), which is limited by optical power available at ONU/ONT.

In general, FTTH can carry voice, video (including TV), and Internet data services to the end users. In AON and PON solutions it is provided by multiplexing the data on a single wavelength. However, with aim to increase the capacity to the end-users with FTTH the next solution is WDM-based optical access in a given direction (WDM is already used for separation of data in downstream and upstream). WDM solution can be classified into coarse WDM (a few wavelengths used on a single fiber) and dense WDM (many wavelengths used on a single fiber). While Dense Wavelength Division Multiplexing (i.e., DWDM) is targeted for usage in the transport networks [which are migrating from SDH/SONET (Synchronous Digital Hierarchy/Synchronous Optical Networking) toward IP/MPLS networks over DWDM], the coarse WDM is suitable for usage in FTTH access networks. Then, there is possibility to multiplex voice, video, and data on the same wavelengths, or to carry different types of traffic on different wavelengths (e.g., voice on one wavelength, TV on another, etc.).

Similar to PON advantages, cost-effective solution for introduction of WDM in FTTH access networks is WDM-PON. In such case the WDM-PON can be implemented as an upgrade of legacy PON (which uses single wavelength per direction, downstream and upstream). For a transition from PON to WDM-PON there should be changed only transceivers at OLT and ONU/ONT.

4.4   Next Generation Passive and Active Optical Networks

The future of access networks is toward all-optical access due to much higher throughputs that can be achieved via the fiber than the copper media (such as twisted-pairs and coaxial cables). However, current level of supported bit rates to end-users is driven by the applications (which demand such bit rates) and end-user equipment (e.g., computers) that can utilize such bit rates (depending on processing capabilities, memory, etc.) via standardized interfaces (e.g., typically the interface of home devices is via Ethernet or WiFi, both standards from the IEEE 802 group).

4.4.1   PON Standards

The main work on the standardization of optical access networks is done by the ITU and the IEEE. There are three major ITU-T standards for PON:

- *ITU Q.834*: Asynchronous Transfer Mode–Passive Optical Network (ATM-PON) [11]. This standard was developed by the Full Services Access Networks (FSAN) and later adopted by ITU-T. It provides maximum downstream bit rates of 622 Mbit/s [compatible with STM-4 (Synchronous Transport Module) of SDH], and upstream bit rates of 155 Mbit/s (compatible with STM-1 of SDH). It uses ATM technology between the CO and user equipment, and the ODN is passive (i.e., PON).
- *ITU G.983*: Broadband optical access systems based on Passive Optical Networks (BPON) [12]. The ODN is PON, with nominal downstream line rates of 155.52, 622.08, and 1244.16 Mbit/s, and nominal upstream line rates of 155.52 and 622.08 Mbit/s.
- *ITU G.984*: Gigabit-capable passive optical networks (GPON) [13]. The ODN is again PON, with nominal line rates of 2.4 Gbit/s in the downstream direction and 1.2 and 2.4 Gbit/s in the upstream direction. Additionally, GPON provides possibility for symmetric bit rates in the upstream and downstream.

The IEEE has contributed to PON standardization by adding its standard based on Gigabit Ethernet access, and that is:

- *IEEE 802.3ah*: Ethernet Passive Optical Network (EPON) [14]. This standard is based on using Ethernet family of protocols between the CO and the end-users. It is also known as Ethernet in the First Mile (EFM). The standard was completed by 2004 and later it was included in overall updated Ethernet standard IEEE 802.3-2008.

Table 4.4 Standardized PON bit rates

	BPON	EPON	GPON
Standard	ITU-T G.983	IEEE 802.3ah	ITU-T G.984
Downstream bit rate (Gbit/s)	Up to 1.244	Up to 1.25	Up to 2.4
Upstream bit rate	Up to 622 Mbit/s	Up to 1.25 Gbit/s	Up to 2.4 Gbit/s
Downstream wavelength (nm)	1490; 1550	1550	1490; 1550
Upstream wavelength (nm)	1310	1310	1310
Transmission	ATM	Ethernet	Ethernet, ATM, TDM

Table 4.4 summarizes main PON standards regarding the access bit rates, wavelengths used in downstream and upstream, as well as transmission technology. All PON technologies use different wavelengths in different direction, in downstream are used wavelengths 1490 and 1550 nm, while in the upstream the used wavelength is 1310 nm in all PON standards. The highest bit rates are provided with GPON, considering legacy PON technologies outlined in Table 4.4.

4.4.1.1   Broadband Passive Optical Networks

Typical usage of BPON is 622 Mbit/s downstream and 155 Mbit/s upstream, although there are standardized higher data rates as given in Table 4.4. It is based on the ATM, which carries data in fixed-length cells (i.e., packets), 53 bytes each. It followed and extended further the ATM-PON (standardized in Q.834 [12]), and therefore it is sometimes referred to as APON/BPON.

However, there is a difference in protocol requirements in downstream and upstream. In downstream it uses stream of time slots where each time slot carries exactly one 53-byte ATM cell. Initially, G.983.1 specified wavelength window from 1480 to 1580 nm, which was later updated to 1480–1500 nm window, in particular, to wavelength centered on 1490 ± 10 nm. Due to good optical amplification possible around 1550 nm it has been specified in BPON as enhancement band for analog video. In the downstream direction only OLT transmits. But, the upstream direction is more challenging because many ONTs transmit on the same wavelength of 1310 nm. Because different ONTs may have different distance from the OLT due to different lengths of the fiber cables between ONTs and passive splitter, G.983.1 specifies so-called ranging protocol.

BPON architecture is shown in Figure 4.9. Split ratio in BPON is up to 1 : 32, meaning one feeder fiber cable can be split to maximum 32 different ONTs.

4.4.1.2   Gigabit-Capable Passive Optical Network

GPON adapts WDM with two wavelengths for separation of downstream and upstream directions. It adopts two multiplexing mechanisms and they are:

- Broadcast in the downstream direction;
- TDMA (Time Division Multiple Access) in the upstream direction.

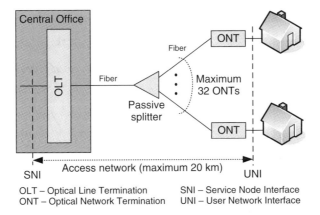

OLT – Optical Line Termination   SNI – Service Node Interface
ONT – Optical Network Termination   UNI – User Network Interface

Figure 4.9   BPON architecture

Regarding the bit rates, generally GPON aims at transmission speed higher or equal to 1.2 Gbit/s. There are two main configuration regarding downstream and upstream bit rates standardized for GPON [13]:

- 2.4 Gbit/s downstream, 1.2 Gbit/s upstream;
- 2.4 Gbit/s downstream, 2.4 Gbit/s upstream.

Regarding the vendor orientation and practical implementations, the most important bit rates for GPON are 2.4 Gbit/s downstream and 1.2 Gbit/s upstream (a legacy implementation).

Reference model of GPON is given in Figure 4.10. It provides maximum logical reach of 60 km and maximum physical reach of 20 km. Logical reach in GPON is the distance between ONU/ONT on the user's side and OLT on the operator's side. Additionally, the distance is limited by maximum signal transfer delay in GPON access network, which should be less than 1.5 ms to be able GPON to provide traditional telecom services such as POTS services, ISDN services, or leased lines (e.g., 2.0 and 1.5 Mbit/s).

Typical split ratios for GPON are 1 : 64, that is, one feeder fiber is shared among up to 64 subscribers. However, split ratios of up to 1 : 128 are also considered for GPON and are dependent upon implementation by the vendors.

4.4.1.3   Ethernet Passive Optical Networks

IEEE standardizes all access networks within its 802 group. Further, fixed access based on Ethernet is 802.3 group, wireless access based on Ethernet principles is within 802.11 (i.e., Wireless Local Area Networks – WLAN), and so on. Regarding the fixed access there are different amendments to the IEEE 802.3 standard, traditionally marked with one or two letters after the standard name. Initially Ethernet was created for access, connecting hosts on distances of couple of hundreds of meters over copper or fiber. In the beginning, Ethernet (such as 10 Mbit/s, 100 Mbit/s, 1 Gbit/s) was based on copper, while Ethernet from 1 Gbit/s onwards (e.g., 10 Gbit/s, standardized in 2002) was based on fiber due to the better transmission characteristics of the fiber (similar to the ITU standards for broadband access).

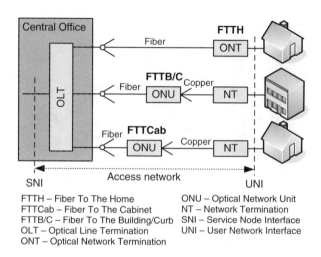

Figure 4.10   GPON reference model

Generally, one may say that Ethernet (together with its wireless version, the IEEE 802.11) is the most spread fixed access networks. That is because Ethernet nowadays is used as an access network (e.g., from a computer to the home gateway node) in homes and in enterprises worldwide. Most of the traffic carried over Internet (with IP packets) starts from an Ethernet and ends in the Ethernet network (including all hosts, either clients or servers). It was initially spread in the business environment (i.e., in enterprises), and in the last decade it also moved into the residential homes, mainly due to the appearance of fixed broadband access to Internet for individual (i.e., residential) users. One may conclude that Ethernet standard has monopolized the LAN.

In 2000 was formed a group titled "Ethernet in the First Mile" within the IEEE with the aim to develop Ethernet-based access networks on longer distances than traditional Ethernet standards. In 2001 was formed a group 802.3ah to work on the standard for EFM, which was completed in 2004. At the same time with the start of the 802.3ah group, in 2001 the Ethernet in the First Mile Alliance (EFMA) was formed by vendors with the aim to promote the technology. In 2005 the EFMA became part of the Metro Ethernet Forum (MEF), an industry consortium dedicated to promotion of Carrier Ethernet networks and services.

EPON uses broadcast mechanism in downstream and TDMA in upstream direction. Transmission is based on Ethernet frames. In downstream each time slot carries frames for all ONTs and each ONT extracts data from its own frame and discards other frames (addressed to other ONTs). In upstream direction each ONT transmits Ethernet frames toward the OLT in the CO by using a transmission time slot that is allocated by the OLT to that ONT. The EPON has a new physical layer compared to Ethernet standards for LAN, but there are little changes in the MAC (Medium Access Control) layer of Ethernet, and the above layers are based on the Internet protocol layering model as in other all-IP networks. On MAC layer EFM introduces in EPON Multi-Point Control Protocol (MPCP). Upstream transmission is controlled by so-called Grants assigned from OLT given to each ONT. This is accomplished in the following way: the ONT caches the data for upstream transmissions and demands bandwidth allocation from OLT (i.e., a Grant). When ONT receives the Grant from the OLT it transmits the data in the specified time slot (obtained from the OLT) with maximum bit rate. However, MPCP does not perform bandwidth assignment or QoS functionalities. Such functions are left to upper layer protocols. The standard, IEEE 802.3ah, does not specify dynamic bandwidth assignments, so vendors should define such algorithms.

Physical network architecture of EPON is P2MP, and each ONT contains MPCP entity which communicates with corresponding MPCP entity of the OLT. Because on the link-layer (i.e., OSI-2 layer) with MPCP each ONT communicates directly with OLT, EPON realizes a P2P emulation sublayer, so physical P2MP architecture of EPON appears as P2P architecture to upper protocol layers, starting from the networks layer (i.e., the IP).

Additionally, EPON does autodiscovery of ONTs without manual interaction by using so-called discovery window (a time-based window) in which ONTs can send register requests to OLT after receiving a message from it for start of that window. If register conflicts (from several ONTs in one discovery window) appear they are solved by adding random delay (within discovery window) before responding or by a response after random overleaping several register grants from the OLT.

Also, EPON does synchronization and ranging, which is needed for the upstream transmissions from ONTs. Because different ONTs may have different distance to the OLT, the RTT (Round Trip Time) between OLT and each ONT is measured and equalization delay is added so all ONTs have the same RTT. Such approach provides possibility to carry TDM-based services (e.g., circuit-switched services such as telephony) over EPON.

4.4.2   Next Generation Passive Optical Networks

The standardization toward the future for PON is also carried by ITU and IEEE. The ITU continues with further standardization toward 10 Gbit/s capable Passive Optical Networks called XG-PON (also referred to as Next Generation Passive Optical Networks 1; i.e., NG-PON1), the G.987 standard [15], and further toward the next generation 40 Gbit/s capable PON (referred to as NG-PON2), the G.989 standard [16].

4.4.2.1   Next Generation PON

G-PON and 1G EPON are deployed world wide after their standardization. Toward the NG-PON, the main tendency is to provide higher bit rates than previous standards. There are two multiplexing factors in the past which will continue to be present in the future as well, and they are: a multiplying factor of four by ITU [e.g., the next generation has a four times higher bit rate from the previous one, which was introduced in 1980s and 1990s, in PDH (Plesiochronous Digital Hierarchy) and SDH technologies], and a multiplying factor of 10 by IEEE (each newer version of Ethernet has 10 times higher bit rates than its predecessor). ITU's NG-PON consists of two PON standards, NG-PON1 (i.e., XG-PON, G.987 series) and NG-PON2 [i.e., 40-Gigabit-capable Passive Optical Networks (XLG-PON), G.989 series].

The XG-PON further extends the standardization of GPON toward higher bit rates in both directions (downstream and upstream) and introduces new wavelengths. XG-PON supports nominal bit rates up to 9.953 Gbit/s (i.e., 10 Gbit/s) downstream and up to 2.488 Gbit/s (i.e., 2.5 Gbit/s) upstream. These bit rates are four times higher in the downstream and two times higher in upstream than legacy G-PON bit rates (legacy bit rates accepted for G-PON by the industry are 2.488 Gbit/s downstream and 1.244 Gbit/s upstream).

XG-PON uses WDM for different directions (downstream and upstream) as other PON standards by the ITU-T. The downstream wavelength is 1578 nm ($\pm$3 nm) and upstream wavelength is 1270 nm ($\pm$10 nm).

The network architecture of XG-PON provides coexistence with a common PON infrastructure and it has full backward compatibility with G-PON (e.g., it can coexist with G-PON in the same ODN). It supports reach up to 60 km.

Regarding the access technologies, XG-PON supports voice, Ethernet access, TDM, xDSL technologies, as well as wireless backhaul. In general, XG-PON provides flexible optical fiber access network for supporting bandwidth requirements for all types of users, that is, individual and business users. Figure 4.11 shows possible XG-PON usage scenarios, as follows:

- *Fiber To The Home (FTTH)*: This scenario is targeted to asymmetric broadband services (e.g., IPTV, VoD, file download, etc.), symmetric broadband services (e.g., e-mail, online

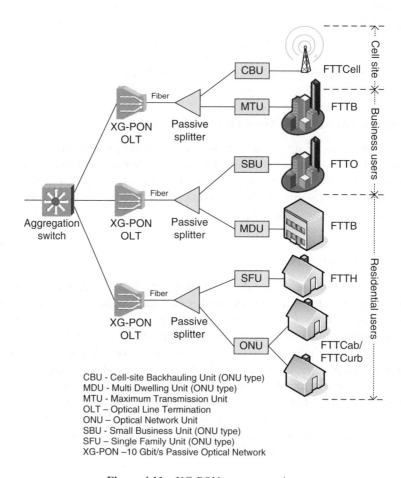

CBU - Cell-site Backhauling Unit (ONU type)
MDU - Multi Dwelling Unit (ONU type)
MTU - Maximum Transmission Unit
OLT – Optical Line Termination
ONU – Optical Network Unit
SBU - Small Business Unit (ONU type)
SFU – Single Family Unit (ONU type)
XG-PON –10 Gbit/s Passive Optical Network

Figure 4.11   XG-PON usage scenarios

gaming, distance learning, file exchange, etc.), and POTS (by either simulation or emulation of narrow-band telephony service).

- *Fiber To The Cell (FTTCell)*: In this scenario the ONU is referred to as Cell-site Backhauling Unit (CBU) and its function is to provide connectivity to base stations of the mobile network, including:
  - Broadband packet-based services, either symmetric or asymmetric, for 3G and 4G mobile networks as mobile backhaul network,
  - Hot spot areas (i.e., WiFi hot spots),
  - Symmetric TDM services (e.g., in 2G mobile networks).
- *Fiber To The Curb or Cabinet (FTTC/FTTCab)*: This scenario provides same services as FTTH such as asymmetric and symmetric broadband services, and POTS. Additionally, this scenario can be used as a backhaul for xDSL access network toward end users.
- *Fiber To The Building (FTTB)*: According to the type of users (e.g., business users, residential users) this scenario can be divided into two scenarios:
  - FTTB for Multi-Dwelling Unit (MDU), which is targeted to residential users for provisioning of asymmetric and symmetric broadband services and POTS (similar to FTTH).
  - FTTB for Multi-Tenant Unit (MTU), which is targeted to business users for provisioning symmetric broadband services (same services as symmetric services for residential users), POTS and private line (at several different bit rates).
- *Fiber To The Office (FTTO)*: In this case the ONU is dedicated to small business environments for provisioning services as those provided for FTTB-MTU scenario, that is, symmetric broadband services, POTS and private line services.

Due to higher bit rates over the same fiber link infrastructure in the access network, G-PON is expected to migrate to NG-PON. The operator that already has a G-PON infrastructure is likely to proceed to a migration scenario to XG-PON in which the two technologies, G-PON and XG-PON, continue to coexist over the same fiber infrastructure (Figure 4.12). In such scenario there is a need for WDM1r as a coexistence wavelength multiplexer, also referred to as a combiner/splitter [15]. This is referred to as brown field migration. When the number of users on G-PON becomes sufficiently low (which is a subjective decision by the operator) then the operator may choose to perform forced migration of remaining G-PON users to XG-PON.

In the cases where legacy G-PON has not been deployed, the deployment of XG-PON is referred to as green field migration. In such a scenario there is no need for the coexistence of XG-PON with G-PONs.

Since telecommunication networks are evolving from traditional telecommunication networks such as PSTN and broadcast networks to the packet-based NGN (which are based on IP on the network layer and Ethernet below the IP, i.e., IP/Ethernet), which can provide different services over a common platform (as elaborated in Chapter 3 of this book). However, existing network infrastructures are not going to disappear in near future, so NGN also provides support for traditional (i.e., legacy) services such as TDM-based access and POTS, either by using emulation (complete replication of traditional service) or simulation (almost the same as the traditional service). Therefore, NG-PON should support POTS and E1/T1 lease lines (i.e., E1 equals to 2.048 Mbit/s, and T1 equals to 1.544 Mbit/s) by either emulation and/or simulation.

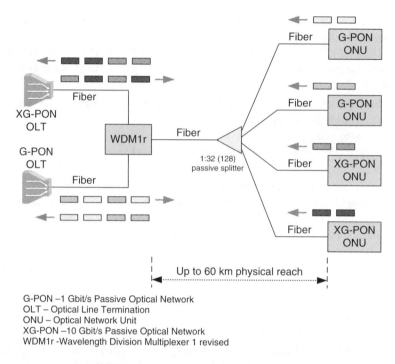

G-PON –1 Gbit/s Passive Optical Network
OLT – Optical Line Termination
ONU – Optical Network Unit
XG-PON –10 Gbit/s Passive Optical Network
WDM1r -Wavelength Division Multiplexer 1 revised

Figure 4.12   Coexistence of G-PON and XG-PON

4.4.2.2   NG-PON2

The follower of XG-PON (i.e., the NG-PON1) is XLG-PON (i.e., NG-PON2), which provides four times higher bit rates than its predecessor. It provides bit rates up to 40 Gbit/s downstream and up to 10 Gbit/s upstream. The given bit rates are achieved by using TDM in downstream direction and TDMA in upstream direction.

As standardizes practice, XLG-PON can coexist with XG-PON as well as G-PON.

4.4.2.3   10G-EPON

IEEE's contribution to 10 Gbit/s passive optical networks, similar to ITU's XG-PON, is the 10 Gbit/s Ethernet Passive Optical Network (10G-EPON). It provides asymmetric access with 10 Gbit/s downstream and 1 Gbit/s upstream, as well as symmetric access with 10 Gbit/s in both directions. Standard is approved by the IEEE in 2009 as IEEE 802.3av [17].

Wavelengths in 10G-PON are chosen to provide coexistence with EPON (i.e., 1G-EPON) and RF video. So, besides asymmetric 10/1 Gbit/s ONUs and symmetric 10 Gbit/s ONUs, the standard enables support for symmetric 1 Gbit/s as well as overlay RF video in the same optical network. For that purpose in downstream 10G-EPON uses three wavelength bands, that is: 1550–1560 nm for RF video, 1480–1500 nm for 1 Gbit/s downstream, and 1575–1580 nm for 10 Gbit/s, all multiplexed on the same fiber. In upstream 10G-EPON OLTs operate in 1260–1360 nm at 1.25 Gbit/s and in the partially overlapping band 1260–1280 nm at 10.3125 Gbit/s (i.e., 10 Gbit/s), thus allowing operation in dual-rate mode.

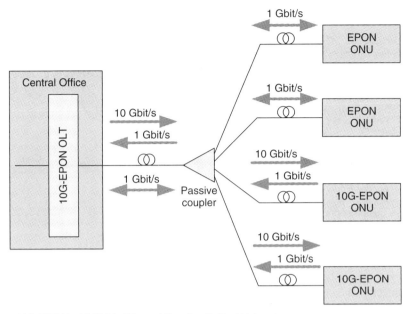

10G-EPON –10 Gbit/s Ethernet Passive Optical Network
EPON – Ethernet Passive Optical Network
OLT – Optical Line Termination
ONU – Optical Network Unit

Figure 4.13   10G-EPON and EPON network architectures

10G-EPON defines a new physical layer while it keeps the MAC layer and layers above it unchanged, so systems for network management and administration (e.g., installed for EPON) can be reused. Also, it uses the MPCP as in EPON, with several changes with aim to provide support for dual-rate operation (1 and 10 Gbit/s), so 1 and 10 Gbit/s ONUs can be implemented in the same network and coexist by sending interleaved bursts at 1 and 10 Gbit/s, as shown in Figure 4.13.

10G-EPON can be deployed at comparable cost to EPON (i.e., 1G-EPON). Low cost coupled with continuously increasing demand for higher bit rates in the fiber access networks accelerates adoption of 10G-EPON as next generation PON, together with the XG-PON.

4.4.3   Next Generation Active Optical Networks

The topologies of AON are very similar to PON. The main difference is in replacement of passive elements (i.e., splitters) in PON with active elements (i.e., switches) in AON.

Today AON are end-to-end Ethernet based networks. AON switches located at the CO in the access network are almost the same as aggregation switches in aggregation part of the network.

Next generation AON are targeted toward mesh-based AON/Ethernet that should result in higher flexibility of the network. Such a Next Generation Access (NGA) network will be connected to home networks via Gateway Sites (GWS) at customer premises. On the other side

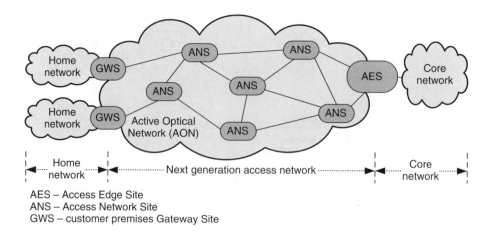

AES – Access Edge Site
ANS – Access Network Site
GWS – customer premises Gateway Site

Figure 4.14   Next generation active optical networks

it will connect to the core network via multi-service Access Edge Sites (AES), as shown in
Figure 4.14. For the purpose of better traffic engineering (e.g., QoS provisioning) and common control plane in access, distribution and core optical networks, next generation AON
may incorporate GMPLS (Generalized Multi-Protocol Label Switching) architecture [18], in
the AON access networks. GMPLS differs from legacy MPLS by adding support for different
types of switching, such as TDM switching, wavelength switching, and fiber link switching.
GMPLS is used as a control plane in wavelength switched optical networks that are initially
used as transport networks. With next generation AON the GMPLS is used in the access
networks as well. It can provide control functions of multi-layered Ethernet networks (e.g.,
Ethernet tunneling), which is covered in the following Metro Ethernet section.

4.5   Metro Ethernet

Most of the infrastructure in operator's core and transport networks was based on SDH in the
past, which is TDM based. Since the Ethernet, as an access network, is the most spread network in business as well as home environments, the MEF has been formed for development of
Ethernet based access in the last mile, or last miles (regarding the metropolitan area orientation
for Ethernet based access). In fact, Ethernet is a natural choice for a technology in core networks and in service domain, when it is already a chosen network by the users and the industry,
mainly due to cost benefits, that is, lower cost of the equipment as well as lower operation and
maintenance costs. On the other side, the end-users which are using Ethernet are demanding
the same level of performance that they had in the past with TDM based wide-area networks.
Such evolution of Ethernet was defined as "Metro Ethernet Network" (MEN) or "Carrier Ethernet" (both names are interchangeable) by the MEF. Hence, the following characteristics are
needed for the Carrier Ethernet [19]:

- *Scalability*: Network operators have requirements to provide broadband access to many subscribers in a given metropolitan or region area.

- *Quality of Service (QoS)*: Carrier Ethernet should provide the same performance guarantees (e.g., regarding the throughput, delay, etc.) as performances received by TDM-based access such as leased lines for business (i.e., enterprise) users.
- *Reliability*: Typical reliability in telecommunications networks (e.g., for provisioning telephony in PSTN and PLMN (Public Land Mobile Networks)) is five nines or 99.999% of the time the network is available for service delivery to end-users. In Ethernet-based environment for carriers is expected to have at least the same reliability as provided in the past by telecom operators via usage of SDH and its ability for link protection (by using redundancy links) and 50 ms link recovery after a failure.
- *Service management*: Service providers have to guarantee the SLAs, therefore it is needed to have tools for performance monitoring of different parameters of the supported services toward the end-users.
- *TDM (Time Division Multiplexing) emulation*: The predecessor of Carrier Ethernet in business environments are the leased lines based on TDM, which are still present in the second decade of the twenty-first century. Hence, Carrier Ethernet should seamlessly interwork with leased line services.

For the Carrier Ethernet purposes the MEF has defined Ethernet-based services in metropolitan areas by using so-called Ethernet Virtual Connections (EVCs). Network (or one may service) providers started using Ethernet connectivity technology to provide Ethernet services. There are defined three types of EVC as follows, called service types [20]:

- *Ethernet Line (E-Line)*: point-to-point EVC;
- *Ethernet Local Area Network (E-LAN)*: multipoint-to-multipoint EVC;
- *Ethernet Tree (E-Tree)*: routed multipoint EVC.

The subscriber network (e.g., computers, switches) attaches to MEN at the User–Network Interface (UNI), which can belong to one of the Ethernet standards such as 10 Mbit/s, 100 Mbit/s, 1 Gbit/s, or 10 Gbit/s. Further, each of the Ethernet services can be provided as a port-based service or virtual service, as shown in Table 4.5.

The EVC-based services are provisioned by using several IEEE standards, from which the most important for Carrier Ethernet is IEEE 802.1Q for virtual LAN [21], IEEE 802.1ad for provider bridges [22], and IEEE 802.1ah for Provider Backbone Bridges (PBB) [23].

Table 4.5   Metro Ethernet services

Type of Ethernet service	Port-based Ethernet service	VLAN-based Ethernet service
E-Line	Ethernet private line (EPL)	Ethernet virtual private line (EVPL)
E-LAN	Ethernet private LAN (EP-LAN)	Ethernet virtual private LAN (EVP-LAN)
E-Tree	Ethernet private tree	Ethernet virtual private tree (EVP-Tree)

4.5.1   Virtual LAN (IEEE 802.1Q)

Provisioning of an E-Line service is based on the IEEE 802.1Q standard for Virtual Local Area Network (VLAN). VLANs are created to provide virtualization of Ethernet infrastructure in enterprises, so the traffic to different departments in the company or different user groups (e.g., ordinary users, network administrators) can be separated into different VLANs. Each VLAN is uniquely identified within the given Ethernet network by a Q-tag (it is also referred to as VLAN ID). The Q-tag field (i.e., VLAN ID) has a length of 12 bits, which results in 4096 different values (from which two are reserved for network administration). Such approach works well in single enterprise. However, when multiple business users (i.e., enterprises) are connected to Internet via a shared Ethernet infrastructure by using virtual LANs, then they should ensure that Q-tags of different VLANs (for different enterprises) do not overlap. So, VLAN is a good solution for a single enterprise. However, to provide higher scalability of Ethernet as a carrier solution for provisioning access to the Internet, there is required mechanism for secure Ethernet connections to individual customers which can further define separate LANs for different departments or groups of users. This is accomplished by provider bridges (IEEE 802.1ad) and provider backbone bridges (802.1ah).

4.5.2   Provider Bridges (IEEE 802.1ad)

Further extension in scalability of Ethernet access networks toward a metropolitan area is provided with provider bridges, introduced by IEEE 802.1ad [22]. Provider bridges add additional tag in the Ethernet frame, called S-tag. Similar to the Q-tag, its length is 12 bits, which results in 4094 possible instances (2 of 4096 values for the S-tag are reserved also for administration). However, the data-link addresses, that is, Ethernet source and destination MAC addresses are the same for provider's networks and for end-user's networks. In this approach VLAN IDs (which identifies different VLANs in a given customer's network) remain unchanged, so provider bridges network appears as one large Ethernet network to the provider's switches. Because there are two tags with the same length (one tag is used as provider ID and the other as VLAN ID), this approach is also referred as Q-in-Q network, as shown in Figure 4.15.

4.5.3   Provider Backbone Bridges (IEEE 802.1ah)

PBB, standardized as IEEE 802.1ah [23], add additional source and destination MAC addresses in Ethernet frames (at UNI) when they enter the PBB network to transfer from one provider's network to another provider's network via a backbone network. This way, switches within PBB network check the added MAC addresses to perform switching according to their forwarding tables. Also, PBB uses an additional tag, called I-tag (it has a length of 24 bits), to identify the services (i.e., Q-in-Q networks), thus separating the PBB network (as a backbone Ethernet network) and service providers' networks. Only switches on the edge of the PBB network should be PBB-enabled, while switches in the core of the PBB network are only provider bridges.

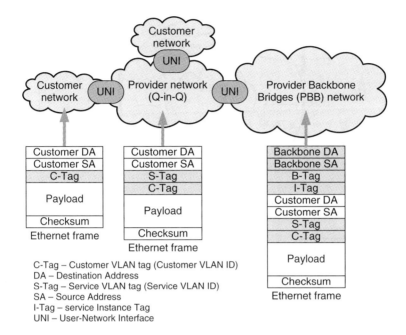

Customer DA
Customer SA
C-Tag
Payload
Checksum

Ethernet frame

Customer DA
Customer SA
S-Tag
C-Tag
Payload
Checksum

Ethernet frame

Backbone DA
Backbone SA
B-Tag
I-Tag
Customer DA
Customer SA
S-Tag
C-Tag
Payload
Checksum

Ethernet frame

C-Tag – Customer VLAN tag (Customer VLAN ID)
DA – Destination Address
S-Tag – Service VLAN tag (Service VLAN ID)
SA – Source Address
I-Tag – service Instance Tag
UNI – User-Network Interface

Figure 4.15   Ethernet architectures with provider bridges and provider backbone bridges

4.5.4   Metro Ethernet for Mobile Backhaul Service

Mobile networks are also going toward all-IP networks, starting from fourth generation (i.e., 4G). However, IP-based access via mobile networks also existed in 2.5G and 3G mobile networks as packet-switched domain (in parallel with circuit-switched domain, mainly dedicated to voice), but transmission between base stations (e.g., referred to as NodeB in 3G radio access networks – RAN) and core network nodes (e.g., a radio network controller in 3G, or gateway nodes in 4G) was designed using TDM based transport (e.g., PDH data structures). The mobile backhaul network can take different forms, something that is dependent upon the required capacity, mobile standards, compatible transport technologies (e.g., with the base stations), preference by the mobile operators, and so on.

When MEN is used by a MEN operator to provide a backhaul network to a mobile operator, then it is done at the UNI reference points, as shown in Figure 4.16. The Radio Access Network Customer Equipment (RAN-CE) is connected to mobile backhaul network via demarcation points. RAN-CE is a generic term to mark the mobile operator's equipment and it can identify a RAN network controller, a gateway, or location with several controllers or gateways, a base station (or NodeB, eNodeB, etc.) or a collection of several base stations of the same or different technologies (including mobile access network as well as transmission interfaces toward the core network).

It is possible RAN CE to have TDM-based interfaces. In such case it can be connected to MEN mobile backhaul by using emulation of TDM services over Ethernet such as Circuit Emulation Service (CES) [24], and PDH circuits (i.e., data structures) [25]. However, if RAN

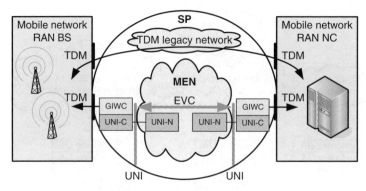

(a) RAN CE with TDM demarcation

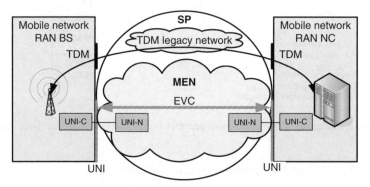

(b) RAN CE with MEF demarcation

EVC – Ethernet Virtual Connection   MEN – Metro Ethernet Network
RAN BS – Radio Access Network Base Stations   SP – Service Provider
RAN CE – Radio Access Network Customer Equipment TDM – Time Division Multiplexing
RAN NC – Radio Access Network Network Controllers UNI-C – User-Network Interface-Customer
MEF – Metro Ethernet Forum   UNI-N – User-Network Interface -Network

Figure 4.16   (a,b) Mobile backhaul network reference model

CE of the mobile operator already has Ethernet interfaces supporting MEN functions, then the mobile operator can obtain all three types of Ethernet services (E-Line, E-LAN, and E-Tree) from the MEN provider (i.e., Ethernet service provider). Following the above discussion, there are two main use-cases for MEN as a mobile backhaul network.

4.5.4.1   Use-Case 1 for Mobile Backhaul: RAN-CE with TDM Demarcation

In the case with TDM-based interfaces at RAN base stations and RAN network controllers/ gateways, there is a need for a Generic Inter-Working Function (GIWF) which has a TDM interface on one side (the RAN-CE's side) and an Ethernet interface with MEN functions on the other side (MEN provider's side). Then, there are further two possible scenarios as follows:

- Scenario with two different mobile backhaul networks – the first over MEN, and the second over legacy TDM-based network (e.g., with PDH data structures, such 2, 34 Mbit/s, etc.). This scenario is convenient is some cases when mobile operator already has established TDM-based infrastructure to RAN-CE (such as base stations) for voice traffic (or higher priority traffic which has low-bandwidth requirements) and wants to offload low priority but high bandwidth traffic (e.g., best-effort Internet traffic such as WWW, e-mail, etc.) from/to RAN-CE, with aim to scale the network with Internet traffic growth due to mobile broadband access.
- Scenario with only MEN, which provides TDM-based services by using the CES of MEN at the TDM demarcation point.

4.5.4.2   Use-Case 2 for Mobile Backhaul: RAN-CE with Ethernet Demarcation

In this use-case the RAN-CE has compliant Metro Ethernet interfaces, referred to as UNI-C (User-Network Interface – Customer), which directly connect to the UNI-N (User-Network Interface – Network) on the side of the MEN provider. For this scenario there is no need for GIWF. However, here are also possible several scenarios, depending upon the type of network used for transfer of the real-time traffic (e.g., voice):

- Offload of low priority traffic high bandwidth traffic onto the MEN, by using the three EVC-based services (E-Line, E-LAN, and E-Tree), while high priority traffic is transferred over a parallel network, independently from MEN.
- All traffic, including real-time and no-real-time traffic is transferred over the MEN. In this scenario there is requirement for synchronization in MEN, which makes this scenario different than previous three scenarios in both use-cases for MEN as mobile backhaul.

Overall, concrete implementation and usage of MEN as a mobile backhaul depends upon the design decisions of the mobile operator.

4.5.4.3   Implementation of Mobile Backhaul with Metro Ethernet Services

Mobile operators most likely are to use VLAN-based services EVPL (Ethernet Virtual Private Line), EVP-LAN (Ethernet Virtual Private Local Area Network), and EVP-Tree. Different types of MEF services are targeted to different connections in the terrestrial segment of RAN. Regarding the technologies in the second decade of the twenty-first century, one should refer to LTE (Long Term Evolution) mobile networks as well as WiMAX networks, which are standardized by 3GPP and IEEE, respectively. In those mobile network, E-Line and EVPL can be used for direct interface between base stations such as X2 for LTE RAN (or WiMAX R8), as well as interfaces between base station and gateways such as S1 for LTE (or WiMAX R6). Further, E-Tree and EVP-Tree with root at the UNI can be used also for X2 interfaces in the aggregation facilities that connect large number of base stations. E-LAN and EP-LAN (Ethernet Private Local Area Network) can be used to support traffic between RAN base stations (eNodeBs in LTE networks) and pooled RAN network controllers/gateways [e.g., Mobility Management Entity (MME) and Serving Gateway (S-GW) in LTE mobile networks].

In many cases of mobile network design, controller/gateway nodes are installed on a single location, which is usually case at network start up. In such case mobile operators have different options to connect RAN base stations with RAN network controllers/gateways, including:

- Port-based implementation with one UNI per RAN base station.
- VLAN-based implementation with EVCs from different RAN base stations multiplexed at one or more RAN network controller/gateway UNIs. However, in the case when multiple EVCs are connected on a single UNI then there is a single point of failure of such connection, so resiliency should be applied (i.e., redundant EVCs) with aim to provide certain level of reliability (e.g., 99.999% availability of the mobile network resources).

In all design approaches mobile networks demand synchronization for mobile applications and services, which is needed to minimize interference in RAN, facilitate handovers in the cellular network (between cells), as well as to satisfy the regulatory requirements. In general, synchronization is distribution of common reference for time and/or frequency to all nodes in a given network with the aim to align their frequency and time scales.

Synchronization is particularly important in all-IP network design scenarios for provisioning of real-time traffic such as voice. It is especially important in mobile environment. Different mobile technologies have different demands for synchronization, demanding that radio signal must be generated in compliance with the frequency, phase, and time accuracy requirements. However, most of the synchronization requirements are related to distribution of a common frequency reference within the mobile network. For example, the deviation of frequency in unit ppb (parts per billion) for 3GPP mobile technologies such as GSM (Global System for Mobile communication) (2G), UMTS (3G), LTE/LTE-Advanced (4G), should be less than ±50 ppb for wide-area base station/eNodeB (e.g., cells with a radius in the range of miles/kilometers), ±100 for local area base station/eNodeB, and ±250 for home base station/eNodeB [26].

Synchronization is very important in the digital telecommunication networks since analog to digital transition in telecommunication at the end of the twentieth century. In fact, the synchronization approach was standardized in digital transport networks in SDH/SONET (the reader may refer to Chapter 1 for more information on SDH/SONET). In a mobile network, in which backhaul between base stations and network controllers/gateways is provided by MEN, there are four main approaches for timing distribution from PRC (Primary Reference Clock) to slave clocks in Radio Access Network base stations (RAN BS), as follows:

- With GPS (Global Positioning System) at the RAN BS.
- Within TDM transport network when legacy TDM (e.g., based on PDH and SDH principles) is also used between RAN BS and core network of the mobile operator.
- With MEN with Ethernet physical layer (i.e., Synchronous Ethernet – SyncE) which is collocated with transport equipment from which the synchronization can be delivered to the RAN BS. This approach is very similar to PDH/SDH approach in legacy TDM networks, and it is specified in ITU-T G.8262 (as Ethernet Equipment slave Clock – EEC).
- With MEN with packet based mechanisms and protocols (e.g., IEEE 1588, Network Time Protocol – NTP specified with RFC 1305, etc.), which are also implemented in a master–slave hierarchy. In this Packet-based Equipment Clock (PEC), positioned at MEN's UNI, is a slave that receives the clock from a PRC (which may be owned either by MEN provider or mobile operator).

4.6 Regulation and Business Aspects

The deployment of broadband technologies worldwide is creating many policy challenges for regulators. The main issues regarding the broadband itself are broadband access and availability. In general, broadband is considered an enabler of economic growth and social inclusion. However, spreading of fixed (i.e., wired) broadband is dependent upon the availability of existing network infrastructure (e.g., twisted pairs or coaxial cables) where it is a shorter time from initial idea to final implementation of the broadband access (e.g., using xDSL over twisted-pairs, or using DOCSIS for broadband access over cable networks). Hence, the penetration of broadband is going with highest speed in developed countries which have present network infrastructure including access and transport networks, while in the less developed parts of the world (which lack a fixed network infrastructure for various reasons) the most prominent way to provide broadband access to the population (i.e., end-users) is based on fixed wireless broadband access (e.g., fixed WiMAX) or mobile broadband access (e.g., UMTS, LTE, mobile WiMAX).

One of the key enablers of broadband access is in fact pricing (when deployment of network infrastructure is not an issue). Pricing at both levels, wholesale level and retail level, can influence all market players including network operators, service providers, and end-users. Because broadband access is important to the society in general (e.g., for online banking, online administration, either local or national, online trade, online shopping, online learning, collaborative work, working on a distance, files and resources sharing, entertainment, richer conversational communication, etc.) regulators and policy-makers should ensure availability and affordability of broadband to the whole population and the same time maintain positive business environment for network providers (e.g., telecom operators) and services providers (e.g., content providers).

However, broadband access is not a single unified product. Broadband technologies are diverse, so their capabilities (e.g., supported bit rates) are not equal. Therefore, for the NGA that certainly will be based on fiber access networks including passive and active ones, it is necessary to distinguish between the already deployed broadband (i.e., existing generation) and NGA broadband. According to certain studies [27], deployed "broadband services" with bit rates between 256 kbit/s and 24 Mbit/s are considered as first generation of broadband access services, and next generation broadband service provides 25 Mbit/s or more.

4.6.1 Regulation of Prices for Broadband Services and Markets

Price regulation for broadband access is one of the tools which may be used to facilitate broadband deployment in a given country. There are two main types of price regulation:

- *Ex-ante regulation*: This is preventive market regulation in a given country by government-specific control, with aim to prevent undesirable outcomes in the broadband markets, or to direct the market development toward higher benefits to the society (e.g., affordable broadband access). This regulation includes measures such as regulation of market concentration level, degree of product differentiation, and so on.
- *Ex-post regulation*: This is enforcement type of regulation which intends to address anti-competitive behavior on the broadband markets or their abuse. This regulation is realized by applying a range of penalties including injunctions, fees, or bans.

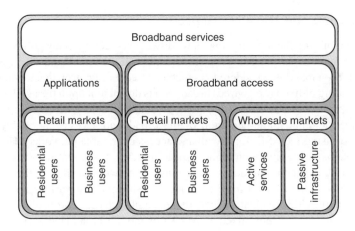

Figure 4.17 Broadband markets and services

With aim to regulate broadband markets and services there is need to differentiate both groups. One such conceptualization is shown in Figure 4.17.

Broadband services can be divided into two main groups [27]: applications and access. Applications belong to the retail market because they are provided downstream to the individual end-users (which can be residential users or business users). So, applications provided to the end-users via broadband access network to Internet are referred to as downstream retail market for broadband. On the other side, access can be classified into two markets: retail markets (consisting of residential retail market and business retail market), and wholesale markets (mainly consisting of passive infrastructure and active services). The wholesale markets for broadband access are typically provided by means of Local Loop Unbundling (LLU), which can be applied either to copper infrastructure in the access network to end users (with xDSL over twisted-pairs) or fiber access networks including both PON and AON. Such markets are provided by leasing or selling access lines toward the end-users to some alternative network operator or service provider, and therefore they are referred to upstream wholesale markets.

Regarding the regulation of different markets in broadband supply chain, one may say that regulation in wholesale markets minimizes the need to regulate the retail markets. Excessive prices in the retail markets (offered to individual end-users) are often a consequence of ineffective competition and dominance in the market of single operator in the upstream wholesale market. Main approach for affordable prices in retail markets is delivering of cost-based pricing by having competitive market. However, in situations where even competition cannot provide desirable pricing, then regulators should use ex-ante regulation methods.

4.6.2 Regulation of Wholesale Prices

Regarding the wholesale broadband markets, the prices naturally go up in cases of uneconomic investments (e.g., investments that have longer cycle of return of investment, or cannot be returned at all) or in cases of protection of the incumbent operator. On the other side, greater efficiency in the use of the existing infrastructure may result in lower prices.

Regulatory pricing control for wholesale broadband markets is usually based on economic costs of the service. If the regulatory pricing is above the economic costs (for both passive and active networks) then in "buy versus build" choice the new entrants on the market will be inappropriately encouraged to build their own network infrastructure. On the other side, if regulatory pricing is below the real economic costs, then the new entrants will be discouraged to build new access network infrastructure and it is more likely that they will lease or buy segments of the existing infrastructure of the dominant operator (in the given wholesale market).

In general, there are two main types of upstream wholesale markets [27]:

- Bitstream access, which can further be distinguished in several types:
  - *Bitstream type 1*: DSLAM access, splitter access for PON;
  - *Bitstream type 2*: Ethernet/ATM level;
  - *Bitstream type 3*: IP level (referred to as managed IP);
  - *Bitstream type 4*: Unmanaged IP.
- Local Loop Unbundling (LLU).

New entrants (i.e., alternative operators) usually enter the broadband markets with bitstream access, starting from unmanaged IP (which does not requires network infrastructure such as routers and links), then managed IP, and so on toward LLU. This is a type of "ladder" for alternative operators entering broadband upstream market, and the regulator should ensure consistent wholesale prices between different types of wholesale markets (e.g., different types of bitstream access). So, when alternative operators reach certain size (in number of users) in certain wholesale market they should be encouraged by the market regulation to go a step up to the upstream wholesale markets, up to LLU and finally toward building its own broadband access infrastructure.

4.6.3   Regulation of Retail Prices

Retail prices for broadband access and services can be controlled by the regulator or by the market itself. Regulator intervention in retail prices may have long-term consequences on the market. So, a practical approach is to have standardized entry-level broadband product that will ensure that broadband is affordable to all residential and business users, and further leave the prices to the market forces (with certain regulation of upstream wholesale markets, where needed).

If the regulator decides to regulate retail prices for broadband, then there are several possible solutions available such as: rate of the return regulation (the regulator specifies maximum rate of return that can be obtained from selling the broadband product to end-users), price-cap regulation (setting up the prices by the regulator), margin (i.e., price) squeeze, benchmarking (with similar markets in similar countries), and so on. However, such regulation of retail prices is usually inappropriate in broadband retail markets, due to heterogeneous nature of the broadband access (which is different than traditional markets related to PSTN and PLMN, or leased lines) regarding the available bit rates as well as provided services and applications.

4.7   Discussion

The world is becoming more and more dependent on broadband access to the Internet. Due to its importance for economic growth and development, globally deployment and greater access to broadband networks is given high priority. For example, the ITU has highlighted the importance of "broadband" by identifying broadband as a key target initiative in the second decade of the twenty-first century.

Broadband access means access to the Internet with high bit rates by using ubiquitous networks, devices, and applications. It is also one of the main drivers for the development of NGN. ITU's ADSL standards offered the first real broadband technology to open up an entirely new experience for a new generation of Internet users. Continuing work in broadband access is focused on standards in fiber optics such as FTTH optical access networks, including PON and AON, as well as MEN which are being widely deployed by carriers.

Finally, regulation and business approaches in broadband are targeted to global universal broadband affordability. Overall, broadband access is a requirement and also main driver toward the NGN.

References

1. ANSI (1998) ANSI T1.413 issue 2. *Network and Customer Installation Interfaces – Asymmetric Digital Subscriber Line (ADSL) Metallic Interface*, ANSI.
2. ITU-T (1999) Asymmetric Digital Subscriber Line (ADSL) Transceivers. ITU-T Recommendation G.992.1, June 1999.
3. ITU-T (1999) Splitterless Asymmetric Digital Subscriber Line (ADSL) Transceivers. ITU-T Recommendation G.992.2, June 1999.
4. ITU-T (1998) A Digital Modem and Analogue Modem Pair for use on the Public Switched Telephone Network (PSTN) at Data Signaling Rates of up to 56 000 bit/s Downstream and up to33 600 bit/s Ppstream. ITU-T Recommendation V.90, September 1998.
5. ITU-T (2009) Asymmetric Digital Subscriber Line Transceivers 2 (ADSL2). ITU-T Recommendation G.992.3, April 2009.
6. ITU-T (2002) Splitterless Asymmetric Digital Subscriber Line Transceivers 2 (splitterless ADSL2). ITU-T Recommendation G.992.4, July 2002.
7. Rigney, C., Willens, S., Rubens, A., and Simpson, W. (2000) Remote Authentication Dial in User Service (RADIUS). RFC 2865, June 2000.
8. Rigney, C. (2000) RADIUS Accounting. RFC 2866, June 2000.
9. CableLabs (2013) Data-Over-Cable Service Interface Specifications – DOCSIS 3.0: MAC and Upper Layer Protocol Interface Specification.
10. CableLabs (2011) PacketCable IMS Delta Specifications: Session Initiation Protocol (SIP) and Session Description Protocol (SDP); Stage 3 Specification. 3GPP TS 24.229.
11. ITU-T (2004) ATM-PON Requirements and Managed Entities for the Network and Network Element Views. ITU-T Recommendation ITU-T Q.834.1, June 2004.
12. ITU-T (2005) Broadband Optical Access Systems Based on Passive Optical Networks (PON). ITU-T Recommendation G.983.1, January 2005.
13. ITU-T (2008) Gigabit-Capable Passive Optical Networks (GPON): General Characteristics. ITU-T Recommendation G.984.1, March 2008.
14. IEEE (2004) IEEE 802.3ah. *Ethernet in the First Mile*, IEEE.
15. ITU-T (2010) 10-Gigabit-Capable Passive Optical Networks (XG-PON): General Requirements. ITU-T Recommendation G.987.1, January 2010.

16. ITU-T (2013) 40-Gigabit-Capable Passive Optical Networks (NG-PON2): General Requirements. ITU-T Recommendation G.989.1, March 2013.
17. IEEE (2009) IEEE 802.3av-2009. *Local and Metropolitan Area Networks – Specific Requirements Part 3: Carrier Sense Multiple Access with Collision Detection (CSMA/CD) Access Method and Physical Layer Specifications Amendment 1: Physical Layer Specifications and Management Parameters for 10 Gbit/s Passive Optical Networks*, IEEE, October 2009.
18. Mannie, E. (ed.) (2004) Generalized Multi-Protocol Label Switching (GMPLS) Architecture. RFC 3945, October 2004.
19. Sanchez, R., Raptis, L. and Vaxevanakis, K. (2008) Ethernet as a carrier grade technology: developments and innovations. *IEEE Communications Magazine*, **46**(9), 88–94.
20. MEF (2008) Ethernet Services Definitions – Phase 2. MEF Technical Specification, MEF 6.1, April 2008.
21. IEEE (2006) IEEE 802.1Q – 2005. *Virtual Bridged Local Area Networks*, IEEE, May 2006.
22. IEEE (2006) IEEE 802.1ad – 2005. *Virtual Bridged Local Area Networks – Amendment 4: Provider Bridges*, IEEE, May 2006.
23. IEEE (2004) IEEE 802.1ah – 2008. *Virtual Bridged Local 14 Area Networks – Amendment 6: Provider Backbone Bridges*, IEEE, August 2008.
24. MEF (2004) Circuit Emulation Service Definitions, Framework and Requirements in Metro Ethernet Networks. MEF Technical Specification, MEF 3, April 2004.
25. MEF (2004) Implementation Agreement for the Emulation of PDH Circuits Over Metro Ethernet Networks. MEF Technical Specification, MEF 8, October 2004.
26. MEF (2012) Mobile Backhaul Phase 2. MEF Technical Specification, MEF 22.1, January 2012.
27. ITU (2012) Regulatory and Market Environment – Regulating Broadband Prices, Broadband Series, Telecommunication Development Sector.

5

Mobile Broadband: Next Generation Mobile Networks

5.1 ITU's IMT-Advanced: the 4G Umbrella

The ITU (International Telecommunication Union) is setting the definition for the next generation mobile networks referred to as 4G (i.e., fourth generation of mobile systems and networks). They are targeted to provide higher data rates to mobile users, and therefore can be referred to as mobile broadband networks and technologies. The requirements for 4G radio interface are specified in ITU-R (International Telecommunication Union Radiocommunication) report M.2134 [1], and are referred to as IMT-Advanced (International Mobile Telecommunications-Advanced). Similar approach was used for the definition of the third generation of mobile networks (3G) which was named IMT-2000. When 3G was already on the ground, with its first implementations, the future development of mobile networks (beyond 3G) was specified in the ITU-R recommendation M.1645 [2].

IMT-Advanced is in fact umbrella specification of all requirements set to a given mobile system with the aim to use a 4G label on it. So, a mobile network is 4G if it satisfies all requirements set in IMT-Advanced [1]. There are several key features required for IMT-Advanced systems:

- High degree of commonality of functionality worldwide while still being flexible to support different applications and services;
- Compatibility of services in IMT environment (i.e., 4G) and fixed networks;
- Interworking with other radio systems, including previous generations of mobile technologies (e.g., 3G);
- Mobile services with high quality;
- Ubiquitous end-user mobile equipment (i.e., mobile terminals);
- User-friendly services and applications;
- Roaming capability on a global scale [something that is present with GSM (Global System for Mobile communication) mobile networks of the second generation];
- Higher bit rates in the radio interface, which in the time of 4G development and implementation (second decade of the twenty-first century) means over 100 Mbit/s for high mobility and over 1 Gbit/s for low mobility of the users.

NGN Architectures, Protocols and Services, First Edition. Toni Janevski.
© 2014 John Wiley & Sons, Ltd. Published 2014 by John Wiley & Sons, Ltd.

In some way to limit different speculations about the expectations from certain technologies, ITU-R states what is high bit rate in the umbrella recommendations such as IMT-Advanced (specified to define the requirements for 4G mobile networks) and previously IMT-2000 (specified to define the 3G mobile technologies). Then, there are two approaches to increase bit rates:

- To increase the spectrum that can be used by the given (4G) technology;
- To increase the spectrum efficiency, that is, to have more bits per second per hertz in the radio interface.

Bit rates that are targeted to individual users in 4G (i.e., Mobile Broadband) are in the range from 1 to 3 bit/s/Hz/cell. For example, targeted average cell spectral efficiency for IMT-Advanced systems is 1.1 bit/s/Hz/cell in downlink and 0.7 bit/s/Hz/cell in uplink in cases of high speed mobility, in urban areas are specified 2.2 bit/s/Hz/cell in downlink and 1.4 bit/s/Hz/cell in uplink, while the spectral efficiency is highest for the indoor cells where it is set to 3.0 bit/s/Hz/cell in downlink and 2.25 bit/s/Hz/cell [1].

To provide more bit/s/Hz, that is, higher spectral efficiency, it is using multiple signals in parallel over the radio interface (between the mobile terminals and the base station-BS) by using multiple antennas (for sending or receiving), something that is known as MIMO (Multiple Input Multiple Output). For example, 4×4 MIMO means four output signals are transmitted in parallel, and four input signals are received in parallel on the other end. The other way for increasing the spectral efficiency is using modulation schemes that can carry more bits per symbol such as 64 QAM (Quadrature Amplitude Modulation). QAM uses different amplitudes and phases of the signal transmitted over the radio network, thus leading to more bits per symbol (i.e., the simplest modulation is based only on different amplitudes of the signal, called BPSK – Binary Phase Shift Keying). For example, with 64 QAM each symbol carries 6 bits (because $64 = 2^6$), while with 16 QAM each symbol carries 4 bits (because $16 = 2^4$), so with 64 QAM one obtains 1.5 times higher spectral efficiency than with 16 QAM. Similar, by 4×4 MIMO we have four signals in parallel (at the same time) over the radio interface, so it increases spectral efficiency four times compared with SISO (Single Input Single Output) antenna systems.

The minimum requirements for peak spectral efficiencies for 4G mobile systems are 15 bit/s/Hz in downlink and 6.75 bit/s/Hz in uplink. Such required spectral efficiencies are given for 4×4 antennas in downlink direction and 2×4 in uplink. Another way for increasing the bit rates is by adding more bandwidth (i.e., frequency spectrum) to be available to a given connection. So, IMT-Advanced requires at least of up to 40 MHz scalable spectrum allocations to be available to mobile users of 4G systems. However, bandwidth up to 100 MHz was also encouraged.

Further, cell edge spectral efficiency is also addressed in the requirements with aim to avoid interference, especially in the cases when same frequency bands are used in adjacent cells in the mobile network, and at the same time the signal on the cell edges has lowest powers in both directions (downlink and uplink). So, the targeted spectral efficiency at the cell edges in IMT-Advanced is around 30 times less than the cell spectral efficiency.

Besides the radio access network (RAN), the IMT-Advanced also defines certain requirements for mobile core networks. Such requirements is the packet-based principle for the mobile network including the access and the core networks, which means that IMT-Advanced are all-IP networks (since packet-based networks today are IP networks).

Another requirement for IMT-Advanced systems is the latency (i.e., delay). There are several types of delay defined, some in the control plane (for signaling) and some in the user plane (for real-time traffic such as VoIP-Voice over IP). Regarding the control plane, the latency is typically measured as time needed for transition from idle to active state, and it should be less than 100 ms. In user plane, the requirements for latency are toward the latencies less than 10 ms in unloaded conditions in the mobile network (i.e., one user with single data stream and small IP packets). All requirements for low latencies in IMT-Advanced networks are mainly targeted to VoIP. Since IMT-Advanced is an all-IP network then it means that telephony should be implemented as VoIP, certainly with QoS (Quality of Service) support (i.e., limited end-to-end delay for VoIP). One may conclude that VoIP is the main driver toward strict latency requirements in IMT-Advanced, due to limited end-to-end delay needed for voice (should be 150 ms for best quality, but higher end-to-end delay than 400 ms is unacceptable). These requirements on the delay also influence the network architecture for 4G mobile networks, which is flat one (only BSs and centralized gateways) to avoid unnecessary delays in intermediate nodes [e.g., Radio Network Controllers (RNCs) in legacy 3G networks]. The importance of VoIP (as the successor of circuit-switched voice) in IMT-Advanced is also pointed by requirements regarding the VoIP capacity in 4G. For example, minimum active VoIP users per sector per MHz is 50 for indoor BSs, 40 VoIP users/sector/MHz for microcellular and urban coverage areas, and 30 VoIP users/sector/MHz for users moving with high speeds. Such VoIP capacities are given as requirements in IMT-Advanced, assuming 12.2 kbit/s VoIP codec, 50% activity factor [average ratio between "on" periods (i.e., talk spurts in active VoIP connection) and total duration of the connection], and 2% outage probability (i.e., less than 2% of VoIP packets is allowed not to be successfully delivered within a one-way radio access delay limit of 50 ms).

Other important issues in mobile networks are mobility and handovers (as a result of the mobility of the users). The IMT-Advanced systems shall be able to provide handover latencies (i.e., handover interruption times) less than 60 ms for handovers between different spectrum bands, less than 40 ms for handovers within the same spectrum band, and less than 27.5 ms for intra-frequency handovers.

In 2008 ITU-R opened an invitation for the submission of candidate mobile technologies for 4G. In October 2009 there were six proposals to ITU-R (and its Working Party 5D), which were centered on the 3GPP (3G Partnership Project) Release 10 and beyond (on one side), and IEEE 802.16m standardization (on the other side). In October 2010 ITU-R announced that it had completed the assessment of six candidates for 4G mobile wireless broadband technologies, that is, IMT-Advanced. Harmonization of the six proposals resulted in two technologies to be accepted as 4G:

- LTE-Advanced (from the 3GPP);
- WirelessMAN-Advanced (i.e., IEEE 802.16m, also known as Mobile WiMAX Release 2 in the WiMAX Forum).

Overall, mobile broadband in the second decade of the twenty-first century means 4G mobile technologies, so the following sections outline the most important aspects of them, the LTE-Advanced and Mobile WiMAX 2.0.

5.2   4G Standard by 3GPP: LTE/LTE-Advanced

The 3GPP evolution, which started with GSM as 2G mobile systems, then continued with UMTS (Universal Mobile Telecommunication System; as a 3G mobile system), has resulted in LTE (Long Term Evolution) and further LTE-Advanced as 4G mobile systems. Different

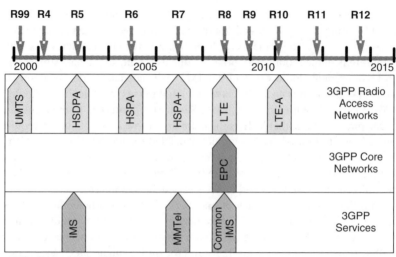

EPC – Evolved Packet Core
HSDPA – High Speed Downlink Packet Access
HSPA – High Speed Packet Access
IMS – IP Multimedia Subsystem
LTE – Long Term Evolution
MMTel – Multimedia Telephony
UMTS – Universal Mobile Telecommunication System

Figure 5.1   3GPP releases timeline

standards for mobile technologies are published by 3GPP as releases, which have started from first 3G standards (they also included continuing development of 2G and 2.5G standards by 3GPP). The first release was Release 99, followed by Release 4, then Release 5, and so on, as shown in Figure 5.1. There are in general three segments that are standardized by the 3GPP:

- *High-speed access*: This includes radio access technology.
- *IP core network*: This includes all controllers/gateways and databases in the networks as well as their interconnection.
- *Services*: This part includes the service overlay network which are implemented over a given mobile network architecture (typical example for services is standardization of the IP Multimedia Subsystem – IMS).

5.2.1   LTE/LTE-Advanced Standardization

The IP based access in mobile networks standardized by 3GPP started with the implementation of GPRS (General Packet Radio System) in the second part of 1990s. In fact, the packet-switching in the core network was introduced in GPRS with two network nodes: Serving GPRS Support Node (SGSN) and Gateway GPRS Support Node (GGSN). Using the same RAN as GSM, based on FDMA (Frequency Division Multiple Access) and TDMA (Time Division Multiple Access), it added another part of the core network, a packet-based core network in parallel with the CS core network of the GSM. Later, EDGE (Enhanced Data Rates for global Evolution) added new modulation scheme in the FDMA/TDMA-based GSM RAN, called GMSK (Gaussian Minimum Shift Keying), which increased the spectral efficiency in the radio network. One may note that initial two nodes added for GPRS will be main nodes for user and control traffic in all 3GPP releases after the GPRS.

The 3G standardization started within 3GPP with Release 99, completed in 2000. It standardized the coexistence of the domains, CS domain and Packet-Switched domain (PS). For the PS domain 3GPP considered two technologies, namely ATM (Asynchronous Transfer Mode) and IP. However, after that, Internet technologies have clearly won the battle with ATM and further 3GPP releases were based on the IP paradigm. The third generation was based on WCDMA (Wideband Code Division Multiple Access) radio access technology, with bit rates up to several megabits per second. The typical carrier width for 3G radio interface is 5 MHz, while in 2G the spacing between frequency carriers was 200 kHz. So, 3G was characterized with carriers with fixed width, something similar to 2G (although 25 times wider bands for frequency carriers are used in 3G than in 2G mobile networks from the 3GPP).

The move toward higher data rates in the radio interface, and next step toward the development of mobile broadband by 3GPP, was 3GPP Release 5 and definition of HSDPA (High Speed Downlink Packet Access). Higher bit rates are easier to introduce in downlink direction due to the fact that BSs (called NodeBs in 3G) have continuous power supply and also they are not limited by the maximum transmitting power unlike mobile terminals. In fact, mobile terminals are limited by their battery life on one side and their maximum transmitted power in the presence of a human body on the other side (in both cases it refers to the uplink direction, that is, when mobile terminal transmits). The enhanced bit rates in the uplink were introduced in 3GPP Release 6 as HSUPA (High Speed Uplink Packet Access), and merged with HSDPA into HSPA (High Speed Packet Access) in Release 7. Additionally, Release 7 introduced more efficient modulation schemes in downlink (i.e., 64 QAM) and uplink (i.e., 16 QAM), as well as MIMO concept for even higher bit rates.

The evolution of all three parts of a mobile network (radio access technology, core network, and services) resulted in standardization of the next step in the evolution of 3GPP in its Release 8 which brought LTE in the radio part, System Architecture Evolution (SAE) for the core network, and IMS for the services. So, regarding the road to 4G, Release 8 has significant importance. The next release (Release 9) standardized the leftovers from Release 8. However, LTE does not satisfy all requirements set by ITU-R for the IMT-Advanced systems, particularly requirements for bit rates above 1 Gbit/s in downlink for nomadic mobile users. Therefore, LTE is usually referred to as 3.9G, but in reality it will be marketed as a 4G technology, due to its similarity with LTE-Advanced (in many aspects).

The road to LTE continued toward LTE-Advanced in 3GPP Release 10, which is followed by releases 11 and 12. Both LTE and LTE-Advanced use OFDMA (Orthogonal Frequency Division Multiple Access) in the radio interface in downlink, and Single Carrier Frequency Division Multiple Access (SC-FDMA) in the uplink. As its name denotes, LTE-Advanced is advanced version of the LTE that also has more similarities than differences with it, such as the same core network, the same radio access technology (on the physical layer), and the same IMS for the services.

5.2.2   System Architecture Evolution

3GPP mobile networks consist of two main overlay network architectures, one for user traffic (i.e., voice, Web, etc.) and other for control traffic (i.e., authentication, signaling for voice, etc.). With the convergence of 3GPP networks toward all-IP, certain changes were needed in

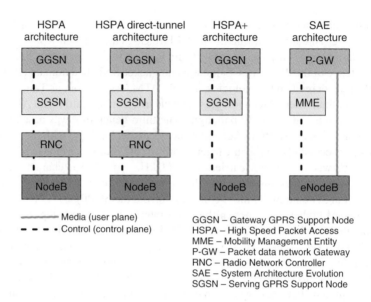

Figure 5.2   3GPP architecture evolution toward SAE (system architecture evolution)

the mobile network architecture with aim to provide lower delay for delay-sensitive traffic such as voice when it is transferred over IP end-to-end. That led to change of hierarchical network architecture (i.e., with RNC until Release 7) to a flat architecture. Lower number of nodes on the path of IP packets are required to decrease the delay budget (i.e., total delay due to various causes) to fit into the end-to-end delay budget for voice when it is carried in a form of VoIP. First move toward flat architecture was in 3GPP release 7 with the introduction of so-called direct tunnel, and finally the flat architecture was standardized completely in Release 8, as shown in Figure 5.2. It is known as System Architecture Evolution and it defines the core network for the LTE mobile networks (also standardized in Release 8).

The main characteristic of SAE is a simplified mobile network architecture, consisting of only two tiers: BSs (i.e., eNodeBs) and centralized gateways. The architecture minimizes the latency through the mobile backhaul network (between the gateways and BSs) and at the same time provides support for higher bit rates which are available to the mobile users via the higher bit rates in LTE radio interface, further increased with LTE-Advanced. Also, SAE provides support for interconnection with heterogeneous access networks, including LTE/LTE-Advanced, UMTS/HSPA, GPRS/EDGE, as well as mobile networks from IEEE and 3GPP2 such as WiMAX, WiFi, and cdma2000. Such a core and backhaul mobile network architecture is also used in further 3GPP releases after Release 8, that is, in Release 9 for LTE and in Releases 10–12 for LTE-Advanced. Hence, SAE architecture suits well the all-IP nature of the LTE/LTE-Advanced. Additionally, in SAE are separated control and user traffic (i.e., control and user plane), meaning that different planes are serviced by different gateways in the core network.

The main part of the SAE is the so-called Evolved Packet Core (EPC), which is the core network for LTE and LTE-Advanced RANs. However, the term "evolution" is not accidental, because the network nodes in the EPC are derived from the network nodes in GPRS (the initial packet-based network architecture by 3GPP, standardized in late 1990s), SGSN, and GGSN.

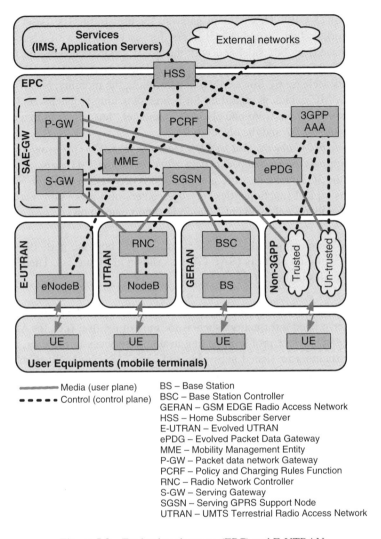

Figure 5.3   Evolved packet core (EPC) and E-UTRAN

However, due to separation of the user plane and control plane, there are three main gateway nodes in the EPC, as shown in Figure 5.3:

- *Mobility Management Entity (MME)*: This is the main control element in the LTE core network (i.e., the EPC), which is responsible for signaling-related mobility management (MM) including user tracking, paging procedures, as well as bearers activation and deactivation. Because mobility is related to roaming between different mobile networks MME has functionality for authentication of the user, and therefore it uses signaling interface toward the users' database HSS (Home Subscriber Server). The Non-Access Stratum (NAS) signaling terminates in MMS (Multimedia Messaging Service) as well. Also, the MME provides interfaces for interconnection with previous 3GPP core networks from 2G and 3G, for handling the mobility between the heterogeneous mobile networks.

- *Serving Gateway (S-GW)*: This is main gateway for user traffic (i.e., user plane) within the core network, which establishes bearers for user traffic with eNodeBs (BSs in LTE/LTE-Advanced) on one side, and with the P-GW (PDN Gateway) on the other side (toward the global Internet). Also, S-GW has defined interface with RNCs of UMTS (i.e., up to 3GPP Release 6). Since MME is controlling the mobility, S-GW has a control interface with MME, which is important for bearer switching at handovers (i.e., a mobile terminal makes a handover from one eNodeB to another, thus resulting in a change of the bearer between the S-GW and the eNodeBs).
- *Packet Data Network Gateway, that is, PDN Gateway (P-GW)*: This provides connectivity from mobile terminals to external packet data networks (PDN). P-GW also acts as an anchor point between the 3GPP core network and non-3GPP networks (i.e., WiFi, WiMAX, etc.). P-GW performs typical functionalities for edge routers of a given core network, such as packet filtering, policies enforcement, charging support, and so on.

S-GW and P-GW are connected via S5/S8 interface, and both are parts of so-called System Architecture Evolution gateway (SAE GW), as shown in Figure 5.3. There are two other important control nodes in the EPC:

- *Home Subscriber Server (HSS)*: This is a centralized database in the mobile network which contains the user-related information, such as location of the user (i.e., the MME to which it connects) and subscriber profile including available services to the user, allowed PDN connections or roaming in visited networks, and so on. User authentication is performed via the Authentication Center (AuC), which is usually a part of the HSS.
- *Policy and charging rules function (PCRF)*: This is a software-based network node in the EPC which is responsible for Policy and Charging Control (PCC) [3]. PCRF detects the service flow and determines policy rules for it in real time. Also, it enforces charging policy in the mobile network. In the SAE architecture the PCRF provides PCC information for bearer setup to the enforcement function located in P-GW, which is named PCEF (Policy and Charging Enforcement Function). When user applications communicate directly with the IMS, the Application Function (AF) in service domain [i.e., service stratum according to the NGN (Next Generation Network) terminology] requires from PCRF to apply PCC rules for dynamic policy or charging control. On the other side PCRF is a central node in the NGN architecture connecting service stratum and transport stratum.

The RAN in LTE and LTE-Advanced is called Evolved UMTS Terrestrial Radio Access Network (E-UTRAN) and it consists of eNodeBs (i.e., BSs) and interfaces on the links to centralized gateway (MME and S-GW) on one side, and to mobile terminals (i.e., User Equipment – UE) on the other side. The radio interface (between mobile terminals and BSs) is referred to as Evolved UMTS Terrestrial Radio Access (E-UTRA). The EPC, E-UTRAN, and E-UTRA form the IP connectivity layer in LTE/LTE-Advanced mobile networks, called Evolved Packet System (EPS).

In E-UTRAN the RNC (which existed in UTRAN in 3G) is completely removed from the network architecture and radio resource management is completely transferred to the BSs, that is, the eNodeBs. With the aim to have lower handover latencies (which is needed for real-time services such as VoIP), E-UTRAN has a defined direct interface between adjacent eNodeBs, called the X2 interface. The interface between eNodeBs and the centralized gateways in the

EPC is S1 interface, which is defined for control traffic (between eNodeBs and MMEs) and for user traffic (between eNodeBs and S-GW). The X2 interface between eNodeBs provides possibility horizontal handover decision (to another eNodeB in the same LTE/LTE-Advanced network) to be made by the eNodeB, and a "temporary" bearer is established between the old eNodeB (before the handover execution) and the new eNodeB (after the handover) for rerouting of all packets that have been addressed to the mobile terminal and has passed the S-GW toward the old eNodeB before the bearer switching between S-GW and eNodeBs (from the old eNodeB to the new one) which is completed by the control signaling from the MME to which the mobile terminal (i.e., UE) is attached.

5.2.2.1   SAE Protocols and Interfaces

The EPC user plane gateways, S-GW and P-GW, use IP packet switching [Open System for Interconnection (OSI) layer 2] and routing (OSI layer 3) technologies. Both nodes perform QoS enforcement, acting as Policy Enforcement Points (PEPs). S-GW performs policy and QoS control at packet level, while P-GW provides policy enforcement at the service level. Similar, S-GW performs charging functionalities at packet level and P-GW generates charging records at service level. S-GW has direct interfaces with previous RANs from 3GPP, such as UTRAN (3G RAN) and GERAN (GSM EDGE Radio Access Network; i.e., 2G RAN), it acts as an anchor point for inter-3GPP RAN mobility. For non-3GPP IP Connectivity Access Network (IP-CAN) [4], mobility anchor point is the P-GW.

In all-IP networks MM is crucial due to Internet addressing principles in which each IP address uniquely identifies the network (network ID, as part of the IP address) and the host (host ID, also part of the IP address), so a change of IP network requires a change of the IP address (this is valid for both IPv4 and IPv6). P-GW is responsible for allocation of IP addresses to mobile terminals, so DHCP (Dynamic Host Configuration Protocol) server usually is collocated or integrated in it. In such case S-GW has DHCP relay functionality to provide transparent allocation of IP addresses to mobile hosts (i.e., UE) in the network. Hence, the P-GW acts as mobility anchor point for mobile terminals moving across non-3GPP IP-CANs (including trusted as well as un-trusted ones). Regarding the user plane (carrying user traffic) P-GW is Local Mobility Anchor (LMA) gateway which terminates Proxy Mobile IPv6 (PMIPv6) [5], signaling in the control plane and IP tunneling (IPv4/IPv6) in the user plane. Such communication is realized via the S2a and S2b interfaces for trusted and un-trusted IP CANs that are non-3GPP, respectively. The home agent in PMIPv6 is located in the P-GW, while the Mobile Access Gateway (MAG) is a function of the access router that manages signaling related to mobility of the mobile terminal that is attached to its access link. The MAG in PMIPv6 is responsible for tracking the movement of the mobile node in the RAN as well as for signaling to the mobility anchor gateway for the given mobile terminal. The non-3GPP IP-CANs usually emulate the MAG functionalities in the PMIPv6 scenario. If S-GW is on the way to the P-GW (i.e., on a home-routed implementation) then S-GW plays the roles of LMA and MAG in a back-to-back manner between the mobile terminal and the P-GW.

Generally, there exist two deployment approaches to address host-based mobility in heterogeneous mobile environment with 3GPP and non-3GPP networks connected via the SAE:

- *Mobile IPv4 (MIPv4)*: S2a and S2b interfaces are based on MIPv4. In such case the P-GW contains MIPv4 Home Agent and the trusted or un-trusted non-3GPP IP-CAN has Foreign

Agent for the mobile terminal (i.e., the UE). The user plane is based on the tunneling of end-to-end IPv4 over IPv4-based transport network.

- *Dual-stack Mobile IPv6 (DSMIPv6)*: Mobile terminal (i.e., UE) has dual stack (i.e., IPv4/IPv6) MIPv6 client, referred to as DSMIPv6 [6], while the P-GW is the DSMIPv6 Home Agent. This model applies to both 3GPP and non-3GPP IP-CANs (where the S2c interface is located between the UE and the P-GW).

Interfaces S5 and S8 are located between the S-GW and P-GW and are similar. In fact, interface between the S-GW and P-GW within the same mobile network is called S5, and interface between the S-GW in visited mobile network and P-GW in home mobile network is called S8 interface (the last one is, in fact, a roaming scenario). Therefore, the interface between the S-GW and P-GW is also referred to as S5/S8. Regarding the protocols that are used on S5/S8 interface there are two options:

- *GTP (GPRS Tunneling Protocol)*: In this case GTP tunnels are established between S-GW and P-GW, and GTP-U (GPRS Tunneling Protocol User plane) is used for the user plane while GTP-C (GPRS Tunneling Protocol Control plane) is used for the control plane. This is typical for 3GPP RANs where S-GW is in fact GTP-U relay between the mobile terminals in the access network (i.e., E-UTRAN) and P-GW. In the case mobile terminals have DSMIPv6 clients then it is possible to use S2c interface over GTP over the connection between S-GW and P-GW.
- *Proxy MIPv6 (PMIPv6)*: This is used as control signaling protocol over the S5/S8 interface in the deployment approaches with DSMIPv6.

In the cases of 3GPP RANs the S-GW terminates the GTP-U tunnels and starts the GTP-U tunnels toward the P-GW. However, both central gateways in the SAE, the S-GW and P-GW, may be physically located in a single unit (i.e., SAE gateway), which is dependent upon the vendors of the equipment.

Overview of main interfaces and protocols in SAE is given in Table 5.1. S-GW and P-GW are connected with the PCRF via control interfaces by using the Diameter protocol. Such interfaces control the PCEF located in the S-GW and P-GW. For access to non-3GPP networks both gateways must implement Diameter interfaces (i.e., S6b and S6c) toward external AAA (Authentication, Authorization, and Accounting) nodes.

Regarding the core network, SAE is all-IP network with flat architecture, and together with the E-UTRAN and the LTE/LTE-Advanced it provides much higher bit rates available to mobile users at lower costs, thus providing higher level of efficiency and flexibility for introduction of new (IP-based) services.

5.2.2.2 SAE Quality of Service

In the heterogeneous service environment and all-IP network in SAE, crucial part becomes the QoS support. Although different so-called QoS classes were defined for 3G mobile networks (i.e., UMTS/HSPA), the main approach in 3G was provisioning of the voice service via the CS domain and delivering of Internet best-effort (BE) services [i.e., WWW (World Wide Web), e-mail, etc.] via the PS domain of the mobile network. However, in SAE all services are based on IP-protocol stack including voice service (i.e., VoIP). So, efficient QoS support

Table 5.1 Overview of main interfaces and protocols in SAE/EPC and E-UTRAN

Interface	Plane	Protocols
S1 (eNodeB–MME)	Control	S1AP/SCTP/IP
S1 (eNodeB–S-GW)	User	GTP-U/UDP/IP
S5/S8 (S-GW–P-GW)	Control	GTP-C/UDP/IP
S5/S8 (S-GW–P-GW)	User	GTP-U/UDP/IP
S11 (MME–S-GW)	Control	GTP-C/UDP/IP
X2 (eNodeB–eNodeB)	Control	X2AP/SCTP/IP
X2 (eNodeB–eNodeB)	User	GTP-U/UDP/IP
S6a (MME–HSS)	Control	Diameter/SCTP/IP
S6b (P-GW–3GPP AAA)	Control	Diameter/SCTP/IP
Gx (PCRF–P-GW)	Control	Diameter/SCTP/IP
Gxc (PCRF–S-GW)	Control	Diameter/SCTP/IP
S9 (home PCRF–visited PCRF)	Control	Diameter/SCTP/IP
SGi (P-GW–external network)	Control	IP, Diameter/SCTP/IP, or RADIUS/UDP/IP
Rx (PCRF–AF)	Control	Diameter/SCTP/IP
S4 (S-GW–SGSN)	User	GTP-U/UDP/IP
S4 (S-GW–SGSN)	Control	GTP-C/UDP/IP
S10 (MME–MME)	Control	GTP-C/UDP/IP
S12 (S-GW–RNC)	User	GTP-U/UDP/IP
S3 (MME–SGSN)	Control	GTP-C/UDP/IP
S2a (P-GW–trusted IP CAN) with PMIPv6	User	IPv4/IPv6 over tunneling layer over IPv4/IPv6
S2a (P-GW–trusted IP CAN) with PMIPv6	Control	PIMPv6 over IPv4/IPv6
S2a (P-GW–trusted IP CAN) with MIPv4	User	IPv4 over tunneling layer over IPv4
S2a (P-GW–trusted IP CAN) with MIPv4	Control	MIPv4/UDP/IP
S2b (P-GW–un-trusted IP CAN) with GTP	User	GTP-U/UDP/(IPv4/IPv6)
S2b (P-GW–un-trusted IP CAN) with GTP	Control	GTP-C/UDP/(IPv4/IPv6)
S2b (P-GW–un-trusted IP CAN) with PMIPv6	User	IPv4/IPv6 over tunneling layer over IPv4/IPv6
S2b (P-GW–un-trusted IP CAN) with PMIPv6	Control	PIMPv6 over IPv4/IPv6
S2c (UE–P-GW)	User	IPv4/IPv6 over tunneling layer over IPv4/IPv6
S2c (UE–P-GW)	Control	DSMIPv6 over IPv4/IPv6

in the IP-based mobile network becomes necessity, especially for real-time services which are sensitive to delay [i.e., VoIP, IPTV (Internet protocol television), gaming, etc.]. The QoS concept in SAE is directly related to bearers. In general, a bearer is a service that allows transmission of information between defined interfaces in the network under defined constraints (i.e., dedicated bit rate, QoS class, etc.). The SAE defined bearers are shown in Figure 5.4.

Every mobile terminal that is connected to the SAE-based mobile network has at least one established bearer for IP connectivity, called default bearer. Further additional bearers may be set up for the mobile terminal. Also, different flows with similar QoS requirements can be mapped on a same bearer.

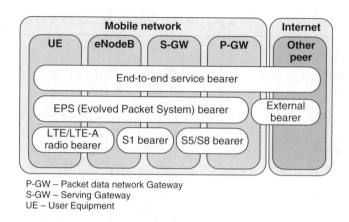

P-GW – Packet data network Gateway
S-GW – Serving Gateway
UE – User Equipment

Figure 5.4   SAE bearers

A bearer is set up when an application in a mobile terminal initiates connections (i.e., Web client connects to a Web server) or when an incoming connection (i.e., voice call) is received. When QoS support is needed the connection is established between the mobile terminal and certain Application Server (AS) on the operator's side (i.e., via the IMS). In the EPS there are three main bearers:

- Radio bearer (on LTE/LTE-Advanced radio interface);
- S1 bearer (on S1 interface, between eNodeB and S-GW);
- S5/S8 bearer (on S5/S8 interface, between S-GW and P-GW).

So, when the AS requests EPS bearer, then it is established with signaling for lower layer bearers, that is, radio bearer, S1 bearer, and S5/S8 bearer. The main goal of bearer set up and existence is minimization of QoS knowledge and configuration in mobile terminals [7].

Regarding the QoS identification SAE uses QoS Class Identifiers (QCIs), which are based on three QoS parameters: priority, loss probability, and delay. There are two resource types defined in SAE for QoS: Guaranteed Bit Rate (GBR) services and non-GBR services. They are further classified into nine QCI values with nine priorities. Logically, highest priority is given to IMS signaling, which therefore can be served with non-GBR resource type (without a signaling none connections can be established via SAE, so its "number one" position on the priority list is deserved). Further, going from higher priorities to lower ones, there are listed QCI for GBR services VoIP, video call, streaming, and gaming, followed by non-GBR services which include TCP/IP-based applications such as interactive gaming, Web browsing, file transfer, e-mail, and so on.

5.2.3   *LTE/LTE-Advanced Radio Access*

LTE-Advanced is an evolution of the LTE designed to meet or even exceed the IMT-Advanced requirements set by ITU-R. The built-in backward compatibility of LTE-Advanced has a direct implication in the way that, for an LTE terminal a network with LTE-Advanced capabilities should appear as an LTE network only, while for LTE-Advanced terminals it will provide full capacity of the LTE-Advanced radio interface. The most important improvements in LTE-Advanced access can be summarized as follows:

- Better flexibility in spectrum management for wideband deployments by using carrier aggregation across different frequency bands.
- Flexible and faster network deployment achieved with the help of heterogeneous networks feature (relay nodes, femto and pico cells, besides typical macro cells in the RAN).
- Improved radio network coverage as well as spectral efficiency (at the cell edges and the average one), which is achieved through robust interference management.
- Higher peak user bit rates by incorporation of higher order MIMO in downlink and uplink (however, MIMO is not exclusively used in LTE-Advanced, but it is also used in LTE and HSPA).

So, there are several important improvements in the radio interface in LTE-Advanced, which are aimed in fulfilling the goals for higher capacity, better cost-efficiency, and greater flexibility.

5.2.3.1 Carrier Aggregation

Carrier aggregation is considered as one of the main features of the LTE-Advanced for provision of higher bit rates than LTE and for fulfilling the IMT-Advanced requirements for 4G set by ITU. LTE introduced flexible spectrum allocations by specification of different spectrum bands for the radio interface, including FDD (Frequency Division Duplex), that is, paired bands, and TDD (Time Division Duplex), that is, unpaired bands. LTE-Advanced provides possibility for spectrum allocations of up to 100 MHz. That is accomplished by carrier aggregation of multiple (separate) component carriers on the physical layer, with allocations of 1.4, 3, 5, 10, 15, and 20 MHz. Maximum carrier aggregation in LTE-Advanced is five bands, each band with maximum 20 MHz, giving overall maximum frequency band allocation of 100 MHz. However, it is not always possible for mobile operators to have large amount of contiguous spectrum allocation. Therefore, LTE-Advanced provides possibility for flexible carrier aggregation of up to five non-continuous component carriers, as shown in Figure 5.5.

For example, one possible spectrum allocation for a mobile operator running LTE-Advanced (4G) will be FDD (i.e., paired bands) allocation of: two components carriers × 20 MHz in 800 MHz spectrum, 10 MHz in 900 MHz (initially used for GSM 900 by 3GPP), 10 MHz in 2.1 GHz (initially allocated to UMTS). In this example, total spectrum allocation is

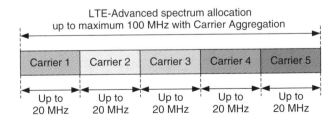

Figure 5.5 LTE-Advanced carrier aggregation

$2 \times 20 + 10 + 10 = 60$ MHz (paired bands), consisted of four non-equal and non-continuous component carriers.

Generally, spectrum allocation in 4G can be characterized as follows:

- Intra-band adjacent component carriers;
- Intra-band non-adjacent component carriers;
- Inter-band component carriers.

At the beginning of the 4G, handheld mobile terminals are not expected to support aggregation over a diverse spectrum. So, lap-top computers are first targeted to exploit carrier aggregation feature via their LTE-Advanced radio interface (i.e., connected via a USB port). In a later phase of 4G development, one may predict that mobile terminals will support carrier aggregation over non-continuous bands.

Regarding the protocol layering model, the carrier aggregation is implemented on OSI layer-2. However, layer-2 consists of two sublayers: MAC (Medium Access Control) and RLC (Radio Link Control), as shown in Figure 5.6. The carrier aggregation is implemented above the MAC sublayer, on the RLC sublayer, which is below the network layer (that is IP in LTE/LTE-Advanced). So, from the network layer (above the radio interface, consisted of lower two OSI layers) the carrier aggregation provides aggregate bit rates obtained from all carriers, while each physical layer and it corresponding MAC sublayer are distinct per frequency band (each band is from 1.4 up to 20 MHz).

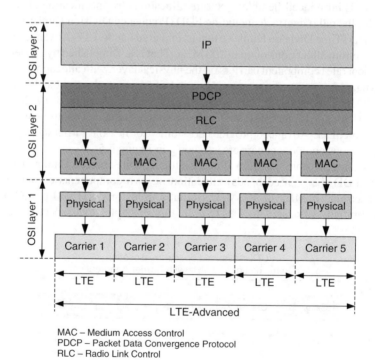

MAC – Medium Access Control
PDCP – Packet Data Convergence Protocol
RLC – Radio Link Control

Figure 5.6 Carrier aggregation and OSI protocol model in LTE-Advanced radio interface

5.2.3.2   Home eNodeB

Radio access to 3G and 4G (i.e., EPS) services may be provided via UTRAN or E-UTRAN cellular BSs that belong to different owners such as home users or business (i.e., commercial) enterprises. This type of access may be provided in 3GPP mobile networks by using Home Node B (HNB) and Home eNodeB (HeNB; in the following part the acronym HeNB is used for both) [8]. The HeNB provides services either only to a Closed Subscriber Group (CSG) or to other mobile subscribers besides its "home" subscribers. The home BSs are connected to the mobile operator core network using IP connectivity via any suitable broadband access technology [i.e., xDSL (Digital Subscriber Line technologies), optical access, cable access, fixed wireless access, etc.]. The cell from the HeNB has smaller coverage (i.e., 10 m) and it is referred to as femtocell. In fact, femtocells provide Fixed Mobile Convergence (FMC) without a need for dual-mode user terminals, because femtocell deployments work with existing mobile terminals in licensed spectrum (i.e., 3G mobile terminals for HNB, LTE mobile terminals for HeNB).

The HeNB deployments also provide two additional features of LTE/LTE-Advanced (as shown in Figure 5.7), given as follows:

- *Local IP Access (LIPA)*: This provides access for IP capable UEs connected via a HeNB to other IP connected devices in the same IP network (residential or enterprise). Data traffic for LIPA is expected to not traverse the mobile operator's network except mobile operator network components in the residential/enterprise premises. However, signaling traffic is necessary to continue to traverse to the mobile operator network by using IP tunneling mechanisms via the access network to the Internet from the LIPA location.

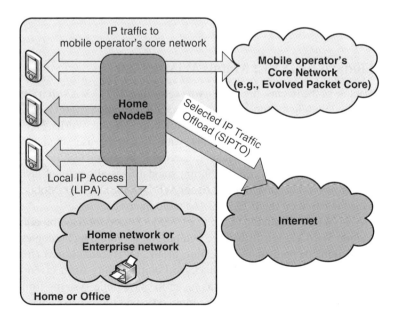

Figure 5.7   Local IP access and IP traffic offload for femtocells

- *Selected IP Traffic Offload (SIPTO)*: This provides access for UE connected via a HeNB toward certain IP network, such as global Internet. There are specific requirements that apply to the HeNB to support SIPTO (including residential and enterprise environments), that are given as follows:
  - SIPTO is possible to be done without traversing the mobile operator network, but such approach should be regulated by the regulator in a given country.
  - The mobile operator and the HeNB host, within the limits set by the mobile operator, shall be able to enable and disable SIPTO for each HeNB.
  - Based on mobile operator SIPTO policies and configured user approval, the mobile network shall be able to offload traffic by using the SIPTO.

5.2.3.3 Self-Organizing Networks

A Self-Organizing Network (SON) is an automation technology which is standardized to provide easier planning, automatic configuration, self-management, self-optimization, and self-healing of mobile RANs, thus making them simpler and faster. The first appearance of SON was in 3GPP Release 8 and it continues in further releases, and the first targeted mobile technologies for SON usage are LTE and LTE-Advanced. For example, some features of E-UTRAN such as Automatic Neighbor Relation (ANR) discovery by eNodeBs are in fact SON features. The idea behind SON in RAN is to have "plug and play" installation of BSs (i.e., eNodeBs in LTE/LTE-Advanced), that is, comparable with typical installation of a WLAN (Wireless Local Area Network; i.e., WiFi) router regarding the required time (i.e., 20 min) and "man power" (i.e., one person for complete installation of an eNodeB). Further, SON is aimed to provide automatic planning of neighborhood relations in the radio network in real time, as well as automatic compensation of cell outage. However, here are mentioned only several features that can be provided by SON, their number will certainly increase in future.

5.3 4G Standard by IEEE: Mobile WiMAX 2.0

In the second decade of the twenty-first century it seems that WiMAX is still trying to catch up with well established mobile technologies, such as 3GPP evolution from GSM/GPRS/EDGE via UMTS/HSPA toward LTE/LTE-Advanced. However, Mobile WiMAX is a "younger" technology compared with 3GPP technologies, with its first release standardized as IEEE 802.16e-2005 [9]. In general, the IEEE standardizes only lowest two OSI layers (i.e., physical layer and data-link layer), while WiMAX Forum standardizes all protocol layers regarding the WiMAX technology. The first release (from WiMAX Forum) for Mobile WiMAX is Release 1.0, which is based on the IEEE 802.16e radio interface.

Mobile WiMAX is a mobile version of the fixed WiMAX used for fixed broadband wireless access. But, Mobile WiMAX is still lacking behind 3GPP mobile technologies. One possible reason is well implemented roaming feature in all 3GPP technologies (from the GSM in the 1990s). Another possible reason is backward compatibility for a given mobile technology. For example, such compatibility was given in 3G mobile systems from 3GPP (i.e., UMTS/HSPA) by having CS domain as in 2G mobile system (2G). Also, such compatibility is given in LTE and LTE-Advanced by having already evolved core architecture and services continuity. Another possible answer is business side of the story regarding the vendors of the equipment used by mobile network operators. Vendors which are stronger in creation of 3GPP equipment and have (or had) larger market share worldwide (especially in the developed countries) are not much interested in Mobile WiMAX, and vice versa, vendors who are lacking behind in the market share for 3GPP mobile technologies show more interest in development of WiMAX solutions.

Mobile WiMAX 2.0 (based on IEEE 802.16m [10]) [11], has many similarities in the radio access part with LTE-Advanced. Also, both technologies are almost at the same time period approved as 4G by ITU. However, this is a different progress compared to the 3G development where Mobile WiMAX entered the 3G umbrella later (in October 2007), and had no significant impact on the 3G mobile networks on a global scale. But, at the same time it was an indication of a "serious intention" of another mobile technology to provide more competition and more possibilities for mobile broadband world. On the other side, the wireless interfaces of Mobile WiMAX 2.0 and LTE-Advanced as well as all-IP core architecture in both standards lead to a convergence of these technologies in some way. That provides possibility to offer both radio access technologies (LTE-Advanced and Mobile WiMAX 2.0, defined over the same frequency bands and having similar radio interfaces) to be integrated in single mobile terminals in cost-effective manner. However, that is dependent upon the regulation (i.e., spectrum management) and business strategies of vendors and mobile operators.

5.3.1 *Mobile WiMAX Network Architecture*

Mobile WiMAX architecture is defined by WiMAX Forum [11]. In general, the architecture for the Mobile WiMAX 1.x releases (based on IEEE 802.16e radio interface) and Mobile WiMAX 2.x releases (based on IEEE 802.16m radio interface) is similar. The network architecture is shown in Figure 5.8. It is consisted of three main parts: Access Service Network (ASN), Connectivity Service Network (CSN), and Mobile Stations (MSs).

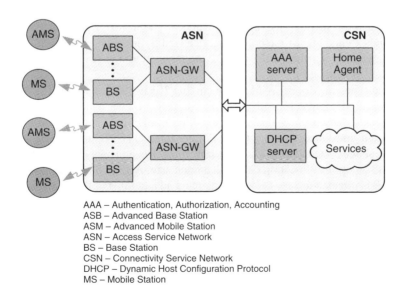

AAA – Authentication, Authorization, Accounting
ASB – Advanced Base Station
ASM – Advanced Mobile Station
ASN – Access Service Network
BS – Base Station
CSN – Connectivity Service Network
DHCP – Dynamic Host Configuration Protocol
MS – Mobile Station

Figure 5.8   Mobile WiMAX network architecture

The ASN is the RAN of Mobile WiMAX, consisted of BSs and Access Service Network Gateway (ASN-GW). The CSN provides the means for IP connectivity between the ASN (and MSs connected to the ASN) and the Internet. The ASN-GW provides two main entities:

- *Data path entity*: This is used for transfer of user data between the MSs (in ASN) and public Internet.
- *Control entity*: This is used for control and context management per MS, including authentication, accounting, key distribution, QoS provisioning, MM with CSN as anchor point, as well as for connection with the CSN and its servers.

The CSN is all-IP based core network which includes different servers for support of necessary functionalities to MSs in the ASN. Main parts of CSN include AAA servers [i.e., RADIUS (Remote Authentication Dial In User Service), Diameter], DHCP server, Home Agent (for MIPv4 and MIPv6) and Foreign Agent (for MIPv4), ASs, and so on.

The main difference in network architectures of Mobile WiMAX release 1.x and 2.x is in the ASN part. Mobile WiMAX Release 2 introduces the Advanced Mobile Stations (AMSs) and Advanced Base Stations (ABSs), which implement the new functionalities of IEEE 802.16m radio interface.

5.3.2   Quality of Service in WiMAX Networks

QoS support in IEEE 802.16e (used for Mobile WiMAX release 1) is based on service flows [12]. They may be based on contention among MSs or contention-free.

Contention-based service flow type is the BE service while contention-free services are Unsolicited Grant Service (UGS) and real-time Polling Service (rtPS). Further, non-real-time Polling Service (nrtPS) can be used in contention-based or contention-free mode. In the case of contention-free service type the BS polls the MS periodically. In contention-based service types BS allocates bandwidth to MS upon Bandwidth Request (BR) message.

UGS is targeted for VoIP service (also rtPS can be used for VoIP with silence suppressions), while rtPS and extended real-time polling service (ertPS) are aimed to be used for video and multimedia streaming. Further, nrtPS and BE are targeted to non-real-time services (i.e., file transfer, WWW, e-mail, etc.).

QoS service types in Mobile WiMAX (including IEEE 802.16e and IEEE 802.16m) are given in Table 5.2. IEEE 802.16m advanced air interface provides a more flexible and efficient QoS framework by adding new service type, which is adaptive Granting and Polling (aGP) service. The aGP introduces primary Grant Polling Interval (GPI) and primary grant size, as well as optional ones: secondary GPI, secondary grant size, and adaptation method. The ABS grants to AMS an uplink allocation GPI with a grant size. ABS polls AMS periodically every GPI. The aGP is targeted to services which demand adaptation and flexibility, such as online gaming, VoIP with Adaptive Multi-Rate (AMR) codecs, as well as TCP-based (Transmission Control Protocol) delay-sensitive services.

5.3.3   Mobile WiMAX 2.0 Radio Interface

The requirements set to Mobile WiMAX 2.0 radio interface are similar to those set to LTE-Advanced. Also, both technologies are using the same spectrum allocations (up to 20 MHz per

Table 5.2 Service flow types in Mobile WiMAX

Service flow type	Traffic type
Unsolicited grant service (UGS)	Real-time traffic with periodic fixed-size packets, for TDM services including VoIP
Real-time polling service (rtPS)	Real-time traffic with periodic variable-size packets; for VoIP with silence suppression, IPTV
Extended rtPS (ertPS)	Real-time traffic with variable-size packets on a periodic basis with active and silence intervals; for video and multimedia streaming
Non-real-time polling service (nrtPS)	Delay-tolerant traffic with minimum reserved rate; for file transfers, Web services
Best-effort (BE)	No guarantees (best-effort Internet concept); for WWW, e-mail, peer-to-peer services, and so on
Adaptive granting and polling (aGP)	More flexible QoS support for both allocation size and inter-arrival time; for online games, VoIP with adaptive multi-rate (AMR), and delay-sensitive TCP-based services

carrier) and can use the same frequency bands, called IMT bands (by ITU-R). In the downlink Mobile WiMAX 2.0 uses OFDMA (as LTE/LTE-Advanced), but it differs in uplink from LTE-Advanced. Mobile WiMAX uses also OFDMA in the uplink (LTE/LTE-Advanced uses SC-FDMA, which is considered more advanced for handheld terminals because it makes easier for a mobile terminal to maintain highly efficient transmission in the uplink and to save power).

Besides similarities in spectrum allocations and available bit rates, Mobile WiMAX 2.0 differs from LTE-Advanced regarding the radio interface structure (i.e., frames structure, resource blocks or units, etc.). Radio resource management in mobile networks is out of the scope of this book. However, certain important features of the radio interface that influence the network and the services (i.e., via the available radio network capacity, i.e., bit rates) are discussed below.

5.3.3.1 Advanced Features of IEEE 802.16m

IEEE 802.16m introduces advanced features in the radio interface that are similar to those introduced by LTE-Advanced, such as carrier aggregation, femtocells, SON, relay, and so on.

Carrier aggregation for IEEE 802.16m is based on the same principles as the one given for LTE-Advanced, with same possible width of component carriers. So, IEEE 802.16m provides carrier aggregation with up to five component carriers, each carrier up to 20 MHz, and maximum spectrum allocation of up to 100 MHz, which satisfies the IMT-Advanced requirements for user bit rates.

Femtocell BSs are small-scale and low-cost devices installed in subscribers' premises which enable high bit rates due to small distance between the mobile devices and the femtocell BSs (similar to HeNB). In such scenario, the control of the radio functionalities and QoS provisioning to the user is enabled by core network connections established over fixed broadband access such as xDSL (Digital Subscriber Line technologies), PON (Passive Optical Network), and so on.

SON defines neighbor discovery, interference mitigation, and load balancing features, which are particularly useful for femtocell deployments due to difficulties for implementation of femtocell parameters in unknown site locations (i.e., carrier frequency, transmitting power, etc.).

Relays increase coverage by closing "blind areas" in unfavorable radio environments. Relay stations are simplified low-cost BSs with limited capabilities which have repeater functionality. IEEE 802.16m supports multi-hop relaying (i.e., relaying over several relay stations).

In general, all advanced features in the radio network are being supported in parallel by both Mobile WiMAX 2.0 and LTE-Advanced.

5.3.3.2 Comparison of IEEE 802.16m and LTE-Advanced

The two 4G technologies, IEEE 802.16m and LTE-Advanced, have similarities and differences. With aim to give an objective view, their main characteristics are compared in Table 5.3, according to the specified requirements at the development of each of them.

5.4 Fixed-Mobile Convergence

FMC is not a new idea. It appeared as an option a couple of decades ago, going back to DECT (Digital Enhanced Cordless Telecommunications) systems in 1990s, then followed by UMA (Unlicensed Mobile Access) and GAN (Generic Access Network) concepts. However, later all ideas about FMC converged. In fact, FMC is based on convergence of different access networks to the same core and transport networks (regarding the transport stratum) and the

Table 5.3 Comparison of IEEE 802.16m and LTE-Advanced

Parameter	IEEE 802.16m	LTE-Advanced
Peak data rates (Mbit/s)	DL: >1000 (low mobility)	DL: >1000
	DL: >100 (high mobility)	UL: >500
	UL: >130	
Spectrum allocation (MHz)	Up to 100	Up to 20–100
Latency (ms)	Control plane: 100	Control plane: 50
	User plane:10	User plane: 10
MIMO technique	Downlink: up to 8×8	Downlink: up to 8×8
	Uplink: up to 4×4	Uplink: up to 4×8
Peak spectral efficiency	DL: 15 (4×4) MIMO	DL: 30 (8×8) MIMO
(bit/s/Hz)	UL: 6.75 (2×4) MIMO	UL: 15 (4×4) MIMO
Mobility support	Maximum data rates (<10 km/h)	Maximum data rates (<15 km/h)
	High performance (<120 km/h)	High performance (<120 km/h)
	Maintain links (<350 km/h)	Maintain links (<350 km/h)
Access scheme	DL: OFDMA	DL: OFDMA
	UL: OFDMA	UL: SC-FDMA
Cell edge spectral efficiency	DL: 0.09 (2×2)	DL: 0.12 (4×4)
(bps/Hz)	UL: 0.05 (1×2)	UL: 0.07 (2×4)

same signaling protocols and overlay networks needed for service provisioning (regarding the services stratum), and finally to same applications and content provided via different (i.e., heterogeneous) mobile and fixed networks. So, one may say that FMC is provisioning of different services/applications to the end-users regardless of the access network, either fixed or wireless/mobile.

On the other side, NGN provides generalized mobility, and FMC requirements are defined in NGN. Generalized mobility requires evolution of the network architectures by separation of transport stratum and service stratum and providing the same services over different wireless and mobile access networks (including horizontal mobility within a given access network and vertical mobility between different access technologies) and fixed networks. This includes convergence of fixed and mobile telecommunications (i.e., FMC).

The fact is that the evolution of core networks toward the generalized mobility as well as FMC is something that is happening with the Fourth Generation (4G) of mobile systems.

Certain objectives for FMC are specified by ITU-T (International Telecommunication Union-Telecommunications), for generalized mobility in NGN. First FMC objective is seamless service operation over heterogeneous fixed networks [i.e., PSTN/ISDN (Public Switched Telephone Network/Integrated Services Digital Network), Local Area Network – LAN such as Ethernet, cable networks, etc.] and mobile networks (i.e., GSM/GPRS, UMTS, cdma2000, Mobile WiMAX, LTE/LTE-Advanced, etc.). Second objective is seamless provisioning of services from the service operator's point of view. Third objective is support to different types of mobility, such as terminal mobility (i.e., one terminal-multiple IP addresses), user mobility (one person, multiple user terminals/devices), and session mobility (meaning one user, multiple terminals in sequence or parallel – for example, audio and video flows from mobile device are transferred to home/office audio system and TV screen after entering home/office, and vice versa). Fourth objective is ubiquity of service availability, that is, the users can enjoy virtually any application or service, anytime, on any end-user terminal, from any location. Of course, in all these objective there are certain limitations from the specific access networks, such as QoS which is strictly defined for 3GPP mobile networks and for the WiMAX (IEEE 802.16 standards), while it is left to contention between stations/terminals in Ethernet (fixed LAN, IEEE 802.3 family of standards) and WiFi (WLAN, IEEE 802.11 family of standards). The last (but not the least important) objective for FMC, as specified for NGN, is support for multiple user identifiers and AAA mechanisms. There are different available identifiers for the end-users such as: E.164 telephone numbers, IMSI (International Mobile Subscriber Identity) which is directly related to MSISDN (mobile subscriber Integrated Services Digital Network, i.e., mobile E.164 number), URI (Uniform Resource Identifier) for different types of protocols [i.e., HTTP (Hypertext Transfer Protocol), Session Initiation Protocol – SIP, etc.], tel URI, SIP URI, and so on. Finally, all these FMC objectives in NGN are addressed via the specification, standardization, and implementation of IMS.

5.5 IP Multimedia Subsystem for NGN

IMS was first standardized within 3GPP in Release 5 as an application development environment. Later, 3GPP Release 7 retargeted the IMS to be used mainly for telephony replacement in IP-based networks, regardless of the access type (i.e., access independent). Finally, 3GPP Release 8 standardized so-called common IMS as a unified standard which implemented different requirements from all other bodies for standardization of mobile and fixed networks,

such as ITU, 3GPP2, TISPAN (Telecommunications and Internet converged Services and Protocols for Advanced Networking), Cablelabs, and so on. The 3GPP standardization of IMS is
included as an integral part in NGN service stratum [13], and also it is considered as legacy
approach for FMC deployments [14]. The 3GPP continues with the development of the IMS
in the releases following Release 8 which defined the commonly known version [15].

IMS uses SIP [16], as signaling and control protocol for different services, including multimedia session services and some non-session services such as presence services and message
exchange services. So, NGN IMS supports SIP-based services, but also PSTN/ISDN simulation services [13].

In Figure 5.9 is shown IMS architecture given with 3GPP functional entities (FEs). However,
it is the same as NGN IMS architecture where the word "Function" (in 3GPP terminology) is
replaced with "Functional Entity" (in NGN terminology). Main FEs in the IMS architecture are
three types of Call Session Control Functions (CSCFs): Proxy-Call Session Control Function
(P-CSCF), Serving-Call Session Control Function (S-CSCF), and Interrogating-Call Session
Control Function (I-CSCF).

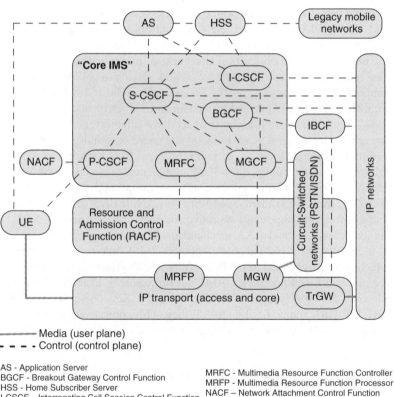

———— Media (user plane)
− − − − Control (control plane)

AS - Application Server
BGCF - Breakout Gateway Control Function
HSS - Home Subscriber Server
I-CSCF – Interrogating Call Session Control Function
IBCF - Interconnection Border Control Function
IMS – IP Multimedia Subsystem
MGCF - Media Gateway Control Function
MGW – Media Gateway

MRFC - Multimedia Resource Function Controller
MRFP - Multimedia Resource Function Processor
NACF – Network Attachment Control Function
P-CSCF – Proxy Call Session Control Function
S-CSCF – Serving Call Session Control Function
TrGW - Transition Gateway
UE – User Equipment

Figure 5.9   IMS architecture

Besides the CSCF entities, the "core" part of IMS [13] includes three additional FEs, given as follows:

- *Breakout Gateway Control Function (BGCF)*: This is used for processing of user requests for routing from an S-CSCF for the situations when the S-CSCF has determined that it cannot use session routing by using DNS (Domain Name System) or ENUM (E.164 Number Mapping)/DNS (e.g., cases with PSTN termination of a given call from an IMS user). So, BGCF determines the next hop for routing of a given SIP message. If a breakout occurs in the same IMS network then BGCF selects a MGCF (Media Gateway Controller Function) for interworking with the PSTN. If the routing in BGCF results in breakout into another network, then BGCF forwards the request to the I-CSCF node in the other IMS network. In the given cases when BGCF is involved it also can generate charging records.
- *Media Gateway Controller Function (MGCF)*: This performs signaling translation between SIP and ISUP (Integrated Services Digital Network User Part, from SS7 signaling system in PSTN). In SAE it interfaces the S-GW by using SCTP/IP protocol (Stream Control Transmission Protocol) stack.
- *Multimedia Resource Function Controller (MRFC)*: This is part of the Multimedia Resource Function (MRF), which is split into two parts – the MRFC and Multimedia Resource Function Processor (MRFP). The MRFC is a signaling node that interprets the information coming from the AS and S-CSCF and using that information for control of the media stream in the MRFP. On the other side, MRFP is a user plane node that provides mixing of media streams (e.g., for multiple receiving parties). Also, it can be a source for media stream (e.g., for multimedia announcements), and provides media streams processing as well as floor control (i.e., management of access rights to shared resources in conferencing environments).

There are also several other entities or nodes for interconnection of the IMS to access networks as well as to user databases such as HSS, given as follows:

- *Home Subscriber Server (HSS)*: This is the main user database in the IMS architecture that contains home subscriber profiles (and subscription information), provides authentication and authorization of the users (via CSCF entities of the IMS) and contains information about the network location of the user (i.e., IP address). In fact, it merges the Home Location Register (HLR) and AuC functions that exist in GSM. It is the same HSS node given in SAE.
- *Subscriber Location Function (SLF)*: This is a resolution function in IMS that provides information on HSS that has information for a given IMS user (upon a query from I-CSCF or S-CSCF, or from an AS). Hence, this function is not required in networks that have single HSS.
- *Application Server (AS)*: This is a server that hosts and executes given service. AS is using SIP for communication with the S-CSCF, which contacts AS in the order supplied by the HSS (in a case of multiple AS). Multiple AS can be any combination of AS, such as SIP AS, OSA (Open Service Access) server, or IP Multimedia Service Switching Function (IM-SSF).
- *Interconnection Border Control Function (IBCF)*: This is used as a gateway to external networks. It is in fact a Session Border Controller (SBC) and provides firewall and NAT (Network Address Translation) functionalities in the IMS architecture.

5.5.1   Proxy CSCF

P-CSCF is the first contact point for the UE (i.e., mobile terminal, fixed terminal) within the IMS. It acts as a SIP proxy and it interacts with the admission control subsystem to provide authorization only of media components that can be provided with appropriate QoS level. However, P-CSCF should not modify SIP messages for session initiation. The IP address of the P-CSCF is provided by the access network during configuration phase of the user terminal (when it obtains access to IMS via a given access network). The P-CSCF may also behave as a user agent. In abnormal conditions it may in fact terminate and independently generate SIP transactions (i.e., to handle the case of a mobile terminal that goes out of the mobile network coverage area). This entity is also an endpoint of security associations between the IMS and the UE (with IMS capabilities), with the aim to maintain confidentiality of SIP sessions. The P-CSCF for a given terminal can be located in a home network or in a visited network (when its IP address is assigned locally in the visited network based on the roaming agreement between the operators).

5.5.2   Serving CSCF

S-CSCF is the central node in the IMS and has SIP server functions. It performs session control services for the UE and maintains the sessions. Regarding the registration, S-CSCF behaves as a SIP registrar (i.e., it accepts registration requests and makes the registration information available via the location server which is HSS). So, it binds the user location (i.e., the IP address) with the SIP address. S-CSCF is always located in the home network of the subscriber and uses Diameter protocol to access HSS database to obtain user profiles (the user profile that is taken from HSS is queried and cached for processing reasons and it cannot be changed by S-CSCF). In the opposite direction S-CSCF uploads user-to-S-CSCF associations to the HSS. As a central node in the IMS architecture the S-CSCF enforces the policies of the operator and provides forwarding of SIP messages between the IMS nodes as well as routing services based on ENUM.

In a given IMS architecture may be available several S-CSCF for load balancing and higher availability of the IMS functions (i.e., higher reliability). In such case HSS assigns one S-CSCF when it is queried for a given user.

An S-CSCF may also behave as a SIP Proxy server or a SIP User Agent (to independently initiate or terminate SIP sessions) and support interaction with service platforms via the SIP-based ISC (IMS Service Control) interface.

5.5.3   Interrogating CSCF

I-CSCF is a SIP location function located at the edge of an IMS network. The IP address of the I-CSCF node is given in DNS, because this node is the first contact for all incoming messages from other IMS-based networks, so DNS records for I-CSCF are needed for such purpose (i.e., for terminating messages from other IMS networks). Even in the case of a session originated by a roaming user who is using a local P-CSCF in some visited IMS network it will be first directed to the I-CSCF of the visited network and routed to its home network S-CSCF before any interaction with the visited network entities (this is because IMS enforces home network control via S-CSCF and HSS nodes). All SIP requests received by an I-CSCF from another network for a given UE are routed toward S-CSCF which is registered for that user.

Such routing function is performed by I-CSCF by using the S-CSCF address obtained from the HSS. I-CSCF also generates charging records that are used in billing arrangements between IMS-based network operators.

5.5.4 Naming and Addressing in IMS

IMS has various identities that can describe users [15]. In general, users may access the IMS via fixed or mobile networks, because IMS is access independent. All user identities in IMS can be grouped into public and private identities (i.e., IDs).

Every IMS user must have one or more Private User Identities. Private User ID (UID) must be stored in HSS, and it is used for registration and AAA, therefore it is not dynamic.

The IP Multimedia Public Identity (IMPU) is UID that can be used by any user for requesting communications to other users (this is similar to telephone numbers in PSTN or e-mail addresses in Internet). Single user may have multiple IMPUs per Private UID. The IMPU can be shared with another phone, so two or more users (i.e., with a single subscription) can be reached with the same identity.

Both of the IDs used in IMS, Private UID and IMPU, are URIs, which typically are in the form of tel URI or SIP URI.

Each Public UID may have one or more Globally Routable User Agent URIs (GRUUs). The GRUU is an identity that identifies a unique combination of IMPU and established UE instance. One may distinguish two types of GRUU:

- *Public-Globally Routable User Agent URI (P-GRUU)*: This makes public the IMPU.
- *Temporary Globally Routable User Agent URI (T-GRUU)*: This is valid until the contact is explicitly de-registered or the current registration expires.

Each pair of a P-GRUU and a T-GRUU is associated with one Public UID and single UE. During subsequent re-registrations of the UE the same P-GRUU will be assigned to the UE, but each time a new and different T-GRUU will be generated and assigned. Also, all previously generated T-GRUUs remains valid after a re-registration. However, the UE may decide to retain or replace (with the new T-GRUU) some or all of the previous T-GRUUs obtained during the initial registration or previous re-registrations. For a given Public UID a current set of the P-GRUU and all T-GRUUs which are currently valid is referred to as the GRUU set. In the case when a UE registers with multiple Public UIDs, then different GRUU set is associated with each Public UID.

IMS introduces standardized presence, messaging, conferencing, and group service capabilities, for which Public Service Identities (PSIs) are needed. Such identities are different from Public UIDs because they identify services, which are hosted by ASs. For example, a chat service may use a PSI (i.e., sip:chat.list@example.com) to which the users establish a session to be able to send and receive messages from other participants in the given session.

5.6 Mobility Management in NGN

MM in NGN is essential for users to communicate anytime and from anywhere. The purpose is to provide possibility to users to use different available wired (i.e., xDSL, PON, etc.) or wireless and mobile access technologies (i.e., WLAN, WiMAX, 3GPP mobile systems, etc.), thus forming heterogeneous network environments. In general, MM can be classified into several types of mobility (depending upon what moves) [17]:

- *Terminal mobility*: This is mobility where the same terminal is moving across different access networks on different locations. It requires the capability of the terminal to access telecommunication services from different access networks and locations, and the capability of the network to identify and locate that terminal.
- *Network mobility*: This is the ability of a network consisting of fixed and/or mobile nodes that are networked to each other, to change, as a whole unit, its point of attachment to the another network upon the network's movement itself.
- *Personal mobility*: This is mobility where the user changes the terminal that is used for network access at different locations. However, the user needs a personal identifier to be able to use the services over different networks. On the other side, networks should have the capability to provide services requested by the user.
- *Service mobility*: This is mobility that is applied for a specific service, that is, the ability of a moving object to use the given service independently of the user's location and the terminal used for access to the service. However, this definition of service mobility is different from the mobility defined for NGN service stratum.

Further, another classification of mobility is based on the service continuity (i.e., having or not interruption of service connection or session at the handover), which is given as follows:

- *Service continuity*: This is the ability for a mobile terminal to maintain its session for the given service, which can be further divide into two types:
  - *Seamless handover*: This is the case when the session is not interrupted at handovers and there is no impact on the agreed service level in Service Level Agreement (SLA).
  - *Handover*: This is the ability to provide service continuity, but with certain impact on the service quality, dependent upon the capabilities of the mobile networks before and after the handover.
- *Service discontinuity*: This is the ability to provide the service regardless of the access network, but without maintaining the session for the service (i.e., interruption happens at handover and session has to be established again between the mobile terminal and corresponding application server). This includes two categories:
  - *Nomadism*: This is the ability for users to change their point of attachment (i.e., to Internet) while moving. However, in such case the session for the service is completely stopped and should be established again over the new point of attachment.
  - *Portability*: This is the ability for a user address (or identifier) to be allocated to different devices or systems when the user is changing their location.

Regarding the types of mobile and wireless networks or fixed networks through which the user is moving there is classification of MM into two types of mobility [2]:

- *Horizontal mobility (i.e., horizontal handovers)*: This refers to mobility when users are moving between different cells or network attachment points (i.e., BSs, eNodeBs) within the same access technology.

- *Vertical mobility (i.e., vertical handovers)*: This refers to mobility when the user is moving between different access technologies (i.e., from Mobile WiMAX to LTE-Advanced).

The NGN has transport and service stratum, and therefore in the following sections are given architectures for MM architectures in both stratums.

5.6.1 Conceptual Framework for MM

Conceptually, MM is a control functionality, which is separated from the Transport Function (TF) in NGN. It is performed by signaling or control operation via the Mobility Management Control Function (MMCF) which is defined in NGN architecture [18].

5.6.1.1 Types of Mobility Management in NGN

Considering the point of attachment of the mobile terminal before the handover and after the handover, that is, whether the old and new points of attachment belong to the same Access Network (AN) or Core Network (CN), there are three cases that can be classified for MM in NGN [19] (shown in Figure 5.10):

- *Inter-CN MM*: Access networks before and after the handover are connected to different core networks.
- *Inter-AN MM*: Access networks before and after the handover are different, but both are connected to the same core network.
- *Intra-AN MM*: The old (before handover) and the new (after the handover) point of network attachment for the mobile user belong to the same access network.

5.6.1.2 Mobility Management Identifiers

For MM provisioning are also needed user and location identifiers. UID belongs to the common identifiers in NGN (as described in Chapter 3), such as IMSI, E.164 number (i.e., for PSTN),

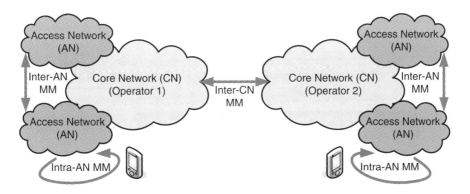

Figure 5.10   Mobility management types in NGN

SIP URI (i.e., for IMS-based services), e-mail address, as well as any other suitable identifier of a user or terminal.

To provide Location Management (LM) (i.e., to locate the end user for terminating calls or sessions), MM framework in NGN requires defining of Location ID (LID). In general, there are two types of LID:

- *Physical/geographical location ID*: This could be Service Set ID (SSID) of Access Point (AP) in WLAN, BS ID in a cellular mobile network, and so on.
- *Logical location ID*: In all-IP world this is a routable IP address (on the network layer). MM in NGN is particularly focused on this type of LID (which is in fact a network ID, i.e., address, allocated to the interface of a host in the Internet).

When used as a LID the IP address may belong to one of the two different types: temporal and persistent. Temporal IP address may change when user moves from one IP network to another, while persistent IP address remains unchanged. A legacy solution for MM in the Internet is Mobile IP (MIP), which uses Home Address as a persistent IP address and Care-of-Address (CoA) as temporal IP address (which is changed every time when users moves from one IP network with a Foreign Agent to another).

So, the MM identifiers are grouped into UIDs and LIDs, where UID is used to identify the mobile user and LID is used to identify the physical or logical location of the user in a given network. Further, LID may be divided into persistent LID or temporal LID (in both cases IP address can be used as LID), as shown in Figure 5.11.

5.6.2 Architecture for Mobility Management in Transport Stratum

MM in the transport stratum addresses device mobility which is provided via MMCFs for the NGN transport stratum. The architecture for MM in transport stratum covers terminal (i.e., device) mobility (from the given types of mobility). It is an IP-based MM approach and provides seamless mobility when moving across different mobile networks, but it does not provide

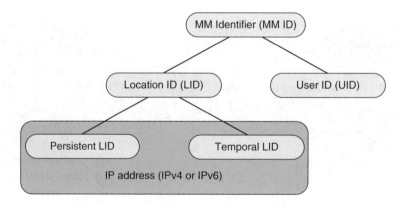

Figure 5.11 Identifiers for mobility management

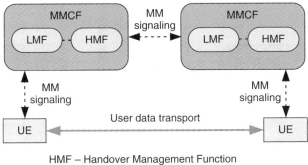

HMF – Handover Management Function
LMF – Location Management Function
MMCF – Mobility Management Control Function
UE – User Equipment

Figure 5.12 A model for MMCF

any mechanism for service adoption for cases when QoS after the handover is degraded from the QoS before the handover.

In all MM schemes mobile terminal (i.e., UE) is registered with one MMCF. However, during movement it may be necessary to change the MMCF with another one. In such case is applied MMCF model as shown in Figure 5.12. The MM signaling operations are performed between the mobile terminal (i.e., UE) and the MMCF, and between MMCFs. MM can be host-based (i.e., mobile terminal initiates handovers) or network-based (mobile network controls all handovers). Then, in host-based MM is possible to have UE to MMCF signaling, while in network-based MM all control operations (i.e., signaling) may be performed only between MMCFs.

Management mobility architecture for NGN transport stratum is shown in Figure 5.13. The main FE is MMCF which is divided into two major functions: Mobility Location Management Functional Entity (MLM-FE) and Handover Control Functions (HCFs).

The MLM-FE is used for LM, which is used to identify the location of the mobile user and to keep track of it while it is moving. LM is most important for terminating calls and sessions, which is a specific problem in all-IP environment where native Internet services are based on the client-server paradigm (where client at the end user is always the one that initiates certain connection or session). With the LM the mobile network to which the user is attached can locate the user and deliver a terminating call or session via appropriate signaling. LM consists of two main functions:

- *Location registration*: This is a procedure to register the current location of the mobile terminal or to update location information.
- *Paging and delivery*: Paging is used to search for the mobile terminal, which is in inactive mode (i.e., offline), in a given network area in which it is registered via the location registration (such areas can be location areas in GSM, routing areas in GPRS and UMTS, or tracking areas in LTE/LTE-Advanced, in all cases they are consisted of bigger or smaller sets of cells of BSs in a given geographical area). Delivery means to deliver the packet to the destined mobile terminal after its location has been determined.

The HCF is used to provide session continuity for established sessions of end users, targeted to minimize the latency and packet losses at handovers. Most of the MM schemes perform

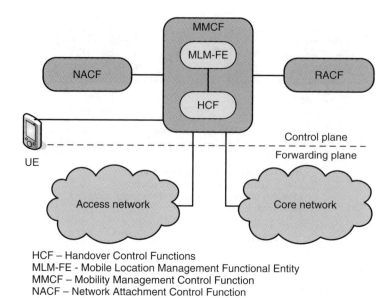

HCF – Handover Control Functions
MLM-FE - Mobile Location Management Functional Entity
MMCF – Mobility Management Control Function
NACF – Network Attachment Control Function
RACF – Resource and Admission Control Function

Figure 5.13 Mobility management architecture for NGN transport stratum

handover management (HM) together with the LM schemes. As discussed before, handovers can be horizontal or vertical, depending upon the type of radio access technology. Also, handovers may be executed on different protocols layers, such as:

- *Handovers on link-layer*: This is usually a standard approach for handovers within the same RAN (i.e., for horizontal handovers), such as Mobile WiMAX, 3GPP mobile networks, and so on.
- *Handovers on network layer*: This is the usual approach for handovers in heterogeneous wireless and mobile environments (i.e., vertical handovers). Typical examples of protocols for handovers on network layers are MIP protocols, such as MIPv4 and MIPv6.
- *Handovers on application layer*: This type of handover is implemented in the service stratum of NGN, and it is typically based on IMS functionalities (for IMS-based NGNs).

5.6.3 Architecture for Mobility Management in Service Stratum

MM in NGN service stratum is based on IMS [20]. In IP-based network (as NGN) a user terminal may change the IP address and it may result in service discontinuity. Hence, link-layer (i.e., OSI-2 layer) information may be used for MM schemes (i.e., IEEE 802.21 standards for Media Independent Handovers – MIHs [21]).

NGN-based mobile networks such as 4G mobile networks generally provide multimedia services. Such standardized system for handling multimedia services is the IMS, especially for real-time multimedia services.

The architecture for MM in a NGN service stratum is shown in Figure 5.14, which is based on interactions between FEs of the Service Control Function (SCF) and TF [20].

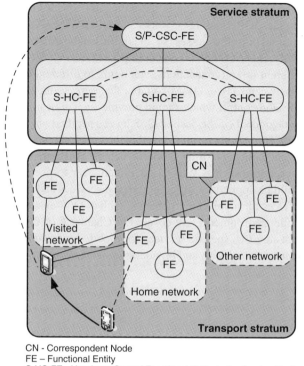

CN - Correspondent Node
FE – Functional Entity
S-HC-FE - Handover Control Functional Entity in the Service Stratum
S/P-CSC-FE – Serving/Proxy Call Session Control Functional Entity

Figure 5.14 Mobility management architecture for NGN service stratum

The SCF may include the following components from the NGN architecture: Call Session Control Functional Entity (CSC-FE), Service Authentication Authorization Functional Entity (SAA-FE), Service User Profile Functional Entity (SUP-FE), and Subscription Locator Functional Entity (SL-FE). CSC-FEs (i.e., CSCFs in 3GPP terminology) are in fact the main IMS entities.

Handover Control Functional Entity in the Service stratum (S-HC-FE) is defined to support the handover of users in the service stratum. Generally S-HC-FE is implemented different from the HCF entities in the transport stratum.

For MM in service stratum are also defined entities for LM and for HM. LM functionality in service stratum is provided by CSC-FE, that is, the IMS call and session control entities. However, for handover execution, the HC-FEs exchange control messages with the entities in the transport stratum (i.e., access or edge gateways, controllers, etc.), which are used for management of handover tunnels and data paths during handover.

5.7 Next Generation Mobile Services

Different wireless technologies have different capabilities and operate in different bands. For instance, WiFi networks have no explicit QoS support because they are not using the TDMA in the RANs as well as they are operating in the unlicensed spectrum (2.4 GHz and around 5 GHz). On the other side, 3GPP mobile networks as well as WiMAX networks have defined QoS support with given set of traffic classes (so, the traffic can be differentiated among several classes, served with different priorities and scheduling mechanisms). However, there should be mapping between different flows (connections) belonging to different traffic classes between different RANs (i.e., between LTE and WiMAX). For that purpose there is a need for an entity for resource and admission control which should provide such allocation in the target mobile network or deny it or degrade it (if the desired QoS level, specified via parameters such as bit rate, losses, delay, jitter, etc., cannot be guaranteed in the target wireless network). Then, there are different services including real-time services such as VoIP, streaming, gaming, and so on, and non-real-time services such as Web-based, e-mail, peer-to-peer, and so on. Therefore, different policies should be applied for different types of services and different RANs.

Important NGN services for mobile environments, such as QoS-enabled VoIP and the Internet of Things in mobile environments, are covered in Chapter 8 of this book. However, the following sections consider two specific services for mobile broadband environment, that is, mobile TV and standardized multicast and broadcast systems, as well as Location-Based Services (LBSs).

5.7.1   Mobile TV

The higher bit rates available to end users in mobile broadband networks are appropriate for bandwidth consuming services such as video streaming and multimedia services (consisted of video, audio, and data components). Provisioning of video and multimedia services can be done on demand or as live streaming or TV. In both cases, one may distinguish between video services offering in BE manner (i.e., without any QoS guarantees) as well as with certain QoS guarantees. Both are possible in mobile broadband networks, such as WiMAX and 3GPP cellular networks, as well as in WiFi (i.e., WLAN). TV is a specific multimedia service (consisted of video and audio streams, and optionally data components) because it should be transmitted to end users with limited time delay (downlink direction is more important in this case), thus it requires certain QoS guarantees in the mobile network. When TV is transmitted using the IP stack (which is the case in NGN and in 4G mobile broadband) it is referred to as IPTV (the provisioning of IPTV services in NGN is covered in Chapter 8). However, Mobile TV has certain specifics also due to mobile nature of the access network, and it results in time-varying wireless channel conditions and thus varying bit rates available to the mobile user.

5.7.1.1   IPTV over WiMAX and WiFi

All wireless access networks standardized by the IEEE are IP-native. Hence, TV is transmitted over WiMAX (including fixed and mobile WiMAX) and WiFi by using the MPEG/RTP/UDP/IP protocol stack (RTP, Real-time Transport Protocol; UDP, User Datagram Protocol). MPEG stands for Moving Pictures Experts Group, where MPEG-2 and MPEG-4 are main standards that can be used for transmission of multimedia streams over networks, from which MPEG-4 is more flexible to the different resolutions of moving pictures, different bit rates for delivering the contents, etc.

To provide TV streaming service in WiMAX, there is a need to use certain QoS class for such service. From the available QoS classes in WiMAX, appropriate class that should be given to TV streams is rtPS or ertPS.

WiFi also can be used for IPTV transmission, especially in home environments. However, WiFi is based on contention-based access to the wireless medium, which reduces the possibilities to provide strict QoS guarantees as in 3GPP mobile networks or WiMAX (both use TDMA-base access schemes in the radio interface, so they can provide explicit QoS support). Also, WiFi operates in unlicensed frequency bands, which may be congested by many users having WiFi networks operational (i.e., in a residential building or corporate offices). However, WiFi also has possibility for QoS support with implementation of IEEE 802.11e standard [22], which uses so-called Access Categories (AC) to provide differentiation between different types of services such as voice, video, BE, and background.

5.7.1.2   Multicast and Broadcast Multimedia System

3GPP has started with standardization of Multicast and Broadcast Multimedia System (MBMS) from Release 6, and it is continued in further Releases [23]. It is aimed to provide the same multimedia content transmitted to multiple users in a same cell or multiple cells. So, MBMS is a point-to-multipoint service in which data is transmitted from a single source entity to multiple recipients. Transmitting the same data to multiple recipients allows network resources to be shared between multicast and broadcast and other services.

Regarding the RANs, MBMS for EPS supports E-UTRAN and UTRAN (it is also referred to as Evolved Multicast and Broadcast Multimedia System, i.e., E-MBMS), while MBMS for GPRS supports UTRAN and GERAN.

The main target of MBMS architecture is efficient usage of radio-network and core-network resources, that is, it saves resources in radio and transport network by not having to send the same packet to multiple nodes individually. However, due to scarce radio resources, it has accent on radio interface efficiency. For introduction of MBMS different core network FEs (i.e., MME, E-UTRAN, and UE for LTE/LTE-Advanced mobile networks) are enhanced to provide the MBMS Bearer Service.

For MBMS in EPS there are added two new FEs, Multicast and Broadcast Multimedia System Gateway (MBMS-GW) and Broadcast Multicast Service Center (BM-SC), as shown in Figure 5.15. BM-SC is responsible for authorization and authentication of content providers, configuration of the data flows through the core network, and charging. On the other side, MBMS-GW is a logical node handling multicast of IP packets from the BM-SC to all eNodeBs in E-UTRAN that are used for transmission of multimedia streams in a given service area. Service area consists of a cell or several cells in which is transmitted the same signal using the same radio resource in the LTE/LTE-Advanced radio network, and it is called MBMS Single Frequency Network (MBSFN). MBMS-GW also handles session control signaling via the MME.

From the BM-SC, the MBMS data is forwarded using IP multicast, a method of sending an IP packet to multiple receiving network nodes in a single transmission, via the MBMS gateway to the cells from which the MBMS transmission is to be carried out.

Typically, a particular instance of the MBMS Bearer Service is identified by an IP Multicast Address and an APN (Access Point Name) Network Identifier.

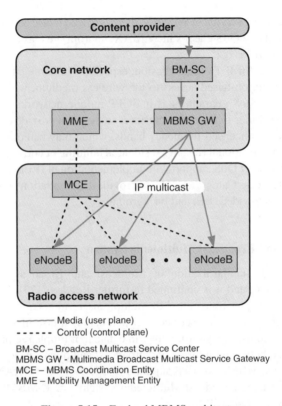

Figure 5.15   Evolved MBMS architecture

Finally, regarding the implementation of MBMS, it is lacking behind the standardization. Although the beginning of MBMS started with HSPA in 3G mobile systems, there were no commercial deployments of MBMS systems during the 3G era. However, one may expect that mobile broadband provided with 4G deployments (i.e., LTE/LTE-Advanced systems) will result in commercial deployments of MBMS in the time period 2014–2020.

5.7.2   Location-Based Services

LBSs provide users of mobile devices personalized services customized to their current location in the mobile network. There are two basic approaches to implement LBS regarding the positioning of the users:

- Process location data in a server and deliver results to the device;
- Obtain location data for a device-based application that uses it directly.

There are two different kinds of Location Services (LCSs) considering whether the information is delivered upon user interaction or delivered to the user based on certain contextual information (i.e., subscription of the user to certain types of location-based information):

- *Pull services*: Deliver information directly requested from the user, similar to the client–server paradigm in the Internet (i.e. the Web).

- *Push services*: Deliver information which is either not or indirectly requested from the user. Such services are more complex to implement because there is no triggering by the user. Also, they are more sensitive, to avoid unwanted information for the mobile user. For example, legacy mobile messaging services (i.e., Short Message Service – SMS and MMS) use using "push" technology to deliver messages to the recipient mobile user.

There are two main architectures for implementation of LCS, that is, it can be implemented using either the user plane or control plane:

- *LBS in control plane*: In this case the positioning data are sent over the control channel and in such case user services (i.e., voice) and location data can occur simultaneously. Such approach can be implemented mainly by vendors of the mobile network elements and mobile terminals, for example, by using 3GPP recommendations for LCS [24].
- *LBS in user plane*: This is an architecture where location related signaling appears as user data to the mobile network. This is end-to-end IP-based approach, which provides quick time to the market and lower deployment costs. The LBS provider can be a third party provider (different from the mobile operator). Standardization of protocols for LBS in user plane is done by the Open Mobile Alliance (OMA) [25].

Possible LBS for mobiles include: emergency services (i.e., 112 in Europe, 911 in the USA, road-side service, etc.), location-based information with push (i.e., reminders, traffic and weather alerts, concierge, etc.), location-based information with pull (i.e., presence-enhanced phone directory, shopping, city guide, weather forecast, etc.), navigation (directions prior to a trip, route planning, etc.), tracking (i.e., workforce and vehicle fleet, stolen/lost tracking, family tracking, etc.), location-enhanced imaging (i.e., location info captured with image or video, sending images with maps, etc.), location sharing (i.e., friend-finder, landmarks, etc.), as well as different combinations of some of the above LBSs.

5.8 Regulation and Business Aspects

Mobile broadband is the driver for development of different and personal telecommunications services. However, its adoption is greatest in those countries where the penetration of fixed broadband is highest.

In the second decade of the twenty-first century most of voice calls are being originated in mobile networks, a trend that is most evident in developing countries where fixed networks have not been widely deployed outside the urban areas. Also, mobile broadband is increasing the importance of mobile video streaming services (i.e., mobile TV) and mobile data services (i.e., Web-based services, peer-to-peer services). However, to have mobile broadband in practice the most important part for mobile networks is spectrum, because more spectrum is needed for higher data rates (i.e., LTE-Advanced and Mobile WiMAX Release 2 may allocate up to 100 MHz of bandwidth). On the other side spectrum is scarce, and usable spectrum for cellular mobile networks is below 5 GHz (with aim to have non-line-of-sight communication between the mobile terminal and the BS or AP). Hence, spectrum management for mobile broadband is increasing its importance, together with the business aspects of mobile broadband services.

5.8.1 Spectrum Management for Mobile Broadband

Spectrum is a limited resource and must be managed to ensure efficient and equitable access for services that use radio communications. On the other side, spectrum is a national resource and hence significant revenue generator (i.e., by giving licenses for spectrum usage to mobile operators). However, radio waves does not respect national borders as people do, thus coordination of access to spectrum depends on international cooperation.

Goals and objectives of the spectrum management system are to facilitate the use of the radio spectrum within the ITU Radio Regulations and in the national interest [26]. Efficient spectrum management approach should provide enough spectrum over both the short and long term for all different services that require radio communications, including public services, private business communications, and broadcasting information to the people. Spectrum management is dependent upon regulation in a given country. Although no two countries are likely to manage the spectrum in exactly the same manner, for spectrum that is important on international level such as mobile broadband spectrum (i.e., for mobility support of users and devices) there is a need for regional and global harmonization of the spectrum usage. Such harmonization of the spectrum management on a global scale is managed and led by the ITU-R.

Global harmonization and recommendations for spectrum management are delivered on ITU-R World Radiocommunication Conference (WRC), held on several years distance. For example, the last WRC is held in 2012, the one before it was in 2007, while the next one is scheduled for 2015. One of the targets of WRC 2012 was spectrum harmonization for IMT, including IMT-2000 (the 3G mobile systems) and IMT-Advanced (the 4G mobile systems). New spectrum are allocated to IMT, as well as existing spectrum allocated to 2G (i.e., GSM) and 3G mobile networks (i.e., UMTS/HSPA, cdma2000) is open for allocation to different standardized mobile systems.

The transition of analog to digital television (i.e., to Digital Video Broadcasting-Terrestrial, i.e., DVB-T), which should be completed in most parts of the world latest by 2015 (and in some developing regions until 2020), there is appearing a new portion of "excellent" spectrum (regarding the radio propagation characteristics) around 800 and 700 MHz, called the digital dividend. In addition to establishing the conditions to use the 800 MHz band (i.e., 790–862 MHz) in some regions (the "first" digital dividend), WRC-12 considered further spectrum allocations to the mobile service to facilitate the development of terrestrial mobile broadband applications. Further extension of the 800 MHz band, that is, 694–790 MHz (called the "second" digital dividend, i.e., the 700 MHz spectrum for mobile broadband services) in some regions in the world will be available from 2015. IMT spectrum allocations for mobile broadband (as a result of ITU-R WRC 2012) are shown in Table 5.4 [27].

Ongoing work at ITU-R is targeted for further harmonization after WRC-12, and it focuses on establishing harmonized channeling arrangements for IMT in band 694–790 MHz, as well as undertaking all the necessary technical studies to ensure coexistence with the networks operated in the new allocation.

In near future ITU-R (at WRC 2015) aims to consider additional spectrum allocations to the mobile service on a primary basis and identification of additional frequency bands (i.e., 470–698 MHz, 3600–3800 MHz, parts of the 4000–4999 MHz spectrum, etc.) for IMT, to facilitate further the development of terrestrial mobile broadband applications.

Table 5.4   IMT spectrum

Spectrum (MHz)	Band name	Paired bands (FDD)				Un-paired bands (TDD) (MHz)
		Uplink – UL (MHz)	Center gap (MHz)	Downlink – DL (MHz)	UL-DL separation (MHz)	
450–470	D1	450–454.8	5.2	460–464.8	10	None
	D2	451.325–455.725	5.6	461.325–465.725	10	None
	D3	452–456.475	5.525	462–466.475	10	None
	D4	452.5–457.475	5.025	462.5–467.475	10	None
	D5	453–457.5	5.5	463–467.5	10	None
	D6	455.25–459.975	5.275	465.250–469.975	10	None
	D7	450–457.5	5.0	462.5–470	12.5	None
	D8	None	–	None	–	450–470
	D9	450–455	10.0	465–470	15	457.5–462.5
	D10	451–458	3.0	461–468	10	None
698–960	A1	824–849	20	869–894	45	None
	A2	880–915	10	925–960	45	None
	A3	832–862	11	791–821	41	None
	A4	698–716	12	728–746	30	716–728
		776–793	13	746–763	30	
	A5	703–748	10	758–803	55	None
	A6	None	–	None	–	698–806
1710–2200	B1	1920–1980	130	2110–2170	190	1880–1920; 2010–2025
	B2	1710–1785	20	1805–1880	95	None
	B3	1850–1910	20	1930–1990	80	1910–1930
	B4	1710–1785	20	1805–1880	95	1880–1920
		1920–1980	130	2110–2170	190	2010–2025
	B5	1850–1910	20	1930–1990	80	1910–1930
		1710–1770	340	2110–2170	400	
2300–2400	E1	None	–	None	–	2300–2400
2500–2690	C1	2500–2570	50	2620–2690	120	2570–2620
	C2	2500–2570	50	2620–2690	120	2570–2620, FDD external DL
	C3	Flexible usage of FDD and TDD				
3400–3600	F1	None	–	None	–	3400–3600
	F2	3410– 3490	20	3510–3590	100	None

5.8.2   Business Aspects for Mobile Broadband

Mobile services have different characteristics than fixed services, because they are targeted to personal communication where each service is typically used by an individual subscriber. Additionally, mobile networks are enhanced by the mobility of the users, something that influences the provided services and their characteristics. In contrast, fixed services are bind to a specific location and usually found in residential (i.e., home) or business environments. The implementation costs of mobile and fixed services are also different (higher complexity is always present in mobile networks due to mobility and less reliable access networks; i.e., radio interface), which influences price plans, price levels, and their structures (i.e., on average, in deployed networks same services have higher costs in mobile networks than in fixed ones) [28].

Business side of the mobile broadband is determined by the pricing, and pricing of services is usually regulated by regulators (including deregulation of certain services in some environments). Also, one should note that mobile and fixed markets are dependent between each other, especially in converged telecommunication world (i.e., FMC). For example, many regulators apply ex-ante retail price controls to fixed telephony. However, that is an approach which is under threat when mobile voice services become substituted for fixed voice services in many business and personal situations (i.e., using fixed broadband access to Internet for voice services, including WiFi access to fixed broadband from mobile terminals, tablets, or lap-top computers).

The telecommunications/ICT (Information and Communication Technology) sector is experiencing a period with major changes driven by the underlying technologies that are being commercially deployed, such as mobile networks and Internet technologies. That results in significant changes in investment, service deployments, convergence on different levels (i.e., network level, service level) and service costs, structure of the service markets and their relationships, as well as user demands which are continuously increasing with the technological development. In such environment legacy pricing regulation may be inappropriate, and therefore it should be more flexible and innovative to follow future growth and development of the markets in mobile (as well as fixed) broadband environments.

5.9   Discussion

The mobile broadband with bit rates comparable with the fixed broadband access is becoming reality with the next generation mobile networks, such as 4G as the following developments in this field. LTE-Advanced and Mobile WiMAX 2.0 are providing higher bit rates to end-users as well as many advances in radio network and in the core network for higher quality of the services. At the same time, 4G in mainly based on NGN principles for separation of the transport entities and service entities. The NGN functionalities are standardized for next generation mobile networks (i.e., PCRF), and IMS is commonly adopted control and signaling platform in all networks for service provisioning. Finally, one may note that with the 4G deployments start the practical implementation of NGN principles in the mobile broadband world. However, in mobile environments this goes along with allocation of more spectrum and mechanisms for even higher spectral efficiency to have higher available bit rates in the future.

References

1. ITU-R (2008) Requirements Related to Technical Performance for IMT-Advanced Radio Interface(s). ITU-R Report M.2134.
2. ITU-R (2003) Framework and Overall Objectives of the Future Development of IMT-2000 and Systems Beyond IMT-2000. ITU-R Recommendation M.1645, June 2003.
3. 3GPP TS (2013) Policy and Charging Control Architecture (Release 12). 3GPP TS 23.203, June 2013.

4. 3GPP TS (2013) Architecture Enhancements for Non-3GPP Accesses (Release 12). 3GPP TS 23.402, June 2013.

5. Gundavelli, S., Leung, K., Devarapalli, V. *et al.* (2008) Proxy Mobile IPv6. RFC 6543, August 2008.

6. Soliman, H. (2009) Mobile IPv6 Support for Dual Stack Hosts and Routers. RFC 5555, June 2009.

7. Holma, H. and Toskala, A. (2011) *LTE for UMTS: Evolution to LTE-Advanced*, 2nd edn, John Wiley & Sons, Ltd, Chichester.

8. 3GPP TS (2012) Service requirements for Home Node B (HNB) and Home eNode B (HeNB) (Release 11). 3GPP TS 22.220, September 2012.

9. IEEE (2006) IEEE 802.16e-2005. *Part 16: Air Interface for Fixed and Mobile Broadband Wireless Access Systems Amendment 2: Physical and Medium Access Control Layers for Combined Fixed and Mobile Operation in Licensed Bands and Corrigendum 1*, IEEE, February 2006.

10. IEEE (2011) IEEE 802.16m-2011. *Part 16: Air Interface for Broadband Wireless Access Systems, Amendment 3: Advanced Air Interface*, IEEE, May 2011.

11. WiMAX Forum (2012) WiMAX Forum Network Architecture: Detailed Protocols and Procedures – Base Specification, 17 April 2012.

12. Alasti, M., Neekzad, B., Hui, J. and Vannithamby, R. (2010) Quality of service in WiMAX and LTE networks. *IEEE Communications Magazine*, **48** (5), 104–111.

13. ITU-T (2006) IMS for Next Generation Networks. ITU-T Recommendation Y.2021, September 2006.

14. ITU-T (2009) Fixed Mobile Convergence with a Common IMS Session Control Domain. ITU-T Recommendation Y.2808, June 2009.

15. 3GPP TS (2013) IP Multimedia Subsystem (IMS); Stage 2 (Release 12). 3GPP TS 23.228, June 2013.

16. Rosenberg, J., Schulzrinne, H., Camarillo, G. *et al.* (2002) SIP: Session Initiation Protocol. RFC 3261, June 2002.

17. ITU-T (2011) Mobility Management Requirements for NGN. ITU-T Recommendation Y.2801, November 2011.

18. ITU-T (2009) Mobility Management and Control Framework and Architecture within the NGN Transport Stratum. ITU-T Recommendation Y.2018, September 2009.

19. ITU-T (2008) Generic Framework of Mobility Management for Next Generation Networks. ITU-T Recommendation Y.2804, February 2008.

20. ITU-T (2011) Framework of Mobility Management in the Service Stratum for Next Generation Networks. ITU-T Recommendation Y.2809, November 2011.

21. IEEE (2008) IEEE 802.21-2008. *Part 21: Media Independent Handover Services*, IEEE.

22. IEEE (2005) IEEE 802.11e-2005. *Part 11: Wireless LAN Medium Access Control (MAC) and Physical Layer (PHY) Specifications, Amendment 8: Medium Access Control (MAC) Quality of Service Enhancements*, IEEE, November 2005.

23. 3GPP TS (2003) Multimedia Broadcast/Multicast Service (MBMS); Architecture and Functional Description (Release 11). 3GPP TS 23.246, March 2003.

24. 3GPP TS (2013) Functional Stage 2 Description of Location Services (LCS) (Release 11). 3GPP TS 23.271, March 2013.

25. OMA (2011) Secure User Plane Location V3.0, September 2011.

26. ITU-R (2010) Economic Aspects of Spectrum Management. ITU-R Report SM.2012-3.

27. ITU-R (2013) Frequency Arrangements for Implementation of the Terrestrial Component of International Mobile Telecommunications (IMT) in the Bands Identified for IMT in the Radio Regulations (RR). ITU-R Recommendation M.1036-4, March 2013.

28. ITU (2012) Regulatory and Market Environment – Regulating Broadband Prices, Broadband Series, Telecommunication Development Sector.

6

Quality of Service and Performance

6.1 Quality of Service and Quality of Experience in NGN

By definition the Next Generation Network (NGN) is a packet-based network [i.e., IP-based (Internet Protocol)] that provides variety of telecommunication services to end users (by using broadband access) by using Quality of Service (QoS) enabled transport technologies, in which service-related functions (in NGN service stratum) are separated from the underlying transport functions (in NGN transport stratum). However, different from traditional telecommunication networks [e.g., PSTN (Public Switched Telephone Network), broadcast networks] which assure the service quality for a particular service type (e.g., PSTN for voice, broadcast networks for television), NGN as a fully IP-based network is aimed to handle heterogeneous services (regarding the QoS requirements, bit rates, etc.) in an integrated manner over the same access and transport networks without failure or downgrading the agreed quality level for a given service. On the other side, the best-effort Internet services [e.g., WWW (World Wide Web), e-mail, etc.] continue to co-exist in NGN with services which have specific QoS requirements [e.g., VoIP (voice over IP), IPTV (Internet Protocol Television), etc.].

6.1.1 What is QoS?

Regarding QoS, there are several definitions. According to ITU-T (International Telecommunication Union–Telecommunications) recommendation E.800 [1], QoS is the totality of the characteristics of a telecommunications service that bear on its ability to satisfy the stated and implied needs of the users of the service. From the network's point of view, QoS can be defined as the ability for segmenting traffic or differentiating between traffic types in order for the network to treat certain traffic flows differently from other. On the other side, QoS also can be defined as a criterion of the degree of user satisfaction of the offered service, which is a more subjective definition of QoS that depends upon the users' perception. However, in traditional telecom traffic engineering (TE) the QoS refers to measurable parameters and techniques to select, control (e.g., via an admission control), measure, and guarantee the required quality for a given service.

Admission control is a typical technique in traditional telecommunication networks, such as PSTN and PLMN (Public Land Mobile Network), where a given call (e.g., a voice call) can be established only if there is available channels end-to-end, otherwise the call is rejected. On

NGN Architectures, Protocols and Services, First Edition. Toni Janevski.
© 2014 John Wiley & Sons, Ltd. Published 2014 by John Wiley & Sons, Ltd.

the other side, in traditional best-effort Internet admission control is absent and the congestion control is provided by the end hosts by using the TCP (Transmission Control Protocol) as a transport protocol (with incorporated congestion control mechanisms). The NGN is a fully converged network which merges the traditional telecommunication networks and the Internet. So, it includes the mechanisms from both "worlds," where QoS comes from the traditional telecommunication worlds (which is important for real-time services, such as voice and television) while separation between services and transport technologies comes from the Internet world (which brings openness for new emerging applications and services that can be independently developed and delivered to users). So, admission control is an intrinsic part of QoS support in any network, including the NGN, although it is not provided for all services. By using already accepted definition in this book, services that do not require admission control (e.g., WWW, e-mail, etc.) are referred to as best-effort services. On the other side, services that do demand admission control (e.g., VoIP, IPTV, etc.) cannot be referred to as best-effort ones. Usually admission control is provided for real-time services where a real-time service is a service that has strict requirements for the end-to-end latency. Further, QoS provisioning in the network requires certain type of signaling prior to given call or session (to check if there are available transport resources end-to-end, such as throughput) as well as during the call or session (with aim to maintain the required QoS level for the given call or session in the access networks and the transport networks). Signaling that is agreed for NGN is Session Initiation Protocol (SIP) and the platform for signaling in NGN is the IMS (IP Multimedia Subsystem). So, regarding the QoS, generally, the two most important techniques that must be implemented in the network are: admission control and signaling. Of course, these two techniques are not enough for QoS provisioning, but they are most important from the networks' perspective. Different mechanisms are available for QoS support on different protocol layers. So, link mechanisms are specified for different access technologies to the resources (including access and transport networks), but they are limited to a given link or layer-2 network segment (e.g., mobile or wireless access, Ethernet, etc.). On the network layer (i.e., the IP layer), there are also defined mechanisms for QoS support end-to-end such as Differentiated Services (DiffServ), Integrated Services (IntServ), and Multi-Protocol Label Switching (MPLS) as legacy solutions for network QoS provisioning in Internet that are also incorporated in the NGN QoS framework.

6.1.2   ITU-T QoS Framework

To provide QoS support in the all-IP based networks (because telecommunication networks are converging to Internet technologies) with heterogeneous user and network equipments (e.g., from different vendors, different access or transport technologies, etc.) and heterogeneous services/applications, standardization of the QoS provisioning is certainly needed end-to-end for those services and applications that have specific QoS requirements. Such standardization is carried out by the ITU-T on a global scale (of course, there are also regional efforts in regional standardization bodies), while Internet technologies are continuing to be standardized by the IETF (Internet Engineering Task Force) and then they are incorporated (where needed) into framework standards (e.g., ITU-T recommendations). To provide QoS support for a given service one needs a definition of the QoS criteria and parameters. Such definitions are given in ITU-T recommendations G.1000 [2], which gives a general QoS framework. It specifies seven QoS criteria: speed (refers to all service functions), accuracy (e.g., speech quality, call success ratio, bill correctness, etc.), availability (e.g., coverage, service availability, etc.), reliability (e.g., dropped calls ratio, number of billing complaints, etc.), security (e.g., fraud prevention), simplicity (e.g., easy of software updates, easy of contract termination, etc.), and flexibility (e.g., easy of change in contract, availability of different billing methods such as online billing, etc.). Also, it differentiates among several service functions, such as service management (e.g., sales and pre-contract, service provision, alteration, service support, repair, and service termination), information transfer (e.g., connection establishment, information transfer, and connection release), billing, as well as network/service management by the end user. However, ITU-T recommendation G.1000 provides too abstract a QoS framework for practical implementation [3], while ITU-T recommendation E.802 [4] explains in more detail the items in the ITU-T QoS framework.

The QoS management can be classified into four viewpoints [2, 4], which are interrelated among each other according to the model given in Figure 6.1, given as follows:

- *Customer's QoS requirements*: This is the QoS level required by the subscriber.
- *QoS offered by the service provider (or planned/targeted QoS)*: This includes QoS criteria or parameters offered by the service provider (which includes network providers that provide access to the Internet as a service), which may be used for several offerings:
  - Service Level Agreement (SLA), as a bilateral agreement between the customer and the service provider.
  - Public offering (i.e., declaration) of the service level that can be expected by the subscribers.
  - Planning and maintaining the service at a given performance level.
  - As a merit for subscribers to make the best choice from the given service provider's offerings.
- *QoS achieved or delivered by the provider*: This is the actual level of QoS achieved or delivered by the service provider, which can be used as a check for delivered QoS (e.g., according to a given SLA) or for basis for any corrective action regarding the QoS.
- *Customer perception of QoS*: This is the QoS level obtained by user ratings of the provided QoS by the service operator, which can be used for comparison purposes among QoS

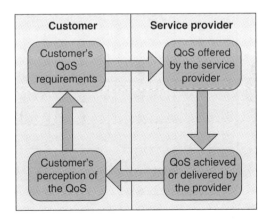

Figure 6.1   Four viewpoints of the QoS by ITU-T

levels provided from different service providers as well as for corrective actions (e.g., when perceived QoS level is below the QoS offering by the provider).

Besides the four viewpoints of the QoS in the ITU-T QoS framework, there is specified a four-market model that is suitable for multimedia services provided by separation of service and transport layers [4], as found in the NGN. It defines the chain of actions from content creation toward service provision, service transport, and customer's equipment. So, there are four elements (content, service provision, service transport, and customer's equipment) that are generally supplied and are working independently of each other, and different parties may be in charge for installation, operation, and maintenance of each of them. The four-market model is shown in Figure 6.2. Different services have different QoS requirements. The given four-market model provides possibility to identify QoS criteria for one or more components in the model. For example, for file download service (e.g., download of music contents) the following criteria can be applied in the four-market model:

- *Content creation*: suitability of the content, its popularity, codec format and its quality, piracy aspects, and so on.
- *Service provision*: ease of navigation to requested music files, fair contract, security of the personal data, pricing, customer care, and so on.
- *Service transport*: bandwidth (in bits/s), latency, (i.e., delay), jitter (i.e., delay variation), error rate, contention, round-trip delay (including delay budget of server, application and network), distortion, and so on.
- *Customer equipment*: quality of playback, required storage capacity (in bytes), ease of selection and playback, ease of navigation and downloading, ergonomic aspects of the user device, and so on.

The given ITU-T QoS framework, including the four viewpoints as well as four-market model, provides basis for definition of NGN regulation which includes Internet regulation,

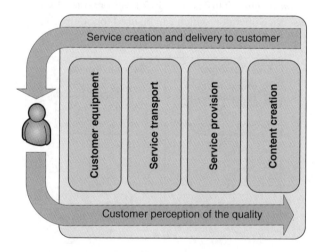

Figure 6.2 Four-market model by ITU-T

since the Internet is not a separate network besides the NGN but becomes its fundamental part in both stratums, the transport stratums (e.g., routers, switches, etc.), and the service stratum (e.g., servers, databases, etc.). However, in the second decade of the twenty-first century QoS regulation regarding the access to the Internet targets mainly Internet access bit rates and particularly is concerned with regulation of the broadband access to Internet [5]. In fact, there are two possible aspects of QoS regulation:

- QoS regulation between operators at interconnection;
- QoS regulation between operators and end users.

However, there is couple of questions regarding the QoS regulation [6]. First question is what QoS issues for certain services (e.g., voice, television) have to be changed when a transition is made from PSTN/PLMN and broadcast networks to all-IP based networks, such as NGN. A second question is what QoS criteria should be included in public regulation (e.g., regulated and monitored by National Regulatory Agencies – NRAs). The situation in the field is different in developed and in developing countries. In some developing countries the quality is low in the PSTN and in the Internet due to various reasons (e.g., low quality of the network infrastructure, not enough installed network capacity or server capacity, human resources, etc.). There are several other key points regarding the QoS regulation. One of them is the dominant best-effort nature of Internet and its architecture, which is not suited well for QoS provision (e.g., VoIP), but it can evolve accordingly by adding certain functions in network nodes and in service platforms, as standardized for NGN. Technical parameters for QoS are also different in packet-switched Internet compared with circuit-switched traditional telecommunication networks. However, the main implication of Internet in the regulation is its multi-service and multi-operator environment as well as its global character. Main area where regulation has a role is SLA in interconnection between dominant players in the given market (e.g., dominant network providers in a given country). However, there is also possibility to have lower QoS than before for certain services such as voice (e.g., latency for QoS-enabled voice delivery in any IP-based network infrastructure cannot be lower than latency in circuit-switched POTS-Plain Old Telephony Service). However, strict regulation regarding the QoS in the Internet is not advised due to dynamic nature of the Internet itself and continuously changing technology solutions. In general, the Internet environment is mostly competitive due to its openness to new services and applications either locally (provided in a given network) or globally (e.g., web sites), hence regulatory intervention should be limited. But, in the cases where markets for Internet services are not sufficiently competitive (e.g., in some developing countries), then there is a need for regulation. However, the criteria that will be used in such cases for competition regulation must be based on market analysis and not the technology itself. Additionally, other possible tool is self-regulation in the interconnection between different market players as well as in the relations between companies (e.g., vendors, network providers, service providers) and end users.

6.1.3   Performance Parameters for IP Services

NGN is all-IP network, where all services are provided by using IP protocol stack, where IP is located on the network layer (i.e., OSI-3 layer). However, the performance of IP-based services (i.e., all NGN services) depends upon the performances of other layers below or above the IP [7]:

- *Lower layers*: These layers belong to certain links (e.g., between two network elements such as routers and switches, or links between network elements and end hosts) which provide connection-oriented or connectionless transport for IP layer (i.e., IP packets). In general links may be based on different types of technologies, such as Ethernet, ATM (asynchronous transfer mode), PDH (plesiochronous digital hierarchy), SDH (synchronous digital hierarchy), ISDN (Integrated Services Digital Network), leased lines, and so on.
- *Higher layers*: These layers include transport layer and above (e.g., presentation, session, and application layers, usually treated in best-effort Internet as single application layer), which include legacy IPs (on those layers) such as TCP, UDP (User Datagram Protocol), RTP (Real-time Transport Protocol) on the transport layer, as well as HTTP (Hypertext Transfer Protocol), FTP (File Transfer Protocol), SMTP (Simple Mail Transfer Protocol), POP3 (Post Office Protocol version 3), and so on, on the application layer. Higher layers may also influence the end-to-end performance (e.g., TCP provides congestion control end-to-end by the end hosts, while RTP provides two-way control information needed for real-time services).

However, in the middle of the protocol "hourglass" is the IP which provides connectionless transport of IP packets end-to-end based on source and destination IP addresses. The layered model of performance for IP services is shown in Figure 6.3. IP packet transfer performance is described by a set of parameters, from which the most important ones are the following:

- IP Packet Transfer Delay *(IPTD)*: This is the time difference between the occurrences of two corresponding IP packet reference events (IP packet reference event is packet transmission via given measurement point in the network). There are several types of IPTD such as: minimum IPTD (the smallest IP packet delay among all IPTDs), median IPTD (the 50th

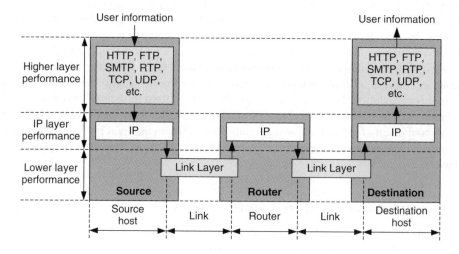

Figure 6.3   Layered model for IP service performance

percentile of the frequency distribution of IPTDs), as well as average IPTD (arithmetic average of IPTDs).

- IP Packet Delay Variation *(IPDV)*: This is the difference between the one-way delay of IP packet and reference IPTD (e.g., average IPTD as a reference delay).
- IP Packet Error Ratio *(IPER)*: This is ratio of total number of errored IP packets to the total number of transmitted IP packets in a given measurement.
- IP Packet Loss Ratio *(IPLR)*: This is ratio of total number of lost IP packets to the total number of transmitted IP packets in a given measurement.

The values of the defined performance parameters vary, depending upon different so-called network QoS classes. Transfer capacity (in bit/s) is a QoS parameter that has highest impact on the performance perceived by the end user, so a higher bit rate (i.e., broadband access and transport) is normally better for all services, including real-time and non-real-time ones. Based on the requirements of the key applications such as conversational telephony, reliable data applications based on TCP (e.g., WWW, e-mail, etc.), and digital television, network QoS classes are specified by ITU-T [8], as given in Table 6.1 which defines upper bounds on key performance parameters (besides the transfer capacity) for end-to-end IP services (end-to-end means from one User–Network Interface – UNI to another UNI, i.e., UNI-UNI performance is defined for each class). For example, class-0 and class-1 are targeted to real-time jitter-sensitive applications (e.g., VoIP, video conferences) where class-0 provides higher interactions due to lower bound on the IPTD parameter. Similarly, class-2 and class-3 are targeted to transaction data, from which class-2 is intended for signaling traffic, while class-3 is for interactive applications. Class-4 is targeted for short transactions, video streaming, or bulk data, while class-5 is unspecified (regarding all performance parameters) and hence it is targeted to traditional best-effort Internet applications. Finally, QoS classes 6 and 7 are provisional in the given table and they do not need to be met until they are revised. However, such QoS classes are needed because some applications such as IPTV (i.e., digital TV over IP networks) cannot be delivered with performance objectives set for QoS classes 0–5, as well as new applications with strict performance parameters are continuously emerging.

Table 6.1 IP traffic classes as defined by ITU

QoS class	Upper bound on IPTD	Upper bound on IPDV (ms)	Upper bound on IPLR	Upper bound on IPER
Class-0	100 ms	50	10^{-3}	10^{-4}
Class-1	400 ms	50	10^{-3}	10^{-4}
Class-2	100 ms	Unspecified	10^{-3}	10^{-4}
Class-3	400 ms	Unspecified	10^{-3}	10^{-4}
Class-4	1 s	Unspecified	10^{-3}	10^{-4}
Class-5	Unspecified	Unspecified	Unspecified	Unspecified
Class-6	100 ms	50	10^{-5}	10^{-6}
Class-7	400 ms	50	10^{-5}	10^{-6}

IPDV, IP packet delay variation; IPER, IP packet error ratio; IPLR, IP packet loss ratio; and IPTD, IP packet transfer delay.

6.1.4   Quality of Experience

Besides the QoS as an objective measure (via performance parameters that can be measured at defined measurements points, such as network interfaces in hosts and network nodes), there is a need for subjective merit for the quality of the given service perceived by the end users, which is even more important in heterogeneous environment such as NGN (based on various Internet technologies including networks and services). Therefore, ITU-T defines so-called Quality of Experience (QoE) as overall acceptability of an application or service, as perceived subjectively by the end user [9]. However, the overall perception of the quality is dependent upon all four parts in the four-market model for QoS (by ITU-T), which on the other side are independent among each other. Using the four viewpoints in the ITU-T QoS framework, the QoE can be considered a viewpoint of the QoS from the user's perspective. However, QoE also includes the user's decision on retaining on the service or giving it up [10]. So, one may consider QoE on the edge between the perceived and assessed service quality by the end user, as it is shown in Figure 6.4.

On one side, it includes the QoS perceived by the customer and QoS that is required by the customer (something that belongs to the ITU-T QoS model). On the other side, QoE also includes qualitative terms that refer to the user satisfaction from the provided service as well as the user attraction with the services. In general, QoE is influenced by all seven QoS criteria. For example, speed influences the available throughput and latencies and it is of crucial importance for the QoE. That is why going toward broadband access and higher access bit rates (including fixed and mobile broadband) the overall QoE improves. Availability and reliability are also very important, which depends upon the capability of the network to recover from a failure [e.g., SON (Self-Organizing Network) solutions in 4G, resilience solutions in

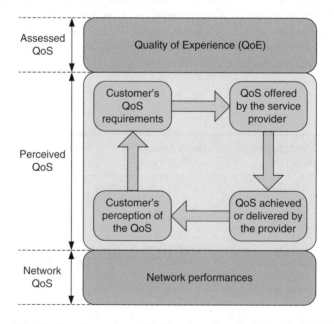

Figure 6.4   Quality of experience (QoE) and quality of service (QoS) by ITU-T

Table 6.2 Mean opinion score (MOS) metrics

Mean opinion score (MOS)	Quality	Impairment
5	Excellent	Imperceptible
4	Good	Perceptible
3	Fair	Slightly annoying
2	Poor	Annoying
1	Bad	Very annoying

optical networks, etc.] as well as appropriate planning and dimensioning of the network (to suit to the expected number of users for a given service or services). For example, typical quality metrics for network availability from the SDH-era onwards are the so-called "five nines," that is, 99.999% of the time service to be available to end users, which request certain survivability mechanisms to be implemented in the network (e.g., re-routing of the traffic in a case of failure over alternative or reserved paths in the network). Also, security aspects, accuracy, simplicity to use the service and flexibility regarding the services influence the QoE.

However, there is needed certain parameter to express the QoE level. The most used measure for QoE is the Mean Opinion Score (MOS). Initially, MOS scale (as given in Table 6.2) referred to voice service only [11], but nowadays it is also used for other services such as video (e.g., IPTV). MOS is expressed as a single number in the range from 1 to 5, where 1 corresponds to the lowest quality experienced by the end user and 5 is the highest quality experienced by the end user. In practice, MOS values of around 4 are also regarded as acceptable quality.

Because NGN provides multiple heterogeneous services over a single converged network it also provides lower operational expenditure (OPEX) due to simplified network operation and management. However, service providers may have a strong competition in such case due to fixed-mobile convergence, and hence may need to benefit by using management and control decisions based on the QoE received from the end users [12]. However, QoE is dependent upon the user itself, application, network conditions and functions, as well as user terminal capabilities. Hence, end-to-end QoE assurance is challenging in NGN due to its multi-network, multi-provider, and multi-vendor environment. However, it is possible to achieve controlled end-to-end QoE in NGN by using adjustable transport functions (e.g., flexible network QoS support) and configurable application parameters (e.g., types of audio and/or video codecs, picture quality, audio quality, number of streams, etc.), which are typically found in NGN.

6.2 Resource and Admission Control Functions

The NGN provides separation between service functions and transport functions. The Resource and Admission Control Functions (RACFs) acts as an arbitrator between Service Control Functions (SCFs) and transport functions for QoS provision. So, RACF has a crucial role for QoS support in the NGN. It performs policy decisions based on contractual information with end users in SLAs, as well as network policy rules, QoS class for the service (e.g., for applying certain priority level), and information regarding the networks and links statuses and utilization.

As crucial QoS functional element in NGN, RACF should satisfy several dozens of requirements for different types of services which demand QoS support from the network and service providers, which include but are not limited to the following [13]:

- Control of QoS-related transport functions in the transport stratum.
- Support for different types of access and core transport technologies as well as different customer equipments while hiding network infrastructure characteristics from the service stratum.
- Arbitration between functions in service and transport stratum.
- Support for absolute and relative QoS control, QoS differentiation, and QoS signaling.
- Support the interactions with standardized Policy and Charging Control (PCC) functions [e.g., 3GPP (3G Partnership Project) PCC].
- Support for resource and admission control for unicast (e.g., VoIP) and multicast traffic (e.g., IPTV).
- Support for resource and admission control for inter-provider SLA for service delivery over multiple NGNs, different types of TV services (e.g., time-shifted TV, IPTV service by invitation, etc.), support for mobility, QoS adaptation (e.g., QoS downgrading), and so on.

The set of requirements for RACF is updated further with each newer release of its recommendation.

6.2.1   RACF Functional Architecture

Functional architecture of RACF is given in Figure 6.5. In general, it consists of two main types of RACFs:

- Policy Decision Functional Entity (PD-FE): This provides a single contact point to the SCF (i.e., the service stratum) and in that manner RACF hides all details of the transport infrastructure and transport functions.
- Transport Resource Control Functional Entity (TRC-FE): This deals with different transport technologies and at the same time it provides the resource-based admission control decision. The TRC-FE is service-independent.

The RACF is connected via its PD-FE with the Network Attachment Control Function (NACF), and optionally with Mobility Management and Control Function (MMCF) when QoS support for mobile nodes is required (e.g., in mobile networks), and with entity for Management for Performance Measurement (MPM). On one side, the PD-FE is connected to SCFs in service stratum, while on the other side it is connected to transport functions in the transport stratum, in particular, to the Policy Enforcement Functional Entity (PE-FE). The PE-FE is a packet-to-packet gateway located at the boundary of different networks, or between the core and access networks. A given transport infrastructure in an NGN can be usually shared among several service providers. In such cases standardized solutions (by using Internet technologies) are used for separation of traffic of different SCF instances [e.g., separation on OSI (Open System for Interconnection) layer-1, layer-2 or layer 3, or usage of Virtual Private Networks (VPNs)].

The TRC-FE (from RACF) is used in transport stratum only. It instructs the Transport Resource Enforcement Functional Entity (TRE-FE) to enforce the transport resource

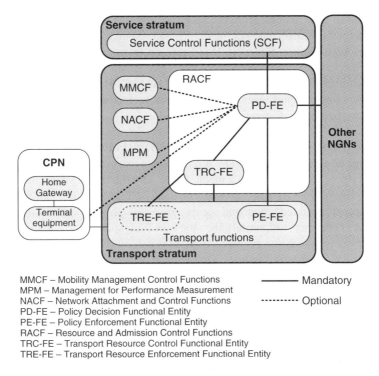

MMCF – Mobility Management Control Functions
MPM – Management for Performance Measurement
NACF – Network Attachment and Control Functions
PD-FE – Policy Decision Functional Entity
PE-FE – Policy Enforcement Functional Entity
RACF – Resource and Admission Control Functions
TRC-FE – Transport Resource Control Functional Entity
TRE-FE – Transport Resource Enforcement Functional Entity

———— Mandatory

---------- Optional

Figure 6.5 RACF functional architecture

policy rules, which is technology-dependent. Hence, TRE-FE differs for different transport technologies including access and core networks.

With aim to transfer different resource control requests between different functional entities in the transport and service stratums, the RACF performs two operations, as follows:

- *Discovery*: This is the discovery of the communicating entities and can be performed with either a static or dynamic mechanism. A static mechanism is mandatory and is based on the configuration of all communicating entities by using location information such as IP addresses and domain names (e.g., retrieved via DNS-Domain Name System). A dynamic mechanism is optional and is used when given functional entity needs to formulate the topology and connectivity information for all communicating entities.
- *Selection*: This refers to a selection of certain discovered instance, which is described with certain information such as Type of Service (ToS), a set of service attributes, and globally unique IP addresses which are bind to the end user's identifier.

When PE-FE in the transport stratum needs to apply certain policy enforcement (e.g., when transport signaling is used to pull the policy decision information from the PD-FE in the RACF), the RACF performs binding of the QoS request by a given media flow with the policy decision information. There are several mechanisms that can be used for such binding of the flow QoS with the applied policies in the transport stratum of the NGN. One possible mechanism is usage of authorization tokens to perform the binding, in which case the PD-FE generates the token upon request from the SCF (i.e., the service stratum functions), where a token may contain the domain name of the PD-FE and resource control session ID to uniquely identify the resource request. Since NGN is based on Internet technologies in the transport stratum, other mechanisms for binding of the flow QoS and the policy is by using network and application identifiers in the Internet environment, such as IP addresses (on network layer) and port numbers (on transport layer). In fact, each flow is uniquely identified by five parameters in Internet: source and destination unique IP addresses [i.e., public IP addresses, not the private ones used via NAPT (Network Address and Port Translation) mechanisms], source and destination ports, and transport protocol (e.g., TCP, UDP). In IPv6 networks where flow label is used, a unique flow can be identified by three parameters: source and destination unique IP addresses, and a flow identifier. When a media flow is identified, its QoS request is mapped to certain session ID. When NAPT is used between the RACF and the user equipment (which is typical usage scenario in fixed broadband access networks with WiFi and/or Ethernet local network for access of end users to Internet), then the flow is identified by the unique IP address of the access gateway [e.g., a home gateway, xDSL (Digital Subscriber Line technologies)/WiFi switch or router, etc.]. Also, for binding a QoS request by a media flow and transport policies, RACF may use the transport subscriber identifier (which is an internal network identifier, not available to the end users) to access subscription profile directly in the NACF.

6.2.2 RACF Deployment Architectures

Regarding the associations between functional entities in RACF and entities in service and transport stratums to which RACF is connected, there are several different architectures for QoS control in NGN, as shown in Figure 6.6.

One may distinguish among three typical RACF architecture implementations, as follows:

- *RACF performs QoS control by using intermediation of SCF*: in this case access and core networks are in separate domains and there is no direct communication between PD-FE (in the RACFs in each of the domains), but only indirect communication via the SCF. Normally, in transport stratum access and core networks are directly connected (via the interface between PE-FEs in access and core networks), only the QoS control by the RACF is going via the service stratum functions.
- *RACF performs QoS control at the RACF level, without intermediation of SCF*: Access and core networks are in separate domains (i.e., each has a separate RACF), but both RACFs communicate directly by using interface between the PD-FEs (i.e., QoS coordination is performed at the RACF level), while SCF communicate in this case only with PD-FE in core networks (there is no communication between SCF and PD-FEs in the access networks).
- *Same RACF is used for access and core networks*: In this case both access and core network are managed by the same network operator. Hence, there is a single PD-FE (that controls both PE-FEs, in access and in core networks) and separate TRC-FEs for access and core networks.

In general, NGN is access independent. However, TRC-FE in the RACF is dependent upon the access technology (e.g., IP networks, IP MPLS, Ethernet, broadband wireless and mobile networks, etc.).

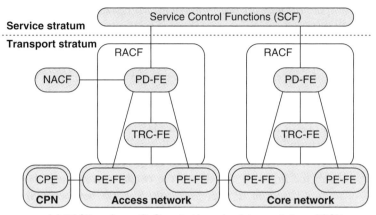

(a) RACF performs QoS control by using intermediation of SCF

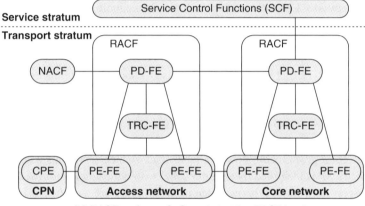

(b) RACF performs QoS control at the RACF level

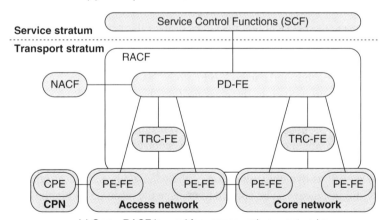

(c) Same RACF is used for access and core networks

Figure 6.6 (a–c) Implementation scenarios for RACF architectures

For example, in "pure IP" network (i.e., without the MPLS) the nodes are routers that handle the packet with IP routing by using some of the standardized routing protocols by IETF [e.g., RIP (Routing Information Protocol), OSPF (Open Shortest Path First), etc.]. In such a case a DiffServ mechanism on the network layer can be used [14], which provides packet differentiation by using different values for packets belonging to different QoS classes in the ToS field in IPv4 headers or Traffic Class field in IPv6 headers, both fields with equal length of one byte (also referred to as Differentiated Services – DS fields). To enable admission control in the network domain one or more TRC-FE are needed to directly manage all physical links in the domain and perform link by link resource allocation and admission control.

In IP transport networks with implemented MPLS (which is typical scenario in transport IP-based networks in the first two decades of the twenty-first century) label switching is used instead of routing based on IP addresses. MPLS works between OSI layers 2 and 3 (below the IP, and above the Medium Access Control – MAC), and adds so-called labels over the IP packet (the position of the label is between the IP header and the MAC header). In IP/MPLS networks Label Switching Path (LSP) technology is used to pre-provision a Virtual MPLS Transport Network (VMTN) for each service type (e.g., VoIP, IPTV, best-effort Internet traffic, corporate VPN, etc.). This can be performed manually (by setting forwarding tables for MPLS) or automatically by using Resource Reservation Protocol for MPLS Traffic Engineering (RSVP-TE), [15], or Constraint-based Label Distribution Protocol (CR-LDP) [16], as well as DiffServ-enabled MPLS Traffic Engineering (DS-TE) for optimization purposes [17]. Hence, admission control and resource allocation for different flows from a single service type are managed within the given VMTN for that service type. In such case one or multiple TRC-FE instances manage all VMTNs. So, in TRC-FE over IP/MPLS networks the QoS route for a given media flow is a label stack (which represents a concatenated LSP set, where each LSP is identified by a unique label within the administrative domain).

When transport network is Ethernet, then network nodes are switches or bridges and only edge nodes are routers (i.e., IP capable). In this case admission control and resource control are applied on link by link resource reservations, where one or multiple TRC-FE instances are deployed to manage the physical link resources of the Ethernet network.

In wireless and mobile broadband access networks [e.g., WiMAX, LTE/LTE-advanced (Long Term Evolution), etc.] mobile terminal handles IP packets via the layer-2 radio interface protocols. Each mobile access technology has QoS signaling in the radio part and certain QoS classification. Such traffic classifications for Mobile WiMAX and 3GPP mobile networks are given in Chapter 5. However, efficient scheduling mechanisms in base stations are needed for efficient QoS provisioning (e.g., scheduling mechanisms in base stations in mobile networks are not standardized and are vendor-dependent). Hence, TRC-FE can be used to provide resource control via priority scheduling (of packets belonging to different QoS classes) as well as dynamic allocation of the bandwidth. Similar to other cases, one or multiple TRC-FE instances can be applied to manage the resource within the mobile network administrative domain.

Regarding the admission control, when Management for Performance Measurement (MPM) is used [18], the TRC-FE admits service requests based on performance information from MPM, such as bandwidth availability, as well as network QoS parameters such as delay, delay variation, and packet loss ratio.

6.2.3 RACF Communication between Different NGN Operators

When a media flow with QoS provision is transported over multiple transport networks, then such action requires inter-operator RACF communication. In such case, for communication between two users, each with its Customer Premises Equipment (CPE), there are three possible scenarios that require inter-operator RACF communication:

- *Only one CPE is in a visited operator network*: In this case both CPEs communicate with the SCF in the home operator (via application-layer signaling), which then sends QoS information to home RACF to reserve resources in the home network. RACF from the home network communicates with the RACF in visited network to reserve resources for the media flow in the visited network.
- *Two CPEs are located in a same visited operator*: This is scenario similar to the first one, with the difference that only resources in the visited network should be reserved, so QoS information is sent from the SCF (upon request from the CPE) to RACF in the home network which forwards it to the RACF entity in the visited network (where the two CPEs are located).
- *Two CPEs are located in different visited operators*: In this case both CPEs communicate directly with the SCF in the home operator network via application-layer signaling. SCF communicates with RACF in the home network of the CPEs and the home RACF communicates directly with RACFs in the visited networks to reserve QoS resources in transport networks.

The three scenarios for inter-operator RACF communication are shown in Figure 6.7.

6.2.4 Example of Admission Control with RACF

NGN includes all different services, so there are different possible examples for admission control applied in the network by the RACF. One example is admission control for VoIP service, other example is for IPTV, and so on. Here is outlined an example of admission control for IPTV service with invitation. This is a service where a user invites one of his/her friends to watch a TV program where the charge for the program is paid by the inviter party.

The QoS reservation procedure for this service is given in Figure 6.8. After the invitation, if invitee accepts it then SCF sends request to invitee's RACF (to its PD-FE entity) for resource initiation. In this case, since the bill will be charged to the inviting party, the RACF does not need to check the subscription information of the invitee. Further, the RACF with its PD-FE makes policy decision (i.e., authorization) by using network policy rules to authorize the required resources (e.g., guaranteed bandwidth for IPTV service to the invitee, QoS class for the service, etc.). Further, within the invitee's network the PD-FE first checks the availability of the required resources for the service with one of the registered TRC-FEs in the network, and then makes the final decision about the admission of the IPTV service by invitation (i.e., final reservation of the required resources). If the media flow (i.e., IPTV flow to the invitee) is not admitted then PD-FE sends a resource initiation response with a rejection reason back to the initiator SCF. On the other side, if the media flow is admitted in the network (upon positive response from a TRC-FE for resource availability), the PD-FE sends resource initiation request to the PE-FE to install the admission decisions. Finally, PE-FE installs the

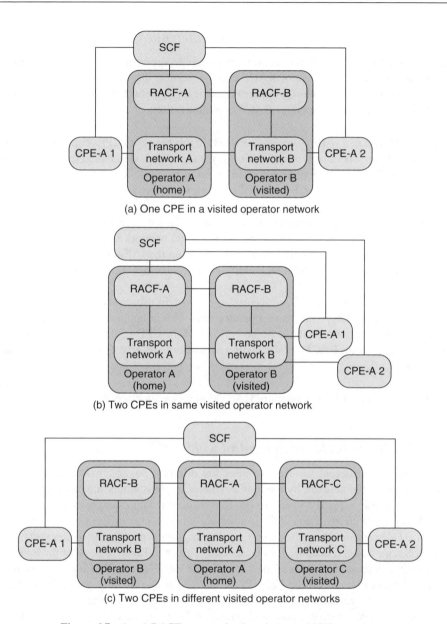

(a) One CPE in a visited operator network

(b) Two CPEs in same visited operator network

(c) Two CPEs in different visited operator networks

Figure 6.7   (a–c) RACF communications between NGN operators

final admission from the PD-FE and sends resource initiation response back to the PD-FE, and PD-FE sends a response back to the initiator SCF.

For services where a calling or called party has to be charged, the Application Function (AF) first requests service information from the Policy and Charging Rule Function (PCRF), which acquires the user profile information from the users database (e.g., HSS-Home Subscriber Server) and makes a policy decision. In a case of positive policy decision, the PCRF sends

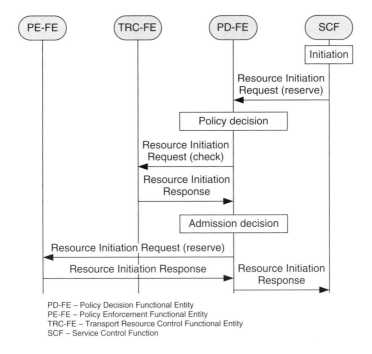

PD-FE – Policy Decision Functional Entity
PE-FE – Policy Enforcement Functional Entity
TRC-FE – Transport Resource Control Functional Entity
SCF – Service Control Function

Figure 6.8 QoS resource reservation procedure for IPTV service with invitation

policy and rules provision to the RACF which further communicates with the policy enforcing functional entity (i.e., the PE-FE) toward the given user (in a case of conversational services, or conference service, this is performed on all sides of a given connection or session, i.e., for all end users). In the case where SCF sends the resource initiation request (e.g., when using IMS for service provisioning) then it sends the request directly to the RACF, and further RACF communicates with the PCRF in the network before it makes a final policy decision for admission of the media flow with the required QoS.

6.3 QoS Architecture for Ethernet-Based NGN

The Ethernet is the dominant access technology in all corporate and residential environments. However, it goes further, from the access to the core and transport networks (e.g., Metro Ethernet, i.e., carrier-grade Ethernet). In such case the all-Ethernet network (on layer 2) with all-IP network (on layer-3) is becoming reality [19]. Therefore, Ethernet-based NGN is an increasingly important network architecture. The main idea in this approach is to use Ethernet technology (i.e., Ethernet frame format) everywhere in NGN transport networks where no conversion of Ethernet frame formats (the "frame" corresponds to OSI layer-2 data structure, where Ethernet belongs) occurs during their delivery end-to-end. On the other side, for QoS support in the Ethernet, it is required Ethernet frames to be able to carry QoS information, that is, to have implemented IEEE 802.1Q standard, as given in Chapter 4 of this book. Overall, Ethernet-based NGN is intended for usage of Ethernet technology over any physical media,

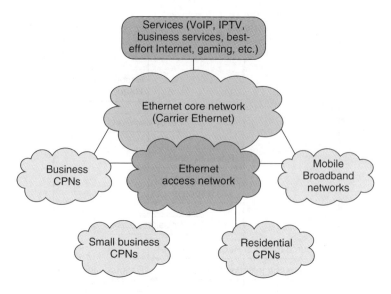

Figure 6.9   Generic architecture model for Ethernet-based NGN

including fixed and wireless environments, in access and core networks. Generic architecture model for Ethernet-based NGN is shown in Figure 6.9.

However, there are certain requirements for user equipment and for the network to provide end-to-end Ethernet approach to be functional. On one side, the user equipments needs to support the same Ethernet format for various underlying physical-layer technologies, auto-discovery via usage of Address Resolution Protocol – ARP (used to convert IP address to Ethernet MAC address of a given network interface attached to Internet) and Reverse Address Resolution Protocol – RARP (does the opposite to ARP; i.e., converts an Ethernet MAC address to the associated IP address for a given network interface), QoS requesting capability (according to the SLA), as well as frame tagging according to IEEE 802.1Q standard for Virtual Local Area Network (VLAN). On the other side, for end-to-end Ethernet the network side needs to support capabilities for Operation and Maintenance (OAM), load sharing, protection and restoration (e.g., similar to SDH transport networks), VPNs, VLAN, auto-configuration capability (e.g., neighbor discovery), security control based on MAC addresses and VLAN IDs, QoS mapping between the core and access networks (and vice versa), as well as to provide traffic management functionalities in Ethernet (to guarantee the requested QoS).

6.3.1   Reference Architecture for Ethernet-Based NGN

Reference architecture for Ethernet-based NGN is shown in Figure 6.10. The interface between the user equipment and the network is Ethernet User-Network Interface (E-UNI) [20]. On the other side, the Ethernet services are provided by concatenation of operator networks. The interface between different operator networks is Ethernet Network-Network Interface (E-NNI). The E-UNI is Ethernet interface which includes the IEEE 802.1Q standard. But, E-UNI frames

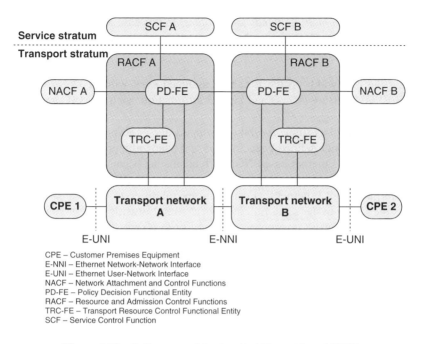

CPE – Customer Premises Equipment
E-NNI – Ethernet Network-Network Interface
E-UNI – Ethernet User-Network Interface
NACF – Network Attachment and Control Functions
PD-FE – Policy Decision Functional Entity
RACF – Resource and Admission Control Functions
TRC-FE – Transport Resource Control Functional Entity
SCF – Service Control Function

Figure 6.10   Reference architecture for Ethernet-based NGN

are not adequate to support OAM and VPN functionalities, and therefore the E-UNI frame is encapsulated in the link-layer payload at the E-NNI.

The Carrier Ethernet (refer to Chapter 4) uses Ethernet Virtual Connections (EVCs) in one of the three possible service types: E-Line, E-LAN, and E-Tree. In Ethernet-based NGN the EVC extends between two E-UNIs. However, there is needed QoS mapping for a service into the Ethernet frame. Because IEEE 802.1Q is mandatory for Ethernet-based NGN, the Class of Service (CoS) is identified by a three-bit field in the 802.1Q frame, also called p-bits. There are eight priority values for CoS (with 3 bits one obtains $2^3 = 8$ different values) which are used for QoS differentiation of network traffic carried in a given EVC (on OSI layer 2). But, network devices (e.g., Ethernet switches) may use or rewrite CoS values in the Ethernet frames (similar to the approach of routers for DS fields in IP headers, used for QoS support on OSI layer-3). In Ethernet-based NGN, the EVC may have one of four possible types regarding the combination of bandwidth profile and CoS, as follows:

- *Single-CoS EVC with a single bandwidth profile*: In this case all frames belonging to the EVC are treated in the same manner regarding the QoS mechanisms, such as scheduling, traffic shaping, and so on.
- *Single-CoS with multiple bandwidth profiles*: In this case all frames are also treated in the same manner regarding the QoS support, but have different bandwidth profiles.
- *Multi-CoS EVC with a single bandwidth profile*: In this case frames are treated differently according to their CoS (different priorities in the frames scheduling), but all frames have equal bandwidth profile.

- *Multi-CoS EVC with multiple bandwidth profiles*: In this case all frames are treated differently according to the CoS assigned, and also all frames have a CoS-specific bandwidth profile.

Regarding the bandwidth profile it can be assigned even without CoS (e.g., cases with single-CoS EVC) by using layer-2 and layer-3 information, such as MAC and IP addresses, respectively.

Single E-UNI as well as single E-NNI can support one or several EVCs. An E-NNI needs to support one of the four defined types for CoS and bandwidth profile.

Support for VPN in Ethernet-based NGN is fundamental because all traffic (including corporate and residential) is transferred by VPNs in transport IP networks. A VPN creates a multipoint-to-multipoint Ethernet that can span several different metro areas providing connectivity as they were in the same Ethernet Local Area Network (LAN). However, it is also possible in practice network operator to connect Ethernet switches with MPLS-enabled networks which use labels instead of VLAN IDs, as well as QoS routing protocols instead of spanning tree protocol that is used in Ethernet networks. With the aim to support VPN services in end-to-end Ethernet NGN architecture, Ethernet frames are encapsulated in the link-layer payload at the E-NNI for their transport across the core networks.

6.3.2   QoS Services in Ethernet-Based NGN

Considering the QoS in Ethernet-based NGN, a user can request or select (from the QoS offering from the operator) certain QoS service with appropriate performance parameters. Each operator is responsible to provide Connection Admission Control (CAC), so it is operator specific. After admission of a connection all parameters including CAC, E-UNI, and E-NNI parameters are set by the operator according to its own network policies.

Ethernet QoS services are grouped in three types as given in Table 6.3. The Ethernet QoS services for NGN include premium service, gold service, and best-effort service. They are created according to the traffic parameters, provider edge rules, transfer capability (regarding the bandwidth for the service) and ITU QoS class for IP-based services.

The traffic management for Ethernet-based NGN is performed by using RACF. The SCF in service stratum determines the QoS requirements for a given flow and then informs the RACF entity. Further, RACF performs policy-based control of transport resources required for the given flow and communicates with transport functions in the transport stratum for applying certain QoS mechanisms for traffic conditioning of the flow. Such mechanisms include classification of the flow and its frames, metering, marking the non-conformant packets (the packets that are received at the excess bit rate or excess bursts) and traffic dropping (e.g., discarding the non-conformant packets at congestion occurrence) or shaping (i.e., delaying some IP packets of the given flow to bring them into compliance with the flow's traffic profile). The edge network nodes (e.g., between the access and core networks of the operator) are applying so-called edge rules on the traffic, that is, traffic is measured by the operator's nodes to check whether it conforms to certain traffic pattern and its traffic parameters, as they are agreed between the operator and the end user (e.g., in the SLA). If customer's traffic exceeds certain traffic parameters, that is, generates an excess traffic in operator's networks (either upstream or downstream), the operator's node (e.g., Ethernet switches) may apply edge rules by marking or dropping the excess frames. However, with aim to identify frames that belong to certain flow

Table 6.3 QoS services for Ethernet-based NGN

Ethernet QoS service type	Traffic parameters	Edge rule	Bandwidth	ITU QoS class for IP traffic
Premium	Constant bit rate and committed burst size, no excess bit rate or bursts	Non-conformant frames are dropped	Dedicated	Classes 0, 1, 6, and 7
Gold	Constant bit rate and committed burst size, excess bit rate, and bursts allowed	Excess frames are admitted with high discard precedence	Delay-sensitive statistical	Classes 2, 3, and 4
Best-effort	No guaranteed bit rate, no guaranteed burst size	All frames are admitted, and first is dropped at congestion	Best-effort	Class 5

and then to correlate it to certain traffic parameters and pattern, one needs traffic classification. In Ethernet-based NGN the classification is usually based on layer-2 (i.e., Ethernet) IDs, but also layer-3 (i.e., IP) identifiers can also be used for end-to-end identification of a given flow. In E-UNI interface the traffic can be tagged (by using IEEE 802.1Q) or untagged (that is traditional Ethernet traffic). When Ethernet frames are tagged (i.e., 802.1Q is used), the flow is classified by using VLAN ID and user priority bits (which determine the CoS). In such a case (with tagged frames), for transport over the operator's core network the VLAN ID is mapped or encapsulated into a label in the MPLS core network (when MPLS is used), or the VLAN ID is mapped into another tag in Ethernet-based transport networks. For example, when an Ethernet-based transport network are used (i.e., Carrier/Metro Ethernet), the VLAN ID (i.e., C-tag in IEEE 802.1Q) is mapped into a S-tag in provider bridges (i.e., Q-in-Q, when IEEE 802.1ad is used in the provider's core network), or into an I-tag or B-tag in provider backbone bridges (when IEEE 802.1ah is used in the provider's core network). These mappings are performed on the network edge nodes (e.g., between access and core networks) by using so-called mapping tables. After the classification, the frames are conditioned according to the agreed traffic profile with the end user. The conformance definition is usually a deterministic algorithm which marks all non-conformant packets considering the Committed Information Rate (CIR), that is, the bit rate, as well as the Excess Information Rate (EIR), the Committed Burst Size (CBS), and the Excess Burst Size (EBS).

For QoS provision end-to-end in Ethernet-based NGN, TE must be supported end-to-end. For that purpose, mapping of different parameters at E-UNIs and E-NNIs should be done in the edge nodes. Traffic management within the operator's core network is performed by scheduling, buffer management, and admission control. Scheduling and buffer management is usually done on a per hop basis in each network node (e.g., Ethernet switch). For such QoS mechanisms there are two main approaches: stateful and stateless. In the stateful approach the QoS mechanisms are based on the connection identifier (e.g., MPLS label in MPLS networks, tag in Metro Ethernet network). The stateless method is performed on the IP layer by using the DS fields in IP headers to determine Per Hop Behavior (PHB), which is a DiffServ QoS approach in Internet. However, similar to DiffServ PHB on network layer, in an end-to-end Ethernet network the DiffServ approach can be applied by using the three user priority bits in the tag control information of the Ethernet header, an approach that is referred to as Ethernet Per Hop Behavior (E-PHB). Further, buffer management in Ethernet switches is applied to handle short-term and long-term congestion. Short-term congestion control is provided by buffering the excess frames certain time (i.e., delaying them), while long-term congestion control is done by discarding (i.e., dropping) the excess Ethernet frames from the buffers.

However, with aim to provide QoS support, it is also needed to provide CAC in Ethernet-based NGN. CAC is based on traffic parameters as well as QoS requirements. Particularly important are traffic parameters and their upper bounds (e.g., maximum committed bit rate to a flow) to be able to predict the traffic behavior and to provide admission control, which means that any certain connection can be rejected if its QoS requirements cannot be met in the given traffic environment in the Ethernet network, something that is different from best-effort only Ethernet where every connection is admitted in the network and then congestion control is left to transport protocols (e.g., TCP) in end hosts as well as buffering mechanisms in network nodes (e.g., dropping of frames or packets).

To provide QoS support in Ethernet-based NGN resource reservation are needed. There are two approaches for resource reservation:

- *Resource reservations triggered by the SCF*: In this case the CPE uses service signaling (e.g., SIP signaling via IMS functional entities) to perform QoS negotiation (e.g., for bandwidth, QoS class, etc.), but the user terminal is unaware of QoS attributes in the transport network. In this approach typically SCF sends resource initiation requests to RACF for authorization and admission control (regarding the transport resources for the requested connection). Finally, RACF pushes the admission control decisions to network nodes such as border gateways, edge nodes, or access controllers.
- *Resource reservations triggered by the CPE*: In this case the CPE supports signaling capabilities (e.g., IETF's RSVP, 3GPP session management signaling, etc.) to request QoS negotiation from the operator's network. The CPE uses dedicated paths for signaling to send QoS request to the network. Such request is intercept by a border node that supports QoS signaling, which further contacts RACF for admission control decision.

Generally, traffic that belongs to a given QoS-enabled connection may traverse multiple domains. In such a case for QoS provision end-to-end, QoS signaling is needed between multiple network or service operators. The inter-operator communication is based on one of the three possible scenarios that are described in the RACF section in this chapter.

In practice, Ethernet-based NGN is realized by tunneling Ethernet frames from one location to another via routed or MPLS-based transport networks. In such case, to support QoS over the Ethernet, traffic grooming is required, that is, Ethernet frames at access router that is connected to MPLS transport network are grouped into virtual connections such as VPNs. So, Ethernet frames are grouped into a given tunnel or tunnels over IP routed network and transported to other network locations of the Ethernet-based NGN, thus connectivity to the end hosts appears to be end-to-end Ethernet network with QoS support.

6.4 Flow-State-Aware Transport

In multimedia networks such as NGN each flow that requires certain QoS support in the network due to given QoS requirements (e.g., bandwidth, end-to-end delay, loss ratio, etc.) needs to be treated separately from its initiation until its termination. However, the flow can be classified in a given QoS class with aim to provide efficient TE in IP transport networks. In terms of QoS support for flow-aware transport, the network may meet performance requirements either statically (e.g., by a network management system) or dynamically (by dynamically requesting a traffic contract for an IP flow in the network). In flow-aware transfer, the QoS requirements are set to packet delay and loss ratio, but also to bandwidth needed by the flow (i.e., the application that uses the given flow) and priority for maintaining the bandwidth at congestion occurrence.

With aim to provide QoS for Flow-State-Aware (FSA) transfer, the network should implement several functions [21]:

- FSA packet forwarding functions in network nodes (e.g., switches and routers);
- Flow-based admission control (e.g., acceptance or rejection of a connection request);
- Functions for signaling needed to reserve resources for QoS support per flow.

An overview of FSA functions in NGN for QoS support per flow is given in Figure 6.11. There are two main methods for signaling: in-band signaling (i.e., signaling messages are within the flow of IP packets) and out-of-band signaling (signaling messages are not part of the same flow and hence may follow different paths than the media flow). However, it is a choice of the network operator or service provider to determine the type of FSA signaling.

In general, there can be different modes of operation for exchange of information between network operators for FSA end-to-end QoS provision. The overall architecture for flow-aware information exchange in NGN includes the following entities:

- *Flow-aware Information Management Functional Entity (FIM-FE)*: This is responsible for management of information per flow basis by using Flow Aggregate (FA) tables for such a purpose.
- *Flow-aware Information Exchange Functional Entity (FIE-FE)*: This is used for exchange of information stored in FA tables, that is, providing flow-aware information to FIG-FE (Flow-aware Information Gathering Functional Entity).
- *Flow-aware Information Gathering Functional Entity (FIG-FE)*: This is the point where gathered information per flow is processed, and it handles requests from applications.

Each flow in the network can be identified by the available mechanisms and addressing principles. For example, in IPv6 networks the flow label octet in the IP header can be used to identify the flow, together with the source and destination IP address. In IPv4 networks, besides the IP addresses, for flow identification are also needed the source and destination port (specified in the transport protocol header, such as TCP header or UDP header) and the transport protocol (e.g., stream transport such as TCP-based, or datagram transport such as UDP-based). For some applications flows need to be identified, such as VoIP connections or IPTV connections. For other applications, such as best-effort Internet applications (e.g., WWW, e-mail, peer-to-peer file sharing, etc.) there is no need for flow identification and the traffic is transported only as aggregate best-effort traffic between end hosts, either client–server

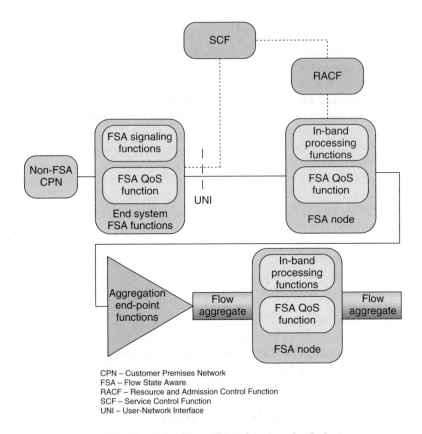

Figure 6.11   Flow-state-aware (FSA) functions for QoS support

or peer-to-peer communication. Hence, the mapping policy of flow IDs to flow-aware network IDs is a complex question [22].

Regarding the bandwidth allocation to a given flow in the network the scheduling policy in network nodes is the one that determines the allocated bandwidth. The main scheduling policy in best-effort Internet is First In First Out (FIFO), that is, IP packets that have first arrived will be first transmitted to an outgoing port in a given node. To provide certain guarantees to certain flows or types of flows (i.e., types of traffic) network nodes use Weighted Fair Queuing (WFQ) scheduling for explicit rate enforcing of certain traffic type (by assigning different weighting coefficients to different flows or traffic types and hence dividing the available bandwidth in explicit manner), as well as priority scheduling for implicit guarantees (i.e., one certain type of traffic to be served before other one, with the aim to provide better QoS for higher priority traffic which will have on average a lower delay and better bit rate sustainability than the lower priority traffic). Of course, these are main scheduling mechanisms, while there are many other scheduling mechanisms which are developed for certain access or transport technology or for certain traffic types in the network.

A given number of flows with the same QoS requirements in the network are aggregated in so-called FA. Each FA is characterized by several parameters including FA identifier (e.g., in MPLS network it is an LSP label), number of flows within a given FA, ratio of maximum packet length and minimum link capacity among all links end-to-end (this is used to estimate the delay, because the different links from one end to the other end of a given flow vary, so the delay is influenced by the link with minimum capacity), number of hops (i.e., links) within the aggregation region (e.g., a domain), as well as the maximum number of aggregation regions. Also, in a given FA, one should consider the maximum sum of sustainable transfer rates of all flows, maximum sum of sustainable burst sizes, scheduling priority, packet discard priority, and service class. However, mapping the flow ID to FA ID is not straightforward and depends upon the operator's policies within the network and the operators' agreement for inter-operator flow-aware communication.

6.4.1 Network Architecture for Flow-Aggregate Information Exchange

Information exchange in FA network concept is not dependent upon the network technology or its architecture. Generic network architecture has several domains, where generally each domain has its specific FA technology. The most spread flow aggregation technology in all-IP transport networks nowadays is the MPLS. However, each MPLS is applied to a given network domain, where edge network nodes add labels to incoming flows and vice versa (extract labels from the outgoing flows from the MPLS domain). For collecting, management and exchange of FA information the three functional entities (FIM-FE, FIE-FE, and FIG-FE) are implemented in hierarchical topology, as shown in Figure 6.12.

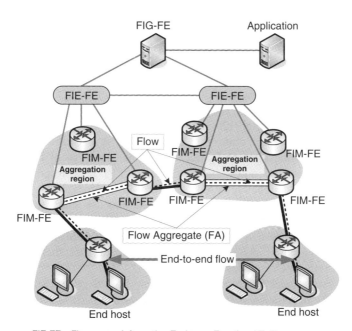

FIE-FE – Flow-aware Information Exchange Functional Entity
FIG-FE – Flow-aware Information Gathering Functional Entity
FIM-FE – Flow-aware Information Management Functional Entity

Figure 6.12   Flow-aggregate (FA) information exchange

The FIM-FE is usually located in the edge switches in the network. Since a network domain usually covers certain geographical area, typically there are multiple FIM-FEs in an aggregate domain. So, FIM-FE is on the lowest hierarchy level among the three entity types for FA. The FIE-FE collects FA information from the FIM-FEs. In each aggregate domain must exist at least one FIE-FE, but multiple such entities are needed for a bigger network. The FIE-FE can coexist in a given edge switch or it can be implemented in a separate server host connected to an edge network node. The top hierarchy level for FA belongs to the FIG-FE. It is responsible for collecting FA information from FIE-FEs. Typically, there is one centralized FIG-FE in the network which is typically implemented on a separate server.

6.4.2   Protocols for FSA Transport

Protocols for FSA transport in fact refer to signaling protocols or modifications of existing protocols with aim to establish FSA transport with required QoS. Such signaling can be in-band or out-of-band, as already indicated in this chapter. Out-of-band signaling in NGN for FSA is performed by using SCFs and RACF in the manner that is elaborated in the RACF section of this chapter. Therefore, in the following section the focus is given on in-band signaling for FSA QoS provision.

Main part of the in-band QoS signaling protocol for FSA transport in NGN represents a set of fields that are added above the header on certain protocol layer, which contain values and indications on requested flow treatment and its QoS requirements from the network as well as network responses to such requests. To perform QoS request with in-band signaling, first packet of the flow carries standardized QoS structure with it [23]. In the case that a QoS request can be supported by the given node on the path through the network, the packet proceeds to the next FSA-capable network node. On the other side, if the FSA network node cannot support the required QoS then it reduces the QoS requirements in the QoS structure to a level that the node can support. However, in the case when FSA network node no longer can support the rates for a given flow (as primary QoS parameter) it may discard packets from the given flow and hence indicate a need for change of the QoS structure of the flow. For FSA QoS provision ITU-T has standardized a set of 10 parameters [23], as given in Table 6.4.

When a sender requests a Maximum Rate Service (MRS) it sends a request packet to the receiver with the following three QoS structure parameters: Type of FSA transport set to MRS, Guaranteed Rate (GR), and Preference Priority (PP). If no response is received, the QoS request with in-band signaling is repeated up to three times. In the case of no answer (the default response timeout value is one second) the data are sent without FSA signaling, while in the case of positive response the data packets are sent to the receiver following the admitted QoS parameters. Also, if no packets are received by the FSA network node for a time period greater than the state-timeout (which is a defined parameter) then the flow state times out and the MRS setup is cleared down.

In the case when a sender requests ARS (Available Rate Service) then it negotiates with the FSA nodes. The negotiation is started with a packet that carries CD (Change/Direction) parameter set to 1, and renegotiation should be renewed after 128 packets or 1 s (the earlier one triggers the renegotiation). Each ARS request packet must have specified the following three parameters: FSA service type set to ARS, requested Available Rate (AR), and requested PP. The ARS service has no starting data rate, so the sender must wait for a response packet before sending data packets. As in the case of MRS, up to three request packets are sent by the

Table 6.4 QoS structure parameters

QoS structure parameter	Description
Available rate (AR)	Rate assigned from network to the flow
Guaranteed rate (GR)	Requested guaranteed rate
Preference priority (PP)	Relative rate priority of the flow
Change/direction (CD)	Change of a request or a response
Type of FSA transport (TP)	Available rate or maximum rate as type of transport
Second QoS structure attached	Two QoS structures are included in the same packet
Security structure attached	Indicates that after QoS structure follows security structure
QoS version	QoS structure version
M (modified) marker	Set to 0 by sender of the request; set to 1 by FSA node that changes the QoS request
Flow sender depth (FSD)	Number of proxies that are entered (but not exited) by the flow

sender before declaring a response timeout. The ARS state is cleared down in an FSA node when no packets are received by the FSA node for a time period equal to or greater than the state-timeout.

In the following sections are given QoS structures (for FSA transport) on different protocol layers.

6.4.2.1 FSA Header for Ethernet

For Ethernet-based transport networks, the QoS structure for FSA transport is placed between Ethernet header and Ethernet payload (in the Ethernet payload typically is stored an IP packet), as shown in Figure 6.13. There are two types of FSA Ethernet packets: data packets and signaling packets.

FSA Ethertype data packets start with a standard 14-byte Ethernet header which is followed by a 4-byte FSA header, which is inserted between the Ethernet header and packet body (i.e., Ethernet frame body). The FSA header has four parameters:

- *Signaling bit (S)*: This field is set to all zeros for all FSA Ethertype data packets, while it has all bits set to one for signaling packets.
- *Packet type (Ptype)*: This field has four bits and denotes the packet type code.
- *First packet ID (ID1)*: This is eight-bit flow identification.
- *QoS offset*: This is equal to the number of bytes from the beginning till the end of the packet body (this field has a length of 16 bits).

FSA Ethernet signaling packets have additional section as shown in Figure 6.13, which is named QoS structure. It is added at the end of the packet and has a length of 20 bytes. The length of the packet section (which is positioned between the FSA header and QoS structure) is specified in the field QoS offset in the FSA header of the given packet.

The QoS structure has two reserved fields that are not used in the Ethernet FSA packet, titled Reserved2 (32 bits) and Reserved3 (16 bits).

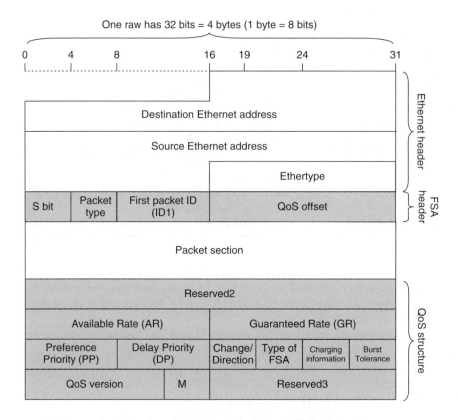

Figure 6.13   Ethernet packet with FSA QoS structure

The first functional fields in the QoS structure are AR field (16 bits) and GR field (16 bits), both containing rates assigned by the network (each of them can have floating point values, i.e., decimal values in the given 16-bit fields).

The PP field (eight bits) indicates a relative priority for Available Rate Service (ARS) as well as acceptance priority level for MRS. Only six bits are used (representing priority levels from 0 to 63) while the lowest two bits are reserved for future use.

Delay Priority (DP) field (eight bits) was initially created for specification of the delay priority, but later it was dropped, so its default value is all zeros.

CD field (four bits) defines whether certain change (i.e., action) is performed. The first bit (bit 0) is always set to zero, while bits 1–3 define whether there is no action required (all bits set to zero) or certain action is required (in the case of values from 1 to 7), such as request for flow to start QoS negotiation (value of 1), response that returns agreed QoS parameters to sender (value of 2), renegotiation (value of 5), close (value of 7), while other values are reserved.

The Type of FSA transmission mode (TP) field (four bits) has bits 0–1 reserved [23], while bits 2–3 define the type of flow: value of 0 denotes ARS, and value of 2 denotes MRS.

The Charging information (CH) field (four bits) has the following definitions [23, 24]: bit 0 defines the type of charging information (0 = forward charging that is paid by sender, 1 = reverse charging that is paid by receiver); bit 1 denotes second QoS structure attached

(0 = single QoS structure, 1 = second QoS structure follows); bit 2 denotes security structure attached (0 = no security structure follows, 1 = security structure follows); and last bit, bit 3, is reserved for further study.

The Burst Tolerance (BT) field (four bits) defines the time at approved bit rate for which a given flow is permitted to exceed its rate (GR + AR) before its packets are discarded [24].

Further, the QoS version field (12 bits) defines the version of the protocol (i.e., QoS structure for FSA transport), and it is set to 2 for Ethernet FSA QoS. Finally, the Modified marker (M) field (four bits) has bit 3 set to 0 by sender on request or renegotiate, and set to 1 by FSA node if any changes occurs (during request or renegotiate). Bits 0–2 of the marker are used to denote the Flow Sender Depth (FSD), initially set to 1 (by the sender of the flow), increasing by 1 when entering a proxy and decreasing by 1 when leaving a proxy.

6.4.2.2 FSA Headers for IP

The QoS structure can also be specified on a network or transport layer in similar manner as the structure specified for Ethernet layer-2 [24]. There are two types of IP addresses, that is, IPv4 and IPv6, and hence the adding of QoS structure to each of them is done on a manner specific to each of the IP versions. One of the main differences between IPv4 and IPv6 is the existence of the Next Header field in IPv6 (it does not exist in IPv4) which provides possibility to have concatenated IP headers in an IPv6 packet, so a QoS structure can be added as a next header to IPv6. So, in IPv4 the QoS structure for FSA transport is added after transport layer header, that is, after TCP or UDP header, as it is shown in Figure 6.14. The fields in the

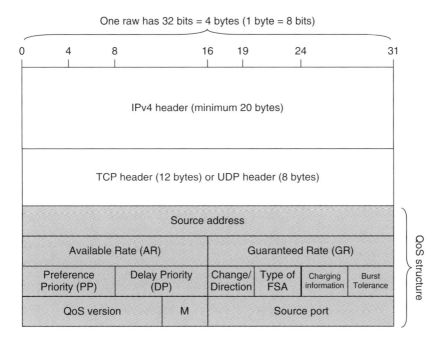

Figure 6.14   IPv4 FSA signaling packet format

QoS structure of IPv4 FSA signaling packet are the same as the same fields in Ethernet FSA signaling packets. The only difference is that the fields named Reserved2 and Reserved3 in Ethernet FSA packets are used in IPv4 FSA signaling packets for specification of source IP address and source port, respectively. In this case the QoS version field in the QoS structure is set to 1. As a reminder, the mentioned QoS structures are used for in-band signaling for FSA QoS provision (the out-of-band signaling is performed by using SCF and RACF, as already described in this chapter).

In IPv6 packets the FSA is implemented by adding FSA QoS structure as next header to IPv6 header. The IPv6 FSA packet format is shown in Figure 6.15. If one compares its structure to the Ethernet FSA packet format, then the main difference is the first field, titled Reserved2 in Ethernet FSA packet, which in IPv6 FSA packets is divided into four fields, as follows: Next Header, Header Extension Length, Option Type, and Option Length. The Next Header (eight bits) is the same field as the one defined in IPv6 header format, and it is used to identify the type of header that follows the given one. Header Extension Length (eight bits) is set to 1 and indicates the QoS extension field. The field Option Type (eight bits) is left for further definition, while the Option Length (eight bits) is set to 12 indicating the 12 bytes that follow in the QoS structure after that field. QoS version is set to 1 (the same value as the one in IPv4 FSA signaling packets).

So, if one summarizes, there is standardized out-of-band FSA signaling that is realized via SCF (e.g., IMS) and RACF, targeted mainly to real-time services such as VoIP and IPTV. Also, there is in-band FSA signaling for QoS provision that is defined for most used protocols on layer-2 (the Ethernet) and layer-3 (the IP, including IP4 and IPv6) and its usage is mainly targeted to streaming and interactive applications that have certain QoS requirements, but generally such requirements are less demanding than those imposed by services VoIP and IPTV.

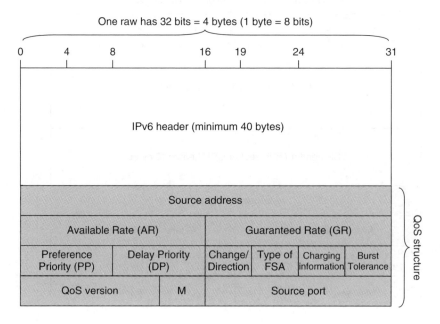

Figure 6.15 IPv6 FSA packet format

6.5   Management of Performance Measurements in NGN

Performance management in NGN is crucial with aim to provide QoS support in heterogeneous network environments consisted of many autonomous domains interconnected between each other, as well as heterogeneous services with different QoS requirements that are provided using the same IP-based transport networks. For that purpose there is defined general reference network model for performance management in an NGN. In that model, along the path end-to-end generally there are involved two Customer Premises Networks (CPNs), two access networks, and one or multiple core networks, as well as zero or multiple transit networks. In such model, access networks are connected to core networks on one side, and to CPNs on the other side. Different core networks are connected via transit networks. A transit network is a network that connects core networks attached to it with other core or transit networks. End users can attach with their CPNs only to access networks. Service provider networks (which provide services to end users) are connected via the core networks. So, regarding the transport stratum (where performance measurements are performed), one may notice a hierarchical network model for performance management in NGN, consisted of customer premises, access, core, and transit networks. With such approach the network end-to-end is partitioned to several network segments and domains, where each of them is responsible for maintaining QoS support in the given segment.

Generally NGN supports three types of assured delivery services regarding the end points:

- *Edge-to-edge*: includes services that extend to the network provider's edge nodes (e.g., gateways).
- *Site-to-site*: for services that extend to CPNs.
- *TE-to-TE* (TE *stands for* terminal equipment): for services that extend to end user terminals within the given CPN.

All three services with assured delivery in NGN have to be supported by a measurement model. The points in the network in which performance measurements are performed are referred to as demarcation points. Typically demarcation points are located on the edge nodes in each of the given network segments in the NGN general model for performance management. For example, edge node in the access network with which it is connected to the core network is a demarcation point. Also, the node with which access network is connected to CPNs is a demarcation point for performance measurements. Similar examples can be derived for other networks in the reference model.

So, an NGN network is partitioned into segments, where in each segment are performed independent performance measurements. Such segments in the NGN are ingress (for incoming traffic) and egress (for outgoing traffic) access segments, core segment, transit segment, and service provider segment. However, demarcation points at the customer ingress and egress segments are dependent upon the traffic type (e.g., VoIP, IPTV, best-effort Internet applications, etc.).

The "big picture" for performance management in NGN is shown in Figure 6.16. There can be defined different pairs of demarcation points for performance measurement. Generally, three models for performance measurements are defined in NGN, for each of the three types for assured delivery services.

In the edge-to-edge model the service delivery is assured between the edge nodes in the access networks to which CPNs are connected, but it is not assured within the customer's

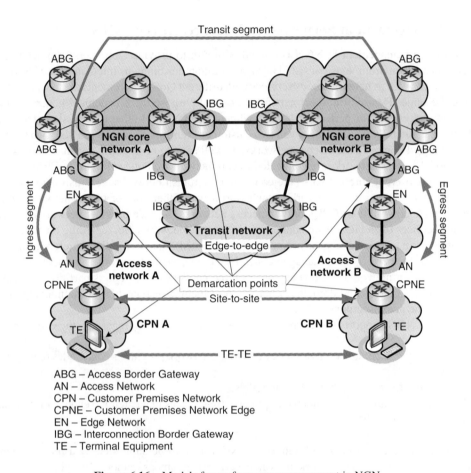

ABG – Access Border Gateway
AN – Access Network
CPN – Customer Premises Network
CPNE – Customer Premises Network Edge
EN – Edge Network
IBG – Interconnection Border Gateway
TE – Terminal Equipment

Figure 6.16   Models for performance management in NGN

network. For this model the measurements provide performance information for all segments between the egress and ingress demarcation points, that is, include performance characteristics of ingress, transit, and egress segments. The transit segment is measured from the demarcation point of the ingress NGN core network to the demarcation point of the egress NGN core network, where the transit segment may include different transit transport providers and can span across large geographical areas (e.g., cities, countries, regions, and continents). In all cases, the demarcation point is a choice of selection by the NGN provider.

Second model for NGN performance measurements is the site-to-site model. In this model the service is assured between the CSN (Connectivity Service Network) edge nodes, while it is not assured by the NGN provider within the CSN (e.g., between user terminals, such as computers, lap-tops, wireless devices, etc., and the CSN edge node, such as home gateway). In this case the ingress and egress segments include the access links as well, such as xDSL (Digital Subscriber Line technologies), Passive Optical Networks (PON), cable networks, WiFi, Ethernet, SDH/PDH leased lines, and so on.

In the third model, the TE-to-TE (i.e., between the TEs), the performance characteristic of the NGN are in fact aggregate of performance characteristic of all segments. Such approach may cover performance measurements for different services such as mobile services, smart homes, corporate LANs, and different end user terminals.

The Management for Performance Measurement (MPM) system is logically separate from the NGN service and transport stratums. However, it is connected to both stratums in the given NGN, to end user functions, as well as to MPM systems in other NGNs. Regarding the functional architecture, an MPM consist of three functional entities: Performance Measurement Execution Functional Entity (PME-FE), Performance Measurement Processing Functional Entity (PMP-FE), and Performance Measurement Reporting Functional Entity (PMR-FE). The PMR-FE is highest in the hierarchy of functional entities within MPM, it collects the measurement information from PMP-FE and reports them to MPM applications (e.g., RACF) or to other MPMs in other NGNs. The PME-FE performs functionalities of active probe initiations (i.e., measurements initiation), active probe termination as well as passive measurements (i.e., measurements by using information from ongoing data and/or signaling traffic, without creating or modifying any traffic in the network). The PMP-FE collects all measurement reports generated by the PME-FEs.

Implementation of MPM in NGN is important with aim to provide desired level of QoS support in the network, either per flow or aggregate, including residential and business users (e.g., to monitor and guarantee the agreed QoS level contracted in the SLA). Also, performance measurements give inputs to the business management system, and directly influences on different aspects such as network planning and dimensioning, service marketing and SLAs, investments (e.g., in more capacity in the network), and so on. Overall, MPM supports business processes to plan, provision, operate, and maintain NGN and its services.

6.6   NGN Architecture for MPLS Core Networks

Most of the transport networks in telecommunications world in the second decade of the twenty-first century are IP/MPLS networks (or simply MPLS). On the other side, NGN provides QoS provision in the network by using RACF between the service and transport stratums. An outcome of such scenario is combination of RACF (for connection between service and transport stratums) and MPLS (in transport stratum) in NGN. Then, there are two possible architectures: centralized RACF architecture [25], and distributed RACF architecture [26].

6.6.1   Centralized RACF Architecture for MPLS Core Networks

In the centralized RACF architecture for MPLS core networks access network aggregate the user traffic to the core network. The core network is MPLS, so the edge routers are Label Edge Routers (LERs) which add labels for ingress traffic to the MPLS core network and deletes the labels for all egress traffic from the core network. The edge nodes of the core network are connected via pre-provisioned LSPs or TE tunnels. In this case there is single centralized RACF in the core network, which typically consists of PD-FE and TRC-FE, as shown in Figure 6.17. Since the architecture is centralized the TRC-FE monitors all traffic in the MPLS core network and accordingly adjusts the bandwidth of the established LSPs, by controlling the aggregate bandwidth in LERs and Label-Switching Routers (LSRs). Admission control is carried by the PD-FE in the centralized RACF and it is implemented in the network via the TRC-FE. Hence,

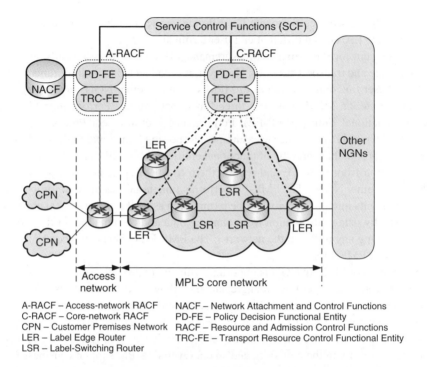

A-RACF – Access-network RACF   NACF – Network Attachment and Control Functions
C-RACF – Core-network RACF   PD-FE – Policy Decision Functional Entity
CPN – Customer Premises Network   RACF – Resource and Admission Control Functions
LER – Label Edge Router   TRC-FE – Transport Resource Control Functional Entity
LSR – Label-Switching Router

Figure 6.17   Centralized RACF architecture for MPLS core networks

the TRC-FE can request resizing of the established LSPs in the MPLS network in certain cases, or rejection at the admission control when there are not enough resources in the network.

The flows are mapped into LSPs at the edge of the MPLS network. Mapping of IPv4 flow to LSP is performed by using the 5-tuple consisted of source and destination IP addresses, source and destination ports, and protocol, as well as the DiffServ field in IP header (i.e., ToS). Additionally, an IPv6 flow can be mapped to an LSP by using the flow label and source and destination IP addresses, as well as the Traffic Class field.

Bandwidth reservation in LSPs or tunnels is performed in two modes, given as follows:

- *Static mode*: This provides reservation of predetermined bandwidth associated with a given LSP or tunnels, which is done independently from the call-by-call admission control that is carried by the PD-FE in the RACF (of course, in this case some calls may be rejected).
- *Dynamic mode*: This provides tight coupling of bandwidth control in mapped LSP and resource request by the given media flow. Aggregate bandwidth in a given LSP is adjusted with each new admitted or released media flow, because bandwidth is reserved per flow in aggregated manner in a given LSP or a tunnel, according to the established policies in the network.

Overall, core network provides QoS-enabled traffic delivery in aggregate manner, where admission control mechanisms should be defined in a scalable manner due to large number of connection or session requests in the core network (from all access networks attached to the given core network). The end-to-end flow control procedure in MPLS core networks starts with CPE service requests targeted to the SCF, which further contacts the destination CPE directly or via a proxy node. After response from terminating CPE the SCF sends a resource request to the RACF, in particular, to its PD-FE. If the received request in the Core-network Resource and Admission Control Function (C-RACF) is acceptable then it sends a request to both Access-network Resource and Admission Control Functions (A-RACFs). In the access segment, A-RACF (on each side of the connection) checks the resource availability with its TRC-FE. Then, A-RACF contacts NACF in the given access network regarding the authorization of the user for the requested QoS (i.e., whether the user has authorization for requested QoS provision). After policy, resource, and user subscription check by the PD-FE in the A-RACF, it sends FSA bandwidth reservation to the edge node in the access network, in particular, to its PE-FE. If everything is confirmed then SCF receives a positive response from the centralized core RACF (i.e., C-RACF), and finally it sends positive response to the initiator CPE. After receiving the response, the CPE starts sending packets in a media flow that is mapped into a selected LSP in the MPLS core network by using the flow mapping configured in the PE-FE.

Generally, using the centralized core RACF architecture the call admission complexity in the network is reduced, while "the stability" of this approach for the network provider is guaranteed by the legacy MPLS transport technology.

6.6.2 Distributed RACF Architecture for MPLS Core Networks

In the distributed RACF architecture for MPLS core networks the two functional entities of the RACF are separated, so there is single centralized PD-FE, and many distributed TRC-FEs at the edges of the MPLS core network. Again, real-time traffic that has strict QoS requirements is aggregated prior to its entrance in the MPLS network. In the case of multiple core networks each network is an autonomous domain. The aggregation of the media flows happens in LERs of the MPLS network. Each flow is mapped into an LSP or a tunnel in the same manner as in a centralized RACF architecture.

The contact point at the edge of the core network for all flows coming from an access network (attached to that core network) is so-defined Border Gateway Functional Entity (BG-FE), which is connected with a LER node. In this architecture the BG-FE is the main node for implementation of the dynamic QoS support in NGN. In general, a BG-FE may be located between an access network and a core network, but also it may be used between two core networks. The implementation of BG-FE between an access network and MPLS core network (with a distributed RACF, as shown in Figure 6.18) is referred to as Session Border Controller (SBC) [26]. All flows from access network are processed by the SBC (i.e., BG-FE) prior to the admission control which is performed by the PD-FE, which communicated with SCF (in the service stratum) to obtain information for incoming calls/sessions about their destination SBC, priority requirements, bandwidth requirements, and DiffServ Code Point (DSCP) based on the priority requirements of the given call or session. The PD-FE checks with TRC-FE (co-located or integrated within the SBC and LER at the core network's edge) the availability of the required bandwidth and possibility to meet the QoS constraints for the given call or session. After QoS requirements have been established by the PD-FE the flows are mapped into LSPs by the LER connected with the SBC that handles the call or session for the given media flow. The SBC is also a PE-FE, which enforces policies for incoming flows into the core network.

Overall, the SBC performs two types of functions, one in the user plane, and other one in the control plane. In the control plane, the SBC executes signaling control functions related

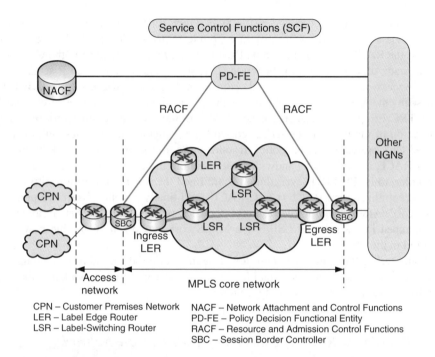

CPN – Customer Premises Network   NACF – Network Attachment and Control Functions
LER – Label Edge Router   PD-FE – Policy Decision Functional Entity
LSR – Label-Switching Router   RACF – Resource and Admission Control Functions
                                   SBC – Session Border Controller

Figure 6.18 Distributed RACF architecture for MPLS core networks

to a given call or session (e.g., authentication, authorization, and accounting, session or call based routing, etc.). In the user plane the SBC performs media path functions (e.g., policing and marking of the non-conformant packets, resource and admission control, etc.). The CAC function performed by the SBC is considered as part of the TRC-FE functionalities. The LER establishes and maintains the state of all LSPs (i.e., traffic engineered tunnels in the MPLS core), which is also part of the functionalities of TRC-FE. So, one may note that TRC-FE functionalities in the distributed RACF architecture for MPLS core networks are implemented in SBC (Session Border Controller) and LER on the given interconnection location between the access network and the core network. Both entities (i.e., SBC and LER) should support RSVP-TE signaling, which is needed for bandwidth reservations in MPLS-TE (MPLS with Traffic Engineering) or DS-TE tunnels for admitted connection by the CAC (admission control is performed by SBC). In all cases, such as RSVP-TE, DS-TE, and MPLS-TE, the TE means existence of mechanisms for traffic demand characterization, setting QoS objectives, traffic control and dimensioning, and performance monitoring in the network.

Finally, it is good to explain the available solutions for QoS provision in NGN core networks. NGN is all-IP network, based on Internet technologies standardized by the IETF. Summarized, the IETF has standardized three main solutions for QoS support in Internet: IntServ based in per flow QoS reservations with RSVP, DiffServ based on the usage of ToS field in IPv4 or DSCP (i.e., Traffic Class) in IPv6 together with appropriate traffic classification (there can be used different classifications in different autonomous domains), and MPLS based on adding stack of labels on IP packets at edge routers in the MPLS domain and performing switching of packets by using labels within the MPLS network.

The initial RSVP for QoS support per flow in Internet (path reservation protocol for IntServ QoS solution in all-IP networks), standardized in the 1990s [27], lacks scalability because each node (i.e., router) in the path of the flow should admit the flow with the given QoS requirements. With the standardization of aggregation of RSVP for IPv4 and IPv6 based flows [28], has been solved the scalability issue with the RSVP. The version of this signaling protocol used in NGN core MPLS networks is RSVP-TE, which is an extension of RSVP for LSP tunnels (in MPLS networks) [29].

Other standardized solution for QoS support in IP networks is DiffServ, which uses DS fields in IP headers in both IP versions (i.e., IPv4 and IPv6) [30]. DiffServ architectures suit well in large scale environments since they support classification of all packets into a limited number of classes based on the DS field in the IP packet header. Elimination of per-flow state and per-flow processing in network nodes has given an advantage to DiffServ over IntServ architectures in the past. However, DiffServ have no mechanisms for communication between individual application and the network. Hence, in the cases where per-flow resource and admission control is required, the IntServ architectures have advantages over DiffServ. With the aggregation of reservations per individual flows at the edge nodes of the core network and their tunneling [31], there is much better scalability of the RSVP-based network architectures for QoS support.

On the other side, from the beginning of the twenty-first century the MPLS became the main approach for implementation of QoS in transport IP networks of the carriers (i.e., telecom operators) since it provides high degree of scalability due to aggregation of traffic (in edge nodes, i.e., LERs) according to its type, QoS requirements as well as business requirements (e.g., type of subscribers, such as residential or business). Hence, MPLS is combined with DiffServ for class-based QoS provision, as well as with RSVP (i.e., IntServ approach) for flow-aware QoS provision. The NGN puts the MPLS, IntServ with RSVP, and DiffServ, in a defined QoS framework within the transport stratum and interconnects it with the SCFs (e.g., IMS) in the service stratum of the NGN via the RACF entity, thus providing practical implementation of NGN core networks by using standardized and well known Internet QoS solutions and technologies.

6.7 Discussion

NGN is a completely heterogeneous network consisting of different networks in all segments, such as customer networks, access networks, core networks, and transit networks. However, all of them are based on Internet technologies, where IP is the default network-layer protocol (including IPv4 and IPv6), while Ethernet is the dominant layer-2 technology. So, there are well known technologies in the NGN transport stratum, which are used for transport end-to-end of heterogeneous services, including real-time services (e.g., VoIP, IPTV, etc.) and non-real-time services (e.g., WWW, e-mail, etc.). Different services have different QoS requirements (e.g., real-time services have stringent requirement on delay) and bandwidth requirements, therefore crucial part of the NGN is QoS provision in all-IP environment with aim to provide desired level of QoS for certain services as well as certain QoE (from the end user's point of view). QoS support in NGN is performed by the definition of the QoS framework by ITU-T, and specification of the RACF entity for performing QoS policy enforcing and admission control by serving as arbitrator between the SCFs in NGN service stratum and transport functions for QoS support in NGN transport stratum. Flow-state-aware transport is required for QoS support in NGN, but it should be done in scalable and standardized manner. Therefore, practical implementation of NGN is based on MPLS-based core networks with RACF-based QoS control and signaling, and appropriate TE by using standardized technologies for QoS support in IP networks (e.g., DiffServ, IntServ with RSVP, and MPLS).

Overall, one may also define NGN as an all-IP network with implemented mechanisms for end-to-end QoS provision.

References

1. ITU-T (2008) Definitions of Terms Related to Quality of Service. ITU-T Recommendation E.800, September 2008.
2. ITU-T (2011) Communications Quality of Service: A Framework and Definitions. ITU-T Recommendation G.1000, November 2011.
3. Ibarrola, E., Liberal, F., Ferro, A. and Xiao, J. (2010) Quality of service management for ISPs: a model and implementation methodology based on the ITU-T recommendation E.802 framework. *IEEE Communications Magazine*, **48** (2), 146–153.
4. ITU-T (2007) Framework and Methodologies for the Determination and Application of QoS Parameters. ITU-T Recommendation E.802, February 2007.
5. Ibarrola, E., Xiao, J., Liberal, F. and Ferro, A. (2011) Internet QoS regulation in future networks: a user-centric approach. *IEEE Communications Magazine*, **49** (10), 148–155.
6. ITU-T and InfoDev (2013) ICT Regulation Toolkit: Key Points and Recommendations on QoS Regulations, http://www.ictregulationtoolkit.org (accessed 08 October 2013).
7. ITU-T (2011) Internet Protocol Data Communication Service – IP Packet Transfer and Availability Performance Parameters. ITU-T Recommendation Y.1540, March 2011.
8. ITU-T (2011) Network Performance Objectives for IP-Based Services. ITU-T Recommendation Y.1541, December 2011.
9. ITU-T (2008) New Definitions for Inclusion in Recommendation ITU-T P.10/G.100. ITU-T Recommendation P.10/G.100 Amendment 2, July 2008.
10. Stankiewicz, R., Cholda, P. and Jajszczyk, A. (2011) QoX: what is it really? *IEEE Communications Magazine*, **49** (4), 148–158.
11. ITU-T (1996) Methods for Subjective Determination of Transmission Quality. ITU-T Recommendation P.800, August 1996.
12. Zhang, J. and Ansari, N. (2011) On assuring end-to-end QoE in next generation networks: challenges and a possible solution. *IEEE Communications Magazine*, **49** (7), 185–191.
13. ITU-T (2011) Resource and Admission Control Functions in Next Generation Networks. ITU-T Recommendation Y.2111, November 2011.
14. Blake, S., Black, D., Carlson, M. *et al.* (1998) An Architecture for Differentiated Services. RFC 2475, December 1998.
15. Awduche, D., Berger, L., Li, T. *et al.* (2001) RSVP-TE: Extensions to RSVP for LSP Tunnels. RFC 3209, December 2001.
16. Jamoussi, B., Andersson, L., Callon, R. *et al.* (2002) Constraint-Based LSP Setup using LDP. RFC 3212, January 2002.
17. Le Faucheur, F. (2005) Protocol Extensions for Support of Diffserv-aware MPLS Traffic Engineering. RFC 4124, June 2005.
18. ITU-T (2008) Management of Performance Measurement for NGN. ITU-T Recommendation Y.2173, September 2008.
19. ITU-T (2007) A QoS Control Architecture for Ethernet-Based IP Access Networks. ITU-T Recommendation Y.2112, June 2007.
20. ITU-T (2009) Ethernet QoS Control for Next Generation Networks. ITU-T Recommendation Y.2113, January 2009.

21. ITU-T (2008) Requirements for the Support of Flow-State-Aware Transport Technology in NGN. ITU-T Recommendation Y.2121, January 2008.

22. ITU-T (2009) Flow Aggregate Information Exchange Functions in NGN. ITU-T Recommendation Y.2122, June 2009.

23. ITU (2012) Signalling Protocols and Procedures Relating to Flow State Aware QoS Control in a Bounded Subnetwork of a Next Generation Network. ITU Recommendation Q.3313, February 2012.

24. ITU (2009) Input of the Draft Recommendation Q.Flowstatesig. ITU Study Group 11 Temporary Document TD 277, September 2009.

25. ITU-T (2008) Centralized RACF Architecture for MPLS Core Networks. ITU-T Recommendation Y.2175, November 2008.

26. ITU-T (2008) Distributed RACF Architecture for MPLS Networks. ITU-T Recommendation Y.2174, June 2008.

27. Braden, R., Zhang, L., Berson, S. *et al.* (1997) Resource ReSerVation Protocol (RSVP) – Version 1 Functional Specification. RFC 2205, September 1997.

28. Baker, F., Iturralde, C., Le Faucheur, F., and Davie, B. (2001) Aggregation of RSVP for IPv4 and IPv6 Reservations. RFC 3175, September 2001.

29. Awduche, D., Berger, L., Gan, D. *et al.* (2001) RSVP-TE: Extensions to RSVP for LSP Tunnels. RFC 3209, December 2001.

30. Nichols, K., Blake, S., Baker, F., and Black, D. (1998) Definition of the Differentiated Services Field (DS Field) in the IPv4 and IPv6 Headers. RFC 2474, December 1998.

31. Le Faucheur, F. (2007) Aggregation of Resource ReSerVation Protocol (RSVP) Reservations over MPLS TE/DS-TE Tunnels. RFC 4804, February 2007.

7

Service Aspects

7.1 Service Architecture in NGN

Service architectures in NGN (Next Generation Network) are based on the network architecture consisting of NGN transport and service stratums. All services that require certain AAA (Authentication, Authorization, and Accounting), and/or admission control, and/or QoS (Quality of Service) support, and so on, require certain functionalities in the NGN environment. Such functionalities within the NGN are located in the service stratum, which is denoted also by the name of the stratum. However, certain functions may be connected with the service stratum and its service control functions (SCFs). To provide different types of services in a common way there is a certain standardized service architecture in NGN, which is referred to as the Open Service Environment (OSE) [1].

In the functional NGN architecture, as presented in Chapter 3, are defined interfaces between the NGN and the applications and service providers, called Application–Network Interface (ANI) and Service–Network Interface (SNI), respectively. ANI is created as a channel for interaction and exchanges between different applications and the network (i.e., the NGN) by using only control plane interactions, while SNI provides interaction between NGN operators and service providers in both planes (i.e., the control plane and the user plane). One may compare the ANI and the SNI (from the perspective of NGN) with the API (Application Programming Interface) in a given host (e.g., client host, server). Similar to API which provides application access to different underlying protocols in the operating system of a host, the ANI and the SNI are intended to provide similar approach in the NGN where the NGN service stratum acts as a "network operating system" (a network through which the services are delivered) while NGN transport stratum (i.e., access networks, transport networks) acts as "network hardware with interfaces."

In general, the requirements set to NGN OSE include the following main characteristics:

- To provide standard API for development and installation of applications in a quick and seamless manner via the NGN to the end users.
- Service level interoperability between heterogeneous networks, operating systems, and platforms as well as different programming languages (e.g., for Web-based services).
- Service development independent of the manufacturers of the network and user equipments (i.e., independent from the underlying network technologies).

NGN Architectures, Protocols and Services, First Edition. Toni Janevski.
© 2014 John Wiley & Sons, Ltd. Published 2014 by John Wiley & Sons, Ltd.

- Location, network, and protocol transparency (i.e., OSE should be technology and terminal agnostic, and should provide standard protocol programming interfaces which hide the network functions from the open service environment).
- Service management capabilities, such as service discovery, service tracking, update management, access control, and so on.
- Possibility for creation and provisioning of services and applications, including service/application development, trial, deployment, and removal.

The position of the OSE in the NGN architecture is shown in Figure 7.1. The functional entity OSE in the service stratum of the NGN architecture provides possibility for applications and services to use NGN capabilities for the given service. Hence, the OSE as a group of functions is positioned within the application support functions (ASFs) and service support functions (SSFs) in the NGN.

The interfaces for OSE, between the application on one side, and the application and SSFs on the other side, do not support the transfer or exchange of media data information [e.g., VoIP (Voice over IP) media, IPTV (Internet Protocol Television) media, WWW (World Wide Web), etc.]. Functional components of the OSE (going from bottom to the top) are the following (as shown in Figure 7.1): policy enforcement, interworking with service creation environments, service development support, service composition, service management, service registration, service discovery, and service coordination.

The open environments in NGN are targeted to speed up and make easier the service creation. Each new service in OSE first needs to be registered, with aim service registration processes to create information for service discovery and management. On the other side, when certain application requests to use a given service, then it has to be authenticated and authorized via the service management (for usage of NGN services). When a given application starts the service discovery process then it gets a list of the offered services via the OSE in the NGN, from which it should make a selection. The service discovery process of the OSE then returns the interface to which the application should connect to use the service. After that step the application is enabled to use the chosen service from the offered list of services in NGN (via the OSE entities). Service providers make the service visible to applications by publishing it in a given registry (e.g., DNS-Domain Name System). The protocol that is used between the service provider and the registry is referred to as Universal Description, Discovery, and Integration (UDDI) protocol. On the other side, user of a given service also uses the so-called UDDI protocol to ask (i.e., query) the registry for the given service. The service description should be created in such manner to allow the end user (i.e., the customer) to interact with the service provider regarding the requested service. Such service descriptions are made in a standardized manner, such as Web Services Description Language (WSDL) for Web-based services or services that provide Web-based interaction (e.g., for service discovery and selection, for the purposes of authentication and authorization of the user, etc.). On the other side, service provider maintains and operates the service via a management interface (e.g., for service configuration, monitoring, etc.). The management interface for service providers can be based on appropriate standardized protocols such as Web Services Distributed Management (WSDM). Examples of distributed services which are used to create overlay service by using certain contexts (e.g., user preferences, location, time, weather, etc.) are already present in best-effort Internet. The goal of the OSE is to provide an environment for standardized approach which will be more efficient regarding all parties, that is, users and their equipments, network operators (fixed and mobile), service developers (e.g., standardized interfaces make a lot easier job

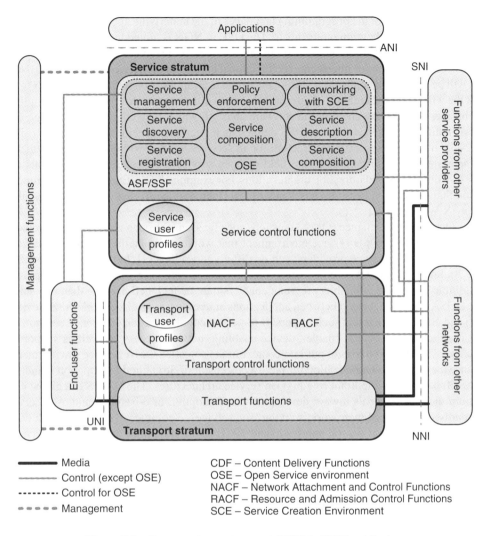

Figure 7.1 Open service environment (OSE) in NGN architecture

for the developers), lower cost for service deployment and management, as well as shorter time to the market.

Besides the OSE in NGN, one may also discuss the Service Oriented Architecture (SOA) approach and its relation to the OSE. In practice, the SOA is a software-based architecture which is based on a structured collection of loosely coupled services to provide a complete functionality of a large software application. In SOA every host has arbitrarily number of programs that it runs, which on the other hand can be connected to other hosts (e.g., computers) to exchange data with programs running on other hosts without human interaction. Hence, the SOA environments are independent services (e.g., for support of business processes) that can be used without explicit knowledge from the underlying software platform implementation. So, the SOA approach provides an infrastructure for the exchange of information in business processes regardless of the programming languages of the underlying applications or operating systems over which they run on the hosts. The OSE in NGN is aimed to provide standardized interfaces to NGN for different applications and services, so it is in line with SOA but requires standardized approach for interfacing the functional entities in the service stratum of the NGN that will be used by all application developers to create an application without the need to "reinvent the wheel" for the creation of interfaces between different services. While SOA simplifies the business process by creating the software "bus" to which they connect to exchange data between different applications, the OSE defines such interfaces specifically to the NGN since the NGN includes transport infrastructure for all services, with or without QoS requirements.

There are also contributions in the field of open service architectures, environments and access, from other standardization organizations. For example, ETSI (European Telecommunication Standardization Institute) has standardized Open Service Access (OSA) with so-called Parlay X Web APIs [2].

The Open Mobile Alliance (OMA) develops specifications for open global service architectures with aim to provide service environment that works over heterogeneous user devices, network operators, service providers, and on different locations [3]. In the OMA Global Service Architecture (OGSA) are defined so-called suites, where each suite is used to represent an arbitrary grouping of enablers (certain enablers may be included in more than one suite). The OGSA suites are located between applications and resources (e.g., networks, user devices, etc.) and consist of enablers such as person-to-person communication enabler, access to content enabler, service access enabler, device enabling, network access, as well as supporting enablers. The OGSA suites are put into context in the so-called OMA Service Environment.

Since the 1990s the Organization for the Advancement of Structured Information Standards (OASIS) has had an important role in open service architectures. The OASIS is a global consortium that has set goals toward development and adoption of e-business and Web service standards. The focus of their work is on the Web services, e-commerce, IP security solutions, applications, document-centric approach, and XML (Extensible Markup Language) processing for interoperability across different platforms, networks, and industry domains. According to the OASIS [4], SOA is a concept organization and utilization of distributed software-based capabilities that may be under the control of different domains, with main purpose of support of business function (to get something done). So, the overall idea in SOAs is to provide access to services via a standardized interface in a manner similar to the way access to electrical networks (i.e., electricity) is provided to different electrical devices.

In general, the NGN Open Service Environment (OSE) includes different efforts from different standardization organizations and provides defined interfaces for all services to the NGN as a global networking and service platform.

7.2   Managed Delivery Services (MDS)

Open service environments in NGN create new business opportunities via integration of applications and telecom infrastructure (which converges to Internet technologies). The dynamic features and comprehensive service delivery controls in NGN are performed via so-called Managed Delivery Services (MDS). The MDS focus is on the online business environment which requires real-time interaction over broadband networks. In that manner, the MDS uses NGN capabilities and functionalities to differentiate between different types of services based on the QoS level provided by the telecom operator or quality experienced by the users.

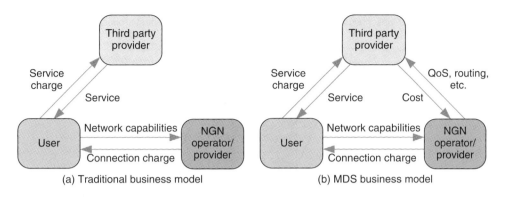

Figure 7.2   (a,b) MDS business models

MDS business model includes three main elements: end user, network operator, and third party service provider, as shown in Figure 7.2. The users are demanding certain capabilities from the network to be able to use the requested services (e.g., from third parties), such as bandwidth provision, IP-based routing, and QoS guarantees (e.g., for VoIP, IPTV, etc.). A third party provider can provide services to end users without any business relation with the NGN operator (e.g., web sites which provide Web-based contents, best-effort streaming, peer-to-peer services such as Skype, Torrent, etc.), but also it may have business relation with the NGN operator (or one may refer to it as NGN provider) regarding the requirements that needs to be satisfied from the NGN network to provide the end users the requested service with needed QoS support.

So, there are two main business models in the Internet-based telecommunication world (i.e., the NGN world), shown in Figure 7.2:

- *Existing business model*: This model includes relations between the end user and NGN operator on one side [for transport resources such as wireless or fixed broadband access network to Internet, and certain functionalities such as IP addressing, NAT (Network Address Translation), routing to/from global Internet, etc.] and relation between end user and third party service provider on the other side. However, in this model there is no relation between the network provider (e.g., telecom provider) and third party service provider. This model refers to best-effort Internet services.
- *MDS business model*: This model establishes business relation between the NGN operator and third party service provider, with which the third party provider can provide more service capabilities by partnering with the NGN operator, as well as different level of services based on the QoS parameters selection. The MDS model is considered a win-win scenario for both, the NGN operator and the third party service provider.

7.2.1   Service Provisioning with MDS

MDS is directly targeted to service provisioning to the end users as well as business relations between the three key players. However, one may distinguish three types for the service provisioning with the MDS, that is, subscription-based type, service request based type, and MDS

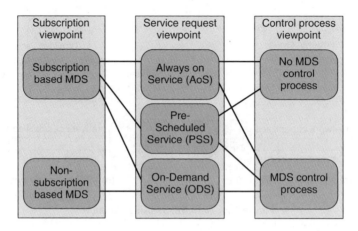

Figure 7.3   MDS service provisioning types and their relations

control process. The three types of MDS service provisioning and their relations are shown in
Figure 7.3.

Regarding the existence of subscriptions between the user and third party provider, two types
of MDS service provisioning may be defined: subscription-based, and non-subscription based.
The first one (i.e., the subscription-based MDS) is convenient for services that have a regular
traffic pattern, such as VPN (Virtual Private Network) business services (i.e., leased line alike),
as well as network-controlled real-time services such as VoIP, streaming services (e.g., IPTV,
e-learning), and so on. Non-subscription-based MDS services do not require existence of a subscription between the user and third party provider prior to service provisioning. This type of
service provisioning is suitable for services without a regular traffic pattern, such as On-Demand
Services (ODSs) which include Video-on-Demand (VoD), Bandwidth-on-Demand (BoD),
video-phone or video-conferencing services on demand, etc.

The second type of MDS service provisioning is based on the service request approach, that
is, whether the user is always online or it demands the service (either in real-time or via a
scheduled demand). There are three subtypes in this approach, as follows:

- *Always on Service (AoS)*: This type refers to the case when MDS is granted to the user
  when it is connected to the network (i.e., it is online), while the user is disconnected from
  the service only when it is offline (i.e., disconnected from the network).
- *Pre-Scheduled Service (PSS)*: This type refers to services provided in a scheduled manner
  that is agreed between the user and the third party service provider.
- *On-Demand Service (ODS)*: This type refers to MDS service provisioning triggered by user
  demand (i.e., on demand).

The third type of MDS provisioning is dependent upon the information whether the third
party service provider takes part in the service process or not. So, this type of MDS is based
on control process participation viewpoint. In this case, there are two subtypes:

- *Third party service provider participates in MDS control process*: In this type the third party
  has an agreement with the NGN operator regarding the service provision. In the case of

AoS and PSS types, the NGN operators informs the third party provider regarding the user attachment to the network, and afterwards the third party provider may request the MDS from the NGN operator. For ODSs, the third party provides MDS requests from the NGN operator upon user demand.

- *No participation of third party provider in MDS control process*: This case covers the AoS and PSS where NGN operator provides MDS control without third party provider participation in the control process. However, in this approach ODS is excluded.

The main purpose of MDS is the possibility of the third party provider to requests certain resources and classification for the services from the NGN operator. Such components that are subject to request by the third party provider from the NGN include traffic handling (e.g., classification of the traffic based on QoS requirements), bandwidth control (e.g., certain amount of bandwidth allocated to the service flows, as well as control of the bandwidth with either in-band or out-of-band signaling), connection admission control, security solutions [e.g., AAA, firewalls, NAPT (Network Address and Port Translation), etc.], as well as other NGN capabilities (e.g., multicast for IPTV service).

For the realization of the MDS, the third party service provider and the NGN operator should exchange and maintain (up-to-date) relevant information and parameters regarding the MDS service provisioning. For example, NGN operator (one may refer to it as NGN provider) maintains the MDS service profiles, third party profiles, user profiles (obtained from the third party), and so on. On the other side, the third party provider maintains the user profiles, IDs of the servers for service provisioning and their status, and so on.

The MDS profile contains Class of Service (CoS) information, admission control information (e.g., always admitted service, admitted only when available, etc.), bandwidth information (e.g., on demand bandwidth, permanent, variable bandwidth, etc.), security information (e.g., AAA, NAPT, privacy requirements, etc.), multicast control if needed for the services, and so on.

The third party profile contains information that is needed to uniquely identify the third party that has MDS business partnership with the NGN operator, which includes IP addresses or URLs (Uniform Resource Locators) of the third party servers, server status information, identifier of the third party provider (e.g., domain name), and so on.

The user profile contains information for unique identification of the user of the third party services. Typically, the user profile includes information on username (or user ID), user's IP addresses (which should be bound with the username in a registry), MDS services type (e.g., subscription based, non-subscription based, etc.), user's ID for NGN access, all information from the MDS service profile (as detailed above), and so on.

The fourth type of profile is MDS control profile which includes information regarding the service and session control, resources control, transport control for MDS. It includes subscriber log session information [e.g., subscriber IP address, MAC (Medium Access Control) address and login ID for NGN network access, IP address and port number information regarding the network access nodes for the given subscriber, service policy, etc.], user session information (e.g., IP address and port number of the MDS source and MDS destination, transport protocol type used for MDS, network information regarding the QoS provision, network path information, etc.), network topology information (e.g., IP addresses, port numbers, and status information of transport nodes in the network path), network resources information (e.g., link bandwidth between transport nodes), as well as network node information (e.g., IP addresses and status information of service control nodes, session control nodes, etc.).

All types of profiles, that is, MDS service profile, third party profile, and user profile, may be upgraded with additional information (in further releases) which will be needed to provide the service with given constraints as well as given demands from the user.

7.2.2   MDS Functional Architecture

MDS functional architecture is based on NGN functional architecture, described in Chapter 3 of this book. Then, main MDS functions are: service provider management, service coordination, session control, and terminal access control. They can be mapped into the NGN functional architecture and its entities according to Table 7.1. According to the table, service provider management MDS functions are provided by ASFs and SSFs in NGN. MDS session control is implemented via SCFs of NGN, while MDS terminal access control is provided by using NGN Transport Control Functions (TCFs).

MDS functional elements in ASF and SSF in the NGN service stratum are used for identification and authentication of the third party service providers via the ANI open interface for receiving MDS requests generated from the third party provider toward the NGN operator. Also, MDS service coordination is provided via these NGN functions.

The SCF entities in NGN provide MDS functionalities regarding the service control, session control functions [in the same manner as P-CSC-FE (Proxy Call Session Control Functional Entity), I-CSC-FE (Interrogating Call Session Control Functional Entity) and S-CSC-FE (Serving Call Session Control Functional Entity) in NGN functional architecture, i.e., the IP Multimedia Subsystem (IMS)], service coordination regarding the policy decisions and path control (e.g., tunneling of service flows), as well as service authentication and authorization in coordination with general service control and session control entities.

MDS functional elements in the TCFs include functional entities for policy decisions, transport resource control (at edge network nodes), terminal access control, then network access configuration (e.g., IP addressing via DHCP-Dynamic Host Configuration Protocol), as well as entities for management transport information by using user profiles (e.g., QoS preferences, session control capabilities, etc.) and location information for user access to the NGN transport stratum (e.g., binding of IP address associated to user equipment and network location information on one side, and user ID on the other side).

However, there is a question about technologies that should be used for managed service delivery. Certainly, such technologies should be standardized and proven to be practical for implementation by NGN operators. The main approach is the open interface between NGN and applications, which is ANI (similar in many aspect to API between applications and operating system in a given host, either client or server). Further, most of the services are using

Table 7.1   Mapping between MDS function and functional entities in NGN functional architecture

MDS function	NGN functions
Service provider management	Application Support Functions (ASFs) and Service Support
Service coordination	Functions (SSFs)
Session control	Service Control Functions (SCFs)
Terminal access control	Transport Control Functions (TCFs)

DNS-based approach for service discovery and Web technologies for service description and different profiles configuration.

On the other side, regarding the SCF in NGN the standardized and globally adopted system is IMS with its three main call session control functions (CSCFs), namely, serving, proxy, and interrogating. So, the ANI in fact is an interface for applications (provided by third party service providers) and IMS in the NGN service stratum. This way, the initially not-clear picture of the NGN open service environment and MDS becomes more obvious and possible to implement in real-world, by telecom operators (which become NGN operators or providers) and service providers (NGN operator can also be a service provider, but in general service provider is a third party besides the end user and the NGN operator).

7.3   IMS-Based Real-Time Multimedia Services

The IMS, as described in Chapter 5 of this book, defines a horizontal architecture in NGN service stratum (over the layered NGN architecture consisted of transport stratum and service stratum) where service enablers and common functions can be reused for multiple applications. Such horizontal architecture of the IMS provides possibilities for operators to move away from traditional vertical implementations of new services.

In non-IMS service implementations (the traditional telecom approach) a new service have to be built with its own functionalities for authentication and authorization, own charging functionalities, service management, dedicated signaling for control traffic as well as specific routing and provisioning for the user traffic. The advantage of IMS-based service implementations is in provisioning (by the common IMS standardized in 3GPP Release 8) of common functions that are generic (i.e., they are not related to some concrete services). So, the IMS improves the approach for service creation and its delivery on two sides. On one side, the IMS provides shorter time to market for the service and simplifies service creation and delivery due to usage of the common functions in IMS (in NGN service stratum). On the other side, such approach provides lower CAPEX (capital expenditure) and OPEX (operational expenditure, i.e., lower costs for operation and maintenance, billing, support, etc.) for service implementation and provisioning.

Overall, the IMS provides functionalities for creation and delivery of multimedia services based on common so-called service enablers, which are reusable building elements for service creation and delivery. In this manner, the IMS fits well into the paradigm for open service architectures (e.g., the OSE in NGN). For example, such enablers in the IMS are presence and group list. The presence enabler provides functions that allows a set of users (i.e., a group of users) to be informed about the means and contractibility of other users in the group or globally. The presence enabler is not invented with the IMS, since it has been used for certain best-effort services such as peer-to-peer best-effort multimedia communication (e.g., Skype, messaging services, etc.). However, IMS presence enabler is a common one, which is also aware of the media types (e.g., VoIP codecs, video codecs) and user preferences. Also, IMS provides the functions for both fixed and wireless environments, and it is therefore the main approach for the Fixed Mobile Convergence (FMC). With the presence functions the user or NGN operator or third party service provider can set different rules about who can view certain information (e.g., availability of the user via a given public ID such as telephone number implemented via tel URI-Uniform Resource Identifier).

IMS multimedia services include conversational as well as streaming IMS-based real-time multimedia services. The NGN (since its release 1) was created to support real-time conversational services based on IMS [5]. The conversational multimedia real-time services include different types of media such as voice, video, real-time text, real-time data transfer, and so on. So, IMS is required to support identical media types in both directions, because different users may be originating or terminating points for a given multimedia service. Also, real-time multimedia services include as a subset single media services (e.g., VoIP). For IMS real-time conversational services NGN needs to provide QoS provision end-to-end as well as codecs negotiation with the end user equipments [to reduce the transcoding which decreases the QoE (Quality of Experience) level]. For control communication between SCFs and the ASF/SSF, the NGN uses Session Initiation Protocol (SIP), as a standardized protocol by IETF (Internet Engineering Task Force) for signaling in all-IP environments [6]. Regarding the real-time multimedia services NGN is required to support also legacy terminals [e.g., PSTN (Public Switched Telephone Network) terminals] as well interconnection with legacy networks (e.g., PSTN, PLMN-Public Land Mobile Network) that support certain real-time services (e.g., voice).

The functional architecture for IMS-based real-time multimedia services is given in Figure 7.4. ASF and SSF contain different service features for real-time multimedia services. Both support functions (ASF and SSF) interact with entities in the service stratum such as S-CSC-FE, Media Resource Control Functional Entity (MRC-FE), Service User Profile Functional Entity (SUP-FE), as well as with end user functions (located in user equipments). The Application Server Functional Entity (AS-FE) requests the media resources by interaction with the MRC-FE, where AS-FE sends announcements (for availability of certain multimedia

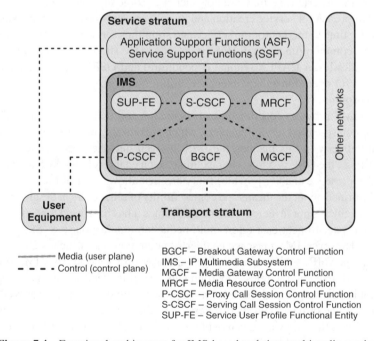

Figure 7.4   Functional architecture for IMS-based real-time multimedia services

service with associated media) and gets response from the MRC-FE about the availability of media resources in the NGN (e.g., codec, type of media content, etc.).

7.3.1   Multimedia Communication Center

Regarding the multimedia services in NGN, the Multimedia Communication Center (MCC) is standardized initially to provide access to information services (or user related services) of an enterprise [7]. In this approach customers communicate with so-called MCC agents (fort manual applications) or with interactive systems, such as Interactive Voice Response (IVR) or Interactive Voice and Video Response (IVVR), in all cases by using multiple media types. When MCC provides services to customers, an agent may serve several customers in parallel (e.g., a chat session). In the case of the MCC the business relationship for the service is between the customer and the enterprise with the MCC. However, there is partnership of the enterprise with the NGN provider as well as relationship between the end users and the NGN for access to the service via the NGN.

Several requirements are specified for MCC deployments [8], such as system extensibility, flexible deployment modes depending upon environments (e.g., large telecom environments, small enterprises, etc.), accessibility via heterogeneous fixed and wireless networks (including legacy networks and user devices), as well as multimedia support for multimedia services based on single or multiple networks simultaneously (i.e., multimedia services to customers from multiple networks). MCC can be deployed in non-IMS network or IMS networks. Since it is adopted that NGN networks are based on common IMS for call and session control functions, the MCC based on IMS (as shown in Figure 7.5) is a long-term approach for deployments.

The main functional block in the MCC architecture is Multimedia Call Distribution Function (MCDF), which provides call control and call distribution toward another entity in the MCC, the Software Control Management Function (SCMF), and with Communication Telephony Integration Function (CTIF). On the other side, MCDF interacts with so-defined Unified Service Broker Function (USBF) to acquire requested service capabilities. When an MCDF has to transfer a call to another MCDF in another NGN network, then it interworks with SSFs. The entity that acts as a service broker in the MCC is the USBF, which can provide interaction with external service capabilities (i.e., outside the given MCC). The Overload Control Function (OCF) is targeted to provide service control for a given service when the CTIF is experiencing an overload (i.e., end users face overload of the MCC). On the other side, the MCC interacts with several application functions (AFs), from which the most important are presence server function, location server function, and Web-enabled service function.

Deployment of MCC in IMS-based NGN, as shown in Figure 7.5, is based on direct logical interaction between the IMS core and the MCC via a single control node (a server) that integrates SCMF, MCDF, and SSF entities. The network interconnection between the IMS core and the MCC is performed by deployment of a Border Gateway (BGW). Control interaction between IMS core and SCMF/MCDF/SSF node is performed by using SIP as a signaling protocol.

7.3.2   IMS-Based IPTV

IPTV service can be delivered in three different ways by service providers: non-NGN IPTV, NGN-based non-IMS IPTV, and NGN-based IMS-based IPTV. While non-NGN IPTV is the

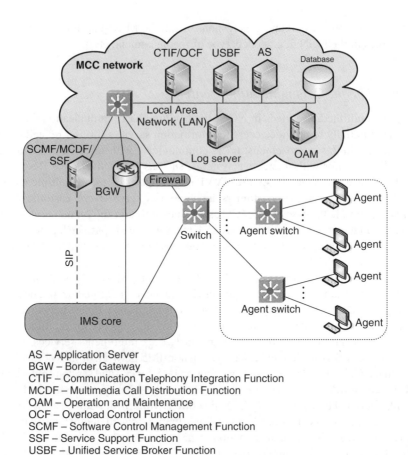

AS – Application Server
BGW – Border Gateway
CTIF – Communication Telephony Integration Function
MCDF – Multimedia Call Distribution Function
OAM – Operation and Maintenance
OCF – Overload Control Function
SCMF – Software Control Management Function
SSF – Service Support Function
USBF – Unified Service Broker Function

Figure 7.5   IMS-based multimedia communication center (MCC) architecture

first wave of IPTV from the first decade of twenty-first century, the NGN-based IPTV is a "natural" continuation of the development. In that manner, the non-IMS NGN implementations of IPTV service are a transitional step toward IMS IPTV deployment. All three types of IPTV implementations can be represented by the IPTV functional architecture shown in Figure 7.6. The main difference in the IMS-based IPTV implementation is the existence of IMS (with its functions) which is located in the service stratum of the NGN.

In NGN-based IMS IPTV realizations the main control protocols are Diameter and SIP, both standards from the IETF. The Diameter protocol is used for communication between functional entities in NGN service stratum or AFs with databases that store application profiles (in AFs) as well as databases that store user service profiles [i.e., HSS (Home Subscriber Server) in NGN service stratum]. The interaction between SCFs in the service stratum with AFs and Content Delivery Functions (CDFs) is realized with SIP. Hence, Diameter and SIP are the two most important control and signaling protocols for NGN, and hence they are covered in the following sections.

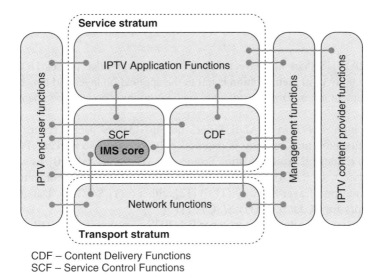

CDF – Content Delivery Functions
SCF – Service Control Functions

Figure 7.6   IPTV functional architecture based on IMS

7.4   Control and Signaling Protocols for NGN

In NGN the signaling is one of the most important advantages over the best effort Internet architecture. However, since NGN is based on Internet technologies, the signaling protocols (for exchange of control information between NGN entities) are also standardized protocols from the IETF. The most important among them are Diameter [9] and SIP [6].

7.4.1   Diameter

The Diameter is a protocol that is standardized by IETF with aim to solve the problems with its predecessor, the RADIUS (Remote Authentication Dial In User Service) protocol. So, Diameter is an AAA framework protocol, which works on the application layer [according to the OSI (Open System for Interconnection) protocol layering model]. It is targeted for usage by applications for network access as well as IP mobility (including local and roaming scenarios).

Diameter uses TCP (Transmission Control Protocol) or SCTP (Stream Control Transmission Protocol) on the transport layer, which provides reliable data transmission between end hosts that are using Diameter protocol. This is different than the RADIUS protocol (as its predecessor) which uses UDP (User Datagram Protocol) on the transport layer, and provides reliability by retransmissions on the application layer. As successor of the RADIUS, the Diameter is created to replace the RADIUS as main AAA protocol in IP-based access networks, including fixed and wireless. The RADIUS was initially created during a time period (in 2000) when single residential users could access the Internet via dial-up modems. Since that time the AAA has become needed by many access technologies, such as xDSL (Digital Subscriber Line technologies), optical access networks, cable networks, mobile broadband access, and so on. So, Diameter introduces several important improvements over

RADIUS, to suit different requirements from heterogeneous fixed and mobile networks, given as follows:

- *Failover*: RADIUS does not have a failover mechanism since it uses UDP as transport protocol, so failover implementation is left to the application layer and hence differs in different implementations. Diameter defines application layer acknowledgments and failover methods.
- *Transmission-level security*: RADIUS does not provide per packet security, application-layer authentication and integrity is required only for response packets. For RADIUS is defined also Extensible Authentication Protocol (EAP) framework [10], as well as optional usage of IPsec with RADIUS [11]. Diameter applies per packet confidentiality by using IPsec and Transport Layer Security (TLS) [12], either as TLS/TCP or Datagram Transport Layer Security for SCTP (i.e., DTLS/SCTP) [13].
- *Reliable transport*: On the transport layer RADIUS uses unreliable UDP, while Diameter uses reliable protocols TCP and SCTP.
- *Agent support*: RADIUS does not have explicit support for agent nodes, such as relays, proxies, and redirect nodes. On the other side, Diameter defines explicitly the behavior of each agent that processes a Diameter message (e.g., a redirect agent does not alter a Diameter message).
- *Server-initiated messages*: Support for this type of messages in Diameter is mandatory, while in RADIUS such messages are defined, but the support is optional and hence it is difficult to function in heterogeneous network environments. These messages are important for implementation of server-to-client initiated unsolicited disconnect or re-authentication/re-authorization on demand.
- *Transition support*: Since the Diameter Protocol Data Unit (PDU) is different than the one from RADIUS, backward compatibility is needed with aim to provide real-world deployments of Diameter, including networks with deployed RADIUS. So, both protocols can coexist in a same network during a transition from RADIUS to Diameter AAA (in parallel with the transition of the legacy IP networks to an NGN).
- *Capability negotiation*: RADIUS does not support capability negotiations, so RADIUS clients and servers are not aware of each others capabilities, hence in some situation they are not able to negotiate an acceptable service. However, Diameter includes error handling support and capability negotiations.
- *Peer discovery and configuration*: While RADIUS requires names or addresses of servers and clients to be manually configured, by using shared secrets. Such approach introduces large amount of administrative work on RADIUS and also limits the reuse of the secret. This is overcome in Diameter by using dynamic discovery of peers via DNS. Diameter peer discovery is performed in two cases: (i) when a Diameter client needs to discover a first-hop Diameter agent and (ii) when a Diameter agent needs to discover another Diameter agent for completion of a Diameter operation.

7.4.1.1 Diameter Message Structure

Diameter is a message-based protocol, where each message is transported is a separate packet. The message structure is shown in Figure 7.7. There are two types of Diameter messages: request messages and answer messages.

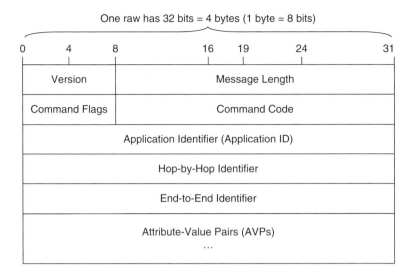

Figure 7.7 Diameter message structure

Each Diameter packet (i.e., messages) contains a packet header followed by one or more Attribute-Value Pairs (AVPs). An AVP is used to encapsulate protocol-specific data, such as routing information, as well as AAA information.

The field Version in the Diameter header denotes the version of the Diameter protocol. In the current version of Diameter it is set to 1 (i.e., denotes version 1 of the protocol). Message length contains the total length of the packet including the header and the AVPs.

The command flags are eight bits, from which first four bits are assigned while the last four bits are reserved. The four bits are defined as follows:

- *"R" (Request) bit*: If set, this denotes a request message. If cleared, the message is an answer.
- *"P" (Proxiable) bit*: If this bit is set the message may be proxied, relayed, or redirected. If cleared, the Diameter message must be locally processed.
- *"E" (Error) bit*: This is set when a Diameter request caused a protocol-related error, otherwise it is cleared. A message with set "E" bit must not be sent as an answer (to a Diameter request message).
- *"T" (potentially re-Transmitted message) bit*: This is used in cases of link failover procedure, and it is set when resending requests are not yet acknowledged (i.e., it serves as indication of a possible packet duplicate due to a link failure). However, this bit must be cleared for sending a request for the first time.

In Diameter each command has assigned a command code, which is stored in the given 24-bit field in the Diameter header. For backward compatibility with the RADIUS as AAA protocol used before the Diameter, the values 0–255 of the command codes are reserved for RADIUS. The command code is used in both types of Diameter messages, that is, requests and answers.

The Diameter is an extensible protocol, a framework for different applications to use it with different approaches. Therefore, the Application ID is a 32-bit field in Diameter packet header that carries the Application ID assigned to a given application by Internet Assigned Numbers

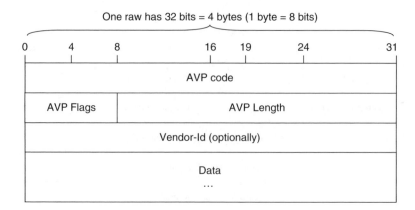

Figure 7.8   Diameter AVP structure

Authority (IANA). However, the Base Diameter protocol does not require an Application ID because its support is mandatory in all entities that use Diameter. Currently, the following application ID values are defined for Diameter: =0 for Diameter common message; =3 for Diameter base accounting; and =0xffffffff (i.e., all bits in the field set to 1) for relay.

Hop-by-Hop Identifier is 32-bits unsigned integer field (always given in network byte order, i.e., most significant byte is transmitted first) that is used for matching requests and answers (i.e., replies). It must be a unique identifier in a request message of a given connection at a given time. The answer message to given request must contain the same Hop-by-Hop Identifier. Its start value is randomly generated and it is monotonically increasing number, by one with each next request. Diameter answer messages without a specified Hop-by-Hop Identifier must be discarded.

Another identifier in Diameter packet header is End-to-End Identifier. It is 32-bit integer field (in network byte order) which is used to target duplicate messages. For that purpose this identifier must stay locally unique for a period of at least 4 minutes, even in the case of reboots of machines running Diameter.

The Diameter packet body consists of AVPs, which structure is shown in Figure 7.8. Each AVP is defined with 32-bit AVP code, where 0–255 codes are the same as RADIUS codes for backward compatibility. The AVP codes from 256 and above are allocated by IANA.

The AVP codes field is followed by the AVP flag, which is 8-bit field. But, only three bits are defined as flags (the other are reserved and sender should set them to zero, while the receiver must ignore them) and they are: "V" bit indicates whether Vendor-Id field is there in AVP or not (if it is set, then Vendor-Id field is present in AVP, otherwise it is not); "P" bit is set when end-to-end security mechanism is used (its default value is 0); and "M" bit (known as Mandatory bit) is set to indicate that Diameter entities such as client, server, proxy, and translation agent, must handle the given AVP, and each Diameter entity that does not understand the AVP (with the "M" bit set) must return an error message.

7.4.1.2   Diameter Communication

Diameter protocol communication is based on connections and sessions, where a connection refers to a transport-level connection between two peers that is used to send and receive

Diameter messages while a session refers to logical connection at the application layer that is established between the Diameter client and server. Each Diameter Session-Id begins with DiameterIdentity of the sender, followed by 64-bit value consisted of high 32 bits and low 32 bits, and optional value (which is implementation specific, i.e., it may be MAC address, timestamp, etc.), in the following format:

Diameter messages
< DiameterIdentity >; < high 32 bits >; < low 32 bits >[; < optional value >]

High 32 bits and low 32 bits are part of monotonically increasing 64-bit value. DiameterIdentity is used to uniquely identify a Diameter node, and in such case it must be the Fully Qualified Domain Name (FQDN) of the Diameter node (e.g., FQDN = access123.example.com). The redirect-host AVP is of type DiameterURI, which follows the URIs syntax rules [14]. In DiameterURI possible values for transport protocol are "tcp," "sctp," and "udp," while possible values for protocol are "diameter," "radius," and "tacacs+." Examples for different Diameter IDs are the following:

Example 7.1: Examples of Diameter IDs (i.e., FQDN):

access1.example.com

Example of Session-Id with no optional value:

access1.example.com;12345;9876

Example of Session-Id with optional value:

access1.example.com;12345;9876;ngn@example.com

Example of DiameterURI with no transport security:

aaa://node.example.com:3868;transport=tcp;protocol=diameter
aaa://node.example.com; transport=tcp (in this case is used the default port 3868 and the
   default protocol is Diameter)
aaa://node.example.com:6666;transport=sctp
aaa://node.example.com:1813;transport=udp;protocol=radius

Example of DiameterURI with transport security:

aaas://node.example.com:5658; transport=tcp;protocol=diameter
aaas://node.example.com:5658; transport=sctp;protocol=diameter

---

In general, there are two types of Diameter sessions:

- *Authorization session*: This is used for authentication and/or authorization.
- *Accounting session*: This is used for accounting.

A given session can be either stateful or stateless, which depend upon the application (whether it requires the session to be maintained for a certain duration, or not). In a stateful Diameter session there are multiple messages exchanged. However, Diameter is an application-layer protocol that uses reliable transport protocols, hence one may virtually distinguish between two connection establishments: (i) transport connection and (ii) diameter connection.

The communication between Diameter peer nodes starts with the establishment of transport connection by using either TCP or SCTP as transport protocol on port 3868. When DTLS is used, the Diameter peer that initiates a connection must establish connection on port 5658. Typically, TLS runs on the top of TCP (when it is used), while DTLS runs on top of SCTP (when it is used).

After proper establishment of the transport connection, the application sender initiates a Diameter connection with Capabilities-Exchange-Request (CER) message as a first sent message to other peer. The other peer sends as a response Capabilities-Exchange-Answer (CEA) message. If the CEA result code is set to Diameter success then the Diameter connection is established and ready for exchange of application messages. When a secure transport is established, then all messages (including CER and CEA) are exchanged on secured transport. However, if no messages are exchanged over an established Diameter connection for a certain time, then either side may send Device-Watchdog-Request (DWR), and in such case the other peer of the Diameter connection must respond with Device-Watchdog-Answer (DWA). So-called watchdog messages are used to probe a given Diameter connection. For termination of a Diameter connection, either side may send a Disconnect-Peer-Request (DPR), which is followed by Disconnect-Peer-Answer (DPA) from the other peer of the connection. After DPR/DPA message exchange the transport connection can be closed.

7.4.1.3   Diameter Nodes and Architecture

The network architecture for Diameter communications consists of the end peers, clients and servers, and Diameter agents. There are four types of Diameter agents given as follows:

- *Relay agent*: This is used to route a message to other Diameter nodes based on routing information in the received message (e.g., Destination-Realm AVP). The relay agent may be connected with multiple IP networks. Diameter relay does not change the message, but it must advertise its Application ID (i.e., 0xffffffff).
- *Proxy agent*: Similar to relay agent, this agent does the routing of the Diameter messages by using the Diameter routing table. So, the proxy agent does the same function as the relay agent, but also it can change Diameter messages. Proxies must maintain the states of their downstream pears (i.e., devices), and must advertise Diameter application they support (since enforcing certain policies by the proxy agent is application dependent).
- *Redirect agent*: This is used for architectures where the Diameter routing information is stored on a central location, so every Diameter node that needs Diameter routing information gets it from the redirect agent. So, redirect agent does not route or forward messages to any

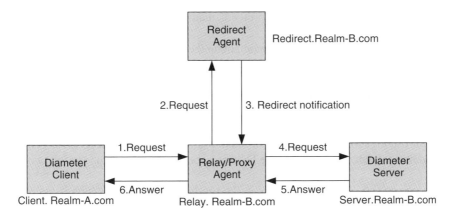

Relay/Proxy Agent (no redirect): messages 1; 4; 5; 6.
Redirect Agent involved: messages 1; 2; 3; 4; 5; 6.

Figure 7.9   Diameter agents architecture

other nodes, it just replies to requests (with the requested routing information included in the response). Redirect agents also must advertise their Application ID, which is the same as the one for relay agents (i.e., 0xffffffff).

• *Translation agent*: This provides translation of RADIUS messages to Diameter messages, and vice versa, for provision of backward compatibility between Diameter and its predecessor RADIUS.

An overview of Diameter architecture is shown in Figure 7.9. The given Diameter architecture include redirect agent as well. If the relay or proxy agent can route the Diameter message based on its own Diameter routing table, then redirect agent is not needed. However, in centralized architectures (with centralized Diameter routing tables at the register agent), when a proxy or relay agent receives the Diameter request message, it asks the redirect agent (sends a request to it) for routing information regarding the given message and the destination information (e.g., destination Diameter realm). In the case of transport failure the relay or proxy agent attempts to reroute a pending request to an alternate Diameter peer (e.g., another relay agent in the network). In such cases (with rerouting) the "T" bit is set. When failed connection is reestablished then failback happens, that is, the message routing is switched back to the original next hop.

The routing of Diameter messages is also a part of the Diameter base standard. Information used for routing of Diameter messages is Application ID (present in the message header) and destination-host or destination-realm AVP. The routing rule is based on the following Diameter routing algorithm:

Algorithm

```
If local identity == destination-host AVP then process locally
else if peer identity == destination-host AVP then send mes-
sage to that peer node
else send unable to deliver answer message back to the sender.
```

Each node that supports Diameter protocol maintains two tables: (i) a peer table and (ii) realm-based routing table.

The Diameter peer-table is used for maintaining the identity of next peer nodes (e.g., relay, proxy) that has direct connection to the given peer, as well as state of the next peer (e.g., idle, open, close, etc.), how the peer was discovered (statically or dynamically), expiration time of the peer entries, and type of transport protocol and associated security with the peer (e.g., whether TLS/TCP and DTLS/SCTP are used). The peer table is also referenced by the Diameter routing table.

Diameter routing tables contain a list of realm routing entries that are used for realm-based routing lookups. Each realm entry in the Diameter rooting table contains the following fields:

- *Realm name*: This is the primary key in the routing table lookup, which is matched with destination-realm AVP.
- *Application ID*: This is the application ID specified in the message headers, and it is secondary key in the routing table lookups.
- *Local action*: This field defines how the message shall be treated by the given peer (i.e., node). Possible actions are: LOCAL (message is processed locally and it is not sent to another peer), RELAY (performed in relay agents), PROXY (performed in proxy agents), and REDIRECT (performed in redirect agents, after receiving message with this action from a relay/proxy agent).
- *Server identifier*: This field contains the identity of one or more servers to which the message is to be routed (when the Local action field is RELAY or PROXY) or redirected (when the Local action field is REDIRECT).
- *Static or dynamic (i.e., discovery type)*: This states whether the routing entry is statically configured or dynamically discovered.
- *Expiration time*: This specifies the time when dynamically discovered route table entry expires.

An example of Diameter routing is shown in Figure 7.9. Since Diameter is request/answer type of protocol, each request message is followed by appropriate answer message in the reverse direction. In the case of Diameter routing management, answer routing path is the of the request message path.

Overall, Diameter base protocol has been developing in a period of a decade, and finally it is ready for its implementation in NGN environments.

7.4.2 Session Initiation Protocol

The SIP is a signaling protocol for IP networks, which is standardized by IETF [6]. SIP is considered as a replacement for SS7 signaling in PSTN and PLMN, which has been reliable and complete signaling for real-time services (e.g., voice) in fixed and mobile environments. The telephony nowadays has the functionalities that are in fact provided by different user parts of the SS7 (e.g., ISDN User Part – ISUP, Telephony User Part – TUP, etc.). SIP appeared in the second half of the 1990s (when the IP dominance in the data world has become evident), and it was finalized at the beginning of the twenty-first century. Although it is already a well-known and mature protocol, its major impact on telecommunication world will become more evident with the transition of different telecommunication networks to all-IP networks, that is, to NGN, which is already started to happen in fixed networks and it will continue to happen in mobile networks with 4G and beyond 4G developments.

As a signaling protocol, SIP works on the application layer (similar to Diameter). Its main role in NGN is to establish, maintain, and terminate multimedia sessions. Hence, it is standardized as a signaling protocol in IMS, which is an essential part of the NGN for real-time multimedia services. In general, SIP can be used for creation of two-party sessions (e.g., VoIP), multi-party sessions (e.g., conference), or multicast sessions (e.g., IPTV). SIP is created to be inherent of the underlying transport protocol, so it can be implemented over TCP, SCTP, or UDP.

SIP supports name mapping and redirection services, which are important in mobile IP-based networks. Also, SIP can be used as a component with other IETF standardized protocols for multimedia, such as Hyper-Text Transfer Protocol (HTTP), Real-Time Streaming Protocol (RTSP) [15], as well as Session Description Protocol (SDP) for describing multimedia sessions [16]. For transmission of real-time multimedia (e.g., voice, video streaming) SIP typically uses Real-time Transport Protocol (RTP) [17].

SIP is a client–server protocol. This means that requests are generated by a single entity (the client) and sent to the receiving entity (the server) that processes them and returns responses to the client. In the case of IP telephony, the caller acts as a client and the called party acts as a server. A call may involve several clients and servers, and a single host can be addressed as a client and as a server for a given call. In general, a caller may directly contact the called party (e.g., by using IP address), which is a peer-to-peer communication approach. So, SIP communication can be client–server, or peer-to-peer.

7.4.2.1 SIP Messages

SIP is a text-based protocol similar to HTTP [and HTTP was similar to SMTP (Simple Mail Transfer Protocol) for e-mail communication]. Similar to HTTP, SIP uses messages in a request/response manner. Each SIP transaction consists of a client request and a response from a SIP server. SIP messages reuse most of the header fields, encoding rules, and status codes (given in response messages) of the format of the HTTP. This allows easy integration of SIP with Web servers and e-mail servers.

SIP calls are identified by the so-called Call Identifiers (Call-IDs), located in the Call-ID field of the SIP message. Call-ID is created by the creator of the call and it is used by all the participants in the given call.

SIP defines six messages used for session setup, management, and termination, and they are given as follows:

- *INVITE*: This message is used by the caller to initialize a session.
- *ACK*: This is a message with which the caller acknowledges the answer of the call (which is initiated with the INVITE message) by the called party.
- *BYE*: This message is used to terminate an established session.
- *CANCEL*: This message is used to cancel already started initialization process (e.g., it is used when a client sends an INVITE and then changes its decision to call).
- *REGISTER*: This message is used by SIP user agent (in user equipment) to register its current IP address and the SIP URIs for which the user will like to receive calls.

- *OPTIONS*: This message is used to request information about the capabilities of a caller, while it does not setup a session (a session is setup only with the INVITE message).

The given messages are used for performing SIP functions (e.g., setup, modification, and termination of sessions) for different services, such as voice, video, messaging, gaming, and so on. Each request SIP message is followed by a response SIP message from the other party in the given session. The response codes for SIP are the same as those defined for HTTP, where a three digit response code is followed by an explanation in a human readable format. All response codes are grouped into six groups, where each group of codes in identified by the first digit, as follows:

- *1xx*: Provisional response, used by SIP servers to indicate progress, but they do not terminate a SIP transaction (e.g., 100 Trying).
- *2xx*: Success response, which means that action was successfully received, understood, and accepted (e.g., 200 OK).
- *3xx*: Redirection response, which indicates that further action is needed for completion of a given request (e.g., 302 Moved Temporarily).
- *4xx*: Client Error, which usually means that the request contains bad syntax or cannot be completed by the server (e.g., 404 Not Found).
- *5xx*: Server Error, which is server-side error on a given valid request from the client-side (e.g., 504 Server Time-out).
- *6xx*: Global Failure, which means that the request cannot be fulfilled by any server (e.g., 603 Decline).

The request and response SIP messages are exchanged between SIP components, that is, network elements and user equipment that supports SIP.

7.4.2.2 SIP Naming and Addressing

To invite a given party to a session, SIP entity must be named. Since the most used addressing form in Internet at the time of SIP standardization (i.e., year 2002) was the form of e-mail addresses (e.g., username@FullyQualifiedDomainName) and URI schemes (e.g., http://www.example.com for WWW), the SIP also adopts the URI scheme based on general standard syntax (the same that is used for WWW and e-mail), as well as user addressing in a form similar to e-mail addressing schemes.

So, each element in a SIP logical network (it is referred to as "logical," because the SIP network is usually an overlay network over an existing IP network infrastructure) is identified by a SIP URI [or a secure SIP (SIPS) URI, when TLS is used, which is similar to the relation between "http" and "https"]. The general form of the SIP URI is the following:

SIP URI

    sip:{user[:password]@}host[:port][;uri−parameters][?headers]

    sips:{user[:password]@}host[:port][;uri−parameters][?headers]

The parts of the SIP URI in the brackets are not mandatory for the SIP URI, and their presence in the URI is dependent upon the type of SIP entity. The format for a SIPS URI is the same as that for a SIP URI, with the difference in the URI scheme (i.e., "sip" is a URI scheme for SIP, while "sips" is a URI scheme for SIPS). The term "user" in the URI refers to a user in the given host, where the host in such a case (i.e., having multiple users) is in fact a domain (e.g., example.com). The password is associated with the given "user," but its presence in SIP URIs is not recommended due to possible security risks (e.g., exposing the password to others). The user and password are parts of the so-called user-info in the SIP URI. However, when the destination host does not have notion of users (i.e., the host itself is the resource that is identified by the URI) then the user and password are excluded. The host name in the URI contains either a FQDN or an IP address (including IPv4 and IPv6 addresses). The port number is used where necessary. If it is omitted then the default SIP port numbers are used, which are 5060 and/or 5061 (for both, TCP and UDP) to connect to SIP servers and other SIP end-peers. The "uri-parameters" contain a list of parameters (in the form: "parameter-name"-"parameter-value"), such as "maddr" field for specification of a proxy that must be traversed on the way to the destination, "ttl" field is used to carry time-to-live value of the UDP multicast packet (when "maddr" is a multicast address), and so on.

Example 7.2: Examples of SIP URI

sip:user123:password123@example.com:6666
sip:user123@example.com
sip:user123@192.168.1.1
sip:server1.example.com
sip:123456789@example.com

The SIP URIs, that is, SIP user addresses, of type "phone-number@gateway" are used to name PSTN telephone numbers to become available through the named gateway node. Also, some users may be able to use e-mail addresses as their SIP addresses, which is dependent upon the SIP implementation in a given domain. E-mail addresses already offered basic location-independent form of address, which is taken later by SIP. But, SIP differs from e-mail because in e-mail communication delivery of messages ends in destination mail server, and later the user obtains the e-mail messages via an e-mail access protocol [e.g., POP3 (Post Office Protocol-version 3), IMAP (Internet Message Access Protocol), Web-based, etc.]. In the case of SIP, the messages are delivered to different end peers (e.g., servers, user equipments, etc.), and in some cases the IP address of the end peer can be changed (e.g., DHCP usage for addressing user equipments) or the user may register different end devices for receiving different types of media in different times of the day (e.g., based on certain contextual information, such as user presence on a given location such as home, office, or a public place).

7.4.2.3   SIP Network Elements

There are two general types of network elements for SIP, and they are User Agents (UA) and servers. Although two user agents can communicate directly between each other (e.g., by using their IP addresses), that is, in peer-to-peer manner, such approach is usually not practical for implementations of public networks such as NGN.

There are two type of User Agents (UA):

- *User Agent Client (UAC)*: This makes SIP requests and sends them to servers.
- *User Agent Server(UAS)*: This receives the requests, processes them, and returns SIP responses.

Regarding the SIP User Agents, a single User Agent can function as both, UAC and UAS. The roles of either UAC or UAS last only for the duration of a given SIP transaction. SIP UA is a logical end-point in the SIP network architecture, which is used to create or receive SIP messages (with which it manages the SIP sessions). On the servers' side, there are the following types of SIP servers:

- *Redirect server*: This is a user agent that generates 3xx responses to redirect the request back to the client indicating that the client needs to try a different route to reach the SIP recipient (e.g., when a recipient has moved to another location). With redirection servers push the routing information for a given request from UAC in a response back to the client, thereby propagating URIs from core network to its edges, which increases network scalability.
- *Proxy server*: This is most common type of server in SIP-based signaling networks. It usually works as both UAC and UAS, with aim to provide recipients IP address to SIP clients. When a request is generated by an UAC, it typically does not know the recipient's IP address (i.e., network location), so the client sends the request to its assigned SIP proxy server, which further forwards the request to another proxy server or to the recipient. Proxy servers are also used to enforce network policy (e.g., checking whether a user is allowed to use given service), and where necessary it may rewrite certain parts of the SIP request message.
- *Registrar server*: This is a logical end-point SIP element that is used for registering current location of the end users (e.g., binding IP address of the client with one or more SIP URIs). A user registers its location by sending REGISTER message to the registrar, which is typically collocated with a SIP proxy server.
- *Location service*: This is a database that stores information about user location, that is, stores bindings between user network location information (e.g., IP addresses) and SIP URIs. Such information in stored by Registrar servers upon receiving REGISTER messages from SIP User Agents.

Figure 7.10 shows three scenarios for using the SIP signaling for VoIP connections. Figure 7.10a shows a general picture of SIP signaling between two VoIP users (i.e., two SIP user agents) when they are in two different domains (usually that means that they are served two different VoIP service providers, where each VoIP service provider has a proxy server in its own domain). Figure 7.10b shows so-called SIP triangle, which is the scenario when both VoIP users (i.e., SIP user agents) are located in the same domain, served by the same proxy server. The third case, shown in Figure 7.10c, presents a scenario for usage of SIP for peer-to-peer telephony, where the SIP network elements are excluded (this is peer-to-peer SIP scenario). Usually Proxy servers are used to connect different network domains with each other, or to connect users (i.e., SIP User Agents) with SIP network nodes (which are SIP servers deployed in the network to which are attached VoIP users in the given example).

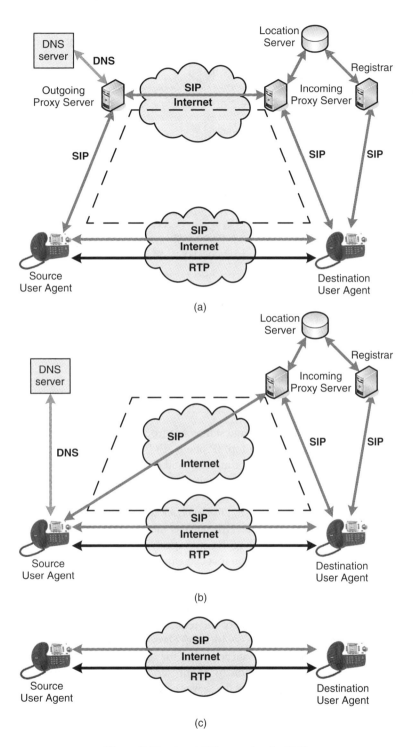

Figure 7.10   (a–c) SIP scenarios for VoIP

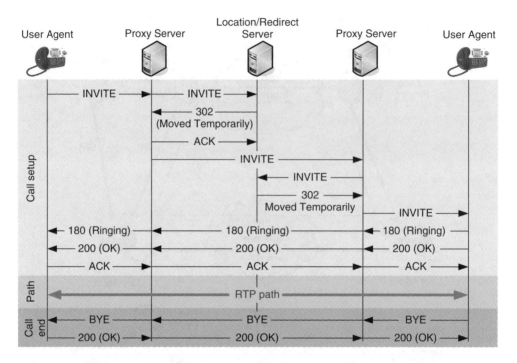

Figure 7.11 SIP diagram for VoIP

SIP signaling typically starts with a request message from a SIP User Agent which is located in a host (e.g., personal computer, lap-top, IP phone, IP home gateway node, mobile device, etc.) and it is used by the end user for originating and receiving (i.e., terminating) calls. The principle of messages in SIP is very similar with HTTP, as it can be seen from the diagram of SIP messages for establishing a VoIP call as shown in Figure 7.11. Thus, each element of a given SIP message sends response (i.e., answer) back to sender. For example, responses for "success" contain only one message, that is the answer message "200 OK," while redirect from a given server is implemented by sending the message "302 Moved Temporarily."

Unlike other applications, the invitation to a call cannot immediately result with a response because locating the one who calls and waiting to answer the call takes a few seconds. The call can be placed in the waiting line if the called party is busy. Responses from class 1xx are used to inform the calling party about the progress of the call, but they are always accompanied by other answers forming thus the outcome of the request. While the responses with 1xx codes are only temporary, the answer messages with other classes of status codes determine the final status of the request, such as 2xx for success, 3xx for redirection or forwarding, as well as 4xx, 5xx, and 6xx in the cases of failures by client, server, and global failures, respectively. For higher reliability of the SIP signaling the server performs forwarding of the final response until the client or server has confirmed receipt by sending ACK message to the sending side.

Finally, SIP is a well standardized protocol that is created to replace SS7 in all-IP environments. However, similar to the SS7 approach, it requires complete standardization to be implemented on a global scale. Such standardization is provided by ITU-T (International

Telecommunication Union-Telecommunications; a similar case as it was done for SS7 several decades before SIP) by accepting SIP globally as the main signaling protocol in NGN, together with the standardized common IMS. Together with the Diameter base protocol, SIP is the most important signaling protocol in NGN and telecommunications in the future.

7.5   Security Mechanisms for NGN

In telecommunication networks security mechanisms have high priority. While in traditional telecommunication networks, such as PSTN and PLMN, the user impact on the security is limited on certain access part of network, in all-IP networks such as NGN, the possible impact on the security is much higher in all network segments due to heterogeneous nature of Internet-based services (including standardized and proprietary ones) as well as higher intelligence of the end user equipment (e.g., computers, mobile devices, etc.).

The selection of the security mechanisms in NGN depends upon the so-called trust model. In NGN are defined many Functional Entities (FEs), which are mainly software-based. In such cases, where it is possible, different FEs are grouped into one physical entity (i.e., a server, a gateway, etc.) in the network, which depends upon the vendor choice as well as NGN operator network design. In general, two trust models can be defined for NGN [18]:

- *Single-network trust model*: This model refers to security in single NGN, and it defines three security zones – trusted, trusted but vulnerable, and un-trusted. Trusted zone includes NGN network elements, trusted but vulnerable zone includes network border elements (e.g., gateways) and un-trusted zone consists of user terminal equipment and provider-controlled equipment on user's side (e.g., home gateway).
- *Peering-network trust model*: This is scenario where NGN is connected to another network, and hence the trust depends upon the physical interconnection (e.g., direct connection in a secure building, or otherwise), the peering model (e.g., NGN operators are directly connected between each other, or via transit transport provider) as well as business relationship among networks [e.g., applying SLA (service level agreement), viewing other NGN providers as un-trusted, etc.].

Regarding the security mechanisms one may distinguish among two main mechanisms: AAA and protection of the data by using encryption. Typically, AAA service is mandatory in NGN, while encryption is optional and depends upon the type of data and the network segment.

7.5.1   Authentication, Authorization, and Accounting in NGN

AAA in NGN and in general consists of three processes: authentication, authorization, and accounting. Authentication is validation of user's identity prior to its access to the network or service resources (e.g., username/password, a security key, a certificate, etc.). Authorization defines the services that the successfully authenticated user is allowed to use (so, authorization follows the authentication, but also they can be merged in some cases). Accounting provides functions for collecting information about user consumption of certain service (e.g., time duration, volume of data, number of messages or transactions, etc.) which can be further processes for billing as well as other purposes (e.g., capacity planning). These three processes are referred to as AAA (as a single word) and are used to provide access control for access networks or services.

Typically, the AAA system consists of AAA client and AAA server, where the server is connected to a database which stores user profiles and configuration information (e.g., HSS). AAA clients are typically located in the access network (e.g., integrated in access control routers), while the AAA server is usually centralized in the network.

Considering NGN reference architecture for network access, both AAA client and AAA server belong to Network Attachment Control Function (NACF) in the transport stratum [19], as shown in Figure 7.12. When an entity in the TCFs detects a connection request from the user terminal, it acts as an AAA client and requests the AAA server to perform user authentication and then authorization. The protocol that is defined for an AAA process in NGN is Diameter, although RADIUS is also a possible solution (e.g., when it is already deployed in the network). The authentication of the user (by the AAA server) can be either explicit (e.g., by using EAP) or implicit (e.g., authentication in the access-line with a username and a password). After successful authorization, the AAA server requests the Resource and Admission Control Function (RACF) for reservation and allocation of NGN transport resources. After obtaining grant for resources from the RACF, the AAA server notifies the AAA client for permission to connect the given user and its equipment.

The AAA mechanism for service access is based on AAA client and AAA server within SCFs in a NGN service stratum (Figure 7.12). The procedure is similar to AAA for network access, that is, the AAA client detects a connection request from a user for a certain service

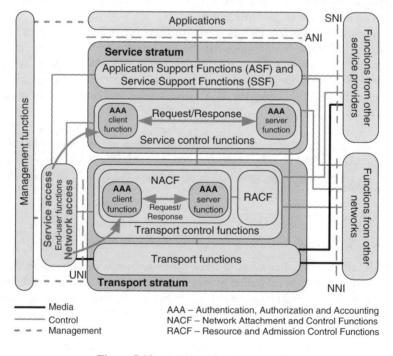

Figure 7.12   AAA architecture for NGN

and sends a request to the AAA server for authentication and authorization. Upon the result of the authentication and authorization the requested service is either provided or rejected. In all cases (AAA for network access or service access) the AAA client sends specified accounting information to the AAA server, which further stores it in an accounting database.

Most of the services in NGN that require AAA are SIP-based services. For SIP-based services the P-CSC-FE (as standardized for IMS and NGN) can be authenticator of subscribers. In such case the user identification and authentication is provided by using credentials between the authenticator and the user equipment.

In general, credentials are used in NGN to identify and authenticate subscriber, end user, or device. Such credentials can have one of the two possible forms, X.509 public key certificate or a shared key [18]. Device credentials may be supplied or burned-in by the vendor of the device (e.g., International Mobile Equipment Identity – IMEI). Such credentials may be used optionally by the NGN provider. Subscriber credentials provide association of the originator of an NGN request with an existing subscriber account. Such credentials can be entered in the device (e.g., in given slots, via download, etc.), where the device should be capable of accepting such credentials. All calls made from the given device with subscriber credentials will be associated with the subscriber owner of the credentials in the device. So, an NGN subscription refers to a "subscriber," which may be consisted of one or more end users (also, an NGN customer may have several NGN subscriptions). End user credentials are used to identify and authenticate individual end users in the NGN (e.g., SIM cards, security tokens, etc.).

7.5.2   Transport Security in NGN

Transport security in NGN is provided with standardized (by IETF) security Internet solutions that work above the network layer (i.e., that is IP layer in Internet and NGN). Such solutions are TLS [20] and IPsec [21].

TLS is used in NGN to secure different types of signaling traffic (e.g., SIP, HTTP, etc.) between various network elements in NGN trusted zone, by using TCP or SCTP as underlying transport protocols. For example, the combination of SIP and TLS gives the SIPS.

Typical solution in IP transport networks for securing the IP packet payload is IPsec. In the NGN, IPsec is used to secure different traffic types (including control plane and user plane) within the trusted zone. The usage of IPsec is typically defined by a policy database. Usually, TLS is not run over IPsec in the NGN. IPsec can be used in transport or tunnel mode. In the tunnel mode, the IPsec header is followed by an IP header of the tunneled packet, which is typical usage in VPNs. In the case of VPN, the end of the IPsec tunnel is not the end destination of the tunneled IP packets, Another IPsec mode, the transport mode, is used for point-to-point communication with IPsec.

Media security (i.e., encryption of the user data) is not mandatory in NGN, but it may be offered to certain types of customers which request protection of their data. A typical protocol for provision of media security for real-timer traffic (e.g., VoIP) in such a case is the Secure Real-time Transport Protocol (SRTP) [22]. This is implemented between the RTP and the transport layer in the protocol stack (RTP typically uses UDP as a transport protocol), and in fact encrypts the payload of the RTP packet on the transmit side and decrypts it on the receiving side of the connection.

7.6   NGN Identity Management

NGN provides Identity Management (IdM) functions according to NGN functional architecture as described in Chapter 3. The IdM framework in NGN utilizes the network information about the user, such as user subscription, location information, policy, presence, as well as NGN entities such as HSS, CSCFs, and Session Border Controller (SBC) [23]. The main goal of IdM services, functions, and capabilities, is to provide support to business and security applications, also including identity-based services in NGN. Hence, scope of IdM includes functions in both NGN stratums, transport, and service stratum.

IdM functions and capabilities provide relationships between business and security applications (including identity-based services) on one side, and identity information in NGN (e.g., identifiers, credential, and attributes) on the other side, as shown in Figure 7.13. The identity information is used in NGN to identify different real-world entities, such as users and subscribers (e.g., organizations, business enterprises, government enterprises), networks, and service peripherals (e.g., user devices), as well as virtual objects (e.g., network elements and objects). Each entity may have one or several identities, which are used for different types of transactions. A person as NGN user may be associated with several digital identities.

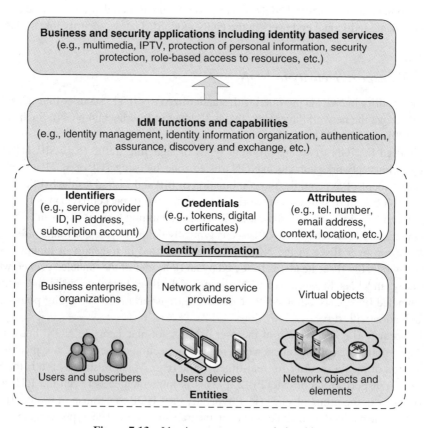

Figure 7.13   Identity management relationships

The identity information that is associated with an entity (e.g., users and subscribers, user devices and network elements, business organizations, network and service providers, and virtual objects) belongs to one of the following types of identities:

- *Identifier*: This can be a subscription account (e.g., for residential users, enterprises), network element address, service provider identifier (e.g., domain name).
- *Attribute*: In NGN this refers to a single user, device, or network interface (e.g., e-mail address, URI, IP address, authentication method, location, etc.).
- *Credential*: This is issued to individual user by a third party, and may include username, password, digital certificate, token, and so on.

Also, NGN provides so-called federated identity information, which is identity information that is shared among multiple NGN providers and/or business and government organizations. For example, identities such as identifiers, attributes, and credentials can be verified by an Identity Service Provider (IdSP) (NGN does not impose limitations regarding such identity providers) which is treated as a trusted party by the relying party (e.g., users, service providers, NGN providers).

Regarding the NGN functional architecture (as it is given in Chapter 3) IdM functions may reside in both NGN stratums (i.e., transport and service stratum) and in different planes (e.g., user, control, and management plane). IdM functions can be implemented either as standalone functions (regarding other NGN entities) or integrated with certain NGN elements. However, implementation of IdM functions also depends upon national and regional regulation (e.g., for privacy, protection of identity data, etc.). The basic data principles that should be implemented with the IdM in NGN are the following:

- Using data binding only for defined purpose (e.g., for a given service);
- Excluding data sharing between applications, at least without end user explicit approval;
- Limitation of the required identity data for a given purpose in a given NGN service;
- Control of the Personally Identifiable Identifiers (PIIs) by the end user.

The identity data in NGN environments is exchanged between different parties (e.g., NGN providers, service providers, etc.) via defined interfaces in NGN, including User-to-Network Interface (UNI), Network-to-Network Interface (NNI), ANI, and SNI. The choice of the NGN interface for IdM communication in NGN depends upon specific application and service requirements, as well as protocol solution (e.g., SIP, Diameter, RADIUS, etc.). For example, Security Assertion Markup Language (SAML) [24], provides standard solution for federation (regarding the IdM in NGN), which is mainly targeted for usage by business environments, government organizations, and their service providers. SAML messages are encoded in XML that provides a set of rules for encoding document data in a human readable and machine readable format (where XML is a specification created by the World Wide Web Consortium – W3C [25]).

Finally, regarding the IdM, the NGN does not put restrictions on IdSPs, but defines the identity discovery, correlation, and binding. Regarding the identity discover there are two main cases: intra-network identity discovery, and inter-network discovery.

For intra-network identity discovery, the NGN functional entity Subscription Locator Functional Entity (SL-FE) provides the address of the SUP-FE, which stores user identity information, subscriber data, and location information. Considering the NGN functional architecture, the SUP-FE can be queried for obtaining user's identity information from several network entities, and they are: Application Support Functional Entity (AS-FE, i.e., AS according to ETSI), Interrogating Call Session Control Functional Entity (I-CSC-FE, i.e., I-CSCF), and Serving Call Session Control Functional Entity (S-CSC-FE, i.e., S-CFCF), where the last two are integral parts of the IMS system (when IMS is deployed in the NGN).

The inter-network identity discovery in NGN is based on pre-established agreements among different parties such as IdSP and relying parties. Mechanisms for inter-network discovery can be based on SAML for exchange of authentication and authorization information between parties, Identity Web Services Framework (ID-WSF) for per user service discovery (e.g., for social networking) [26], OpenID that allows service authentication by relying parties instead of the third service providers [27], and so on.

In general, intra-network identity discovery by using IMS entities (including HSS), connected via SIP and Diameter protocols, may be considered in long-term as a main mechanism for identity discovery and management in NGN networks. In such case inter-network identity discovery can be accomplished by using DNS (e.g., for discovery of the IP address of the IMS functional entity in other domain) and peer-to-peer communication between IMS functional entities in different NGNs.

7.7   Service Continuity

Mobile environments are specific to the end users due to necessity to maintain service continuity for intra access network and inter access network scenarios. In that context, service continuity includes continuity of a given service on the same terminal while moving as well as service continuity on different terminals. Service continuity generally is defined as the ability of a moving object (e.g., a mobile terminal) to maintain an ongoing service including the current states, the network environment for the user, as well as established session for the service [28].

The service continuity on NGN is implemented by FMC solutions, such as IMS and its predecessor Unlicensed Mobile Access (UMA). Since the IP networks, as Packet Switching (PS) networks, are successors of Circuit Switching (CS) networks, the service continuity is also important to be supported between CS and PS networks (besides PS-only networks), during the transition period from traditional CS to IP-based PS networks. Such services are primarily real-time legacy services, including voice (i.e., telephony) and TV (or real-time multimedia).

Support for service continuity can be provided in transport stratum and service stratum. Legacy standards for providing transport level service continuity in IP environments such as NGN are based on Mobile IP standard, that is, Mobile IPv4 for IPv4 networks, and Mobile IPv6 for IPv6 networks. However, IETF has developed experimental specifications around a novel Host Identity Protocol (HIP) which introduces additional protocol layer between the IP layer (i.e., the network layer) and the transport layer (i.e., TCP, UDP) to separate the network locator role of IP addresses (i.e., IP addresses are used to identify an interface attached to Internet, for routing of IP packets to a given destination) and identifier role of IP addresses from application point of view (i.e., IP address is also used to define every open socket from a given application that uses the interface associated with that IP address). This way HIP introduces a new name space, based on 128-bit Host Id Tags (HITs) as host identities, but it is not certain how it will affect the global Internet which is currently based on two existing name spaces: IP addresses, and domain names. Also, solutions below IP layer (i.e., between MAC layer and IP layer), such as IEEE 802.21 for Media Independent Handovers (MIH) can provide seamless handovers between wireless networks [e.g., WiFi, WiMAX, UMTS (Universal Mobile Telecommunications System), LTE/LTE-Advanced] and/or fixed networks (e.g., Ethernet), and hence to provide service continuity.

Although transport-level mobility management technologies can provide continuity of network attachment (e.g., continuation of the session by keeping the open socket between the application and IP protocol stack in the end host), such technologies lack capabilities to dynamically adapt the service to time-varying and dynamic conditions in different access networks (e.g., different traffic load in different access networks, different capabilities for QoS support, time-varying capacity, etc.). Also, user preferences and inter-operator SLA have influence on the service provisioning. Therefore, service-level convergence is needed for service continuity (besides transport-level support for seamless mobility).

In general, one can differentiate the service continuity in NGN into two types [29]:

- *Multimedia Service Continuity (MMSC)*: Multimedia services consist of voice, video, and/or data services, so service continuity of multimedia service means continuity of all respective services and synchronization among them.
- *Voice Service Continuity (VSC)*: This is seamless voice services provided to end users across different access networks.

Further, service continuity, either MMSC or VSC, can include different scenarios, such as service continuity on the same user device, on different multimedia user terminals or different voice terminals (for VSC), as shown in Figure 7.14.

For example, service continuity on the same user device will be seamless vertical handover from mobile network to a fixed home network [e.g., from LTE (Long-Term Evolution) or UMTS interface on the device to Ethernet interface on the same device and vice versa]. The rationale for such handover can be better in-house coverage, conserving mobile bandwidth, better user experience with higher bit rates via fixed access, and so on. The same rationale is applied for service continuity between different user terminals. Special attention regarding the service continuity is given to the voice service due to its importance as a real-time conversational service with strict requirements for end-to-end delay.

Overall, FMC is the main target toward service continuity. In NGN, service continuity in service stratum is supported by IMS while in transport stratum existing solutions are Mobile IP as well as IEEE 802.21 framework for Media Independent Handovers. In the case of Packet-Switching to Packet-Switching (i.e., PS to PS) service continuity both NGN domains are supposed to have IMS, so the communication is between the IMS entities. In the case of Packet-Switching to Circuit-Switching (PS to CS) service continuity, or vice versa, the most important is Voice Call Continuity (VCC) which is an IMS centric approach that is applicable to cases when multi-mode mobile terminals (typically WLAN-SIP/2G-3G-4G terminal) changes its point of attachment. In such case the VCC application is implemented and deployed as IMS application server, and it controls the transfer the call between CS and PS domains. However, single mobile terminal (with a possibility to connect via both PS and CS networks) has to perform registration to both IMS (in the PS domain) and CS domain, with the aim to use the VCC.

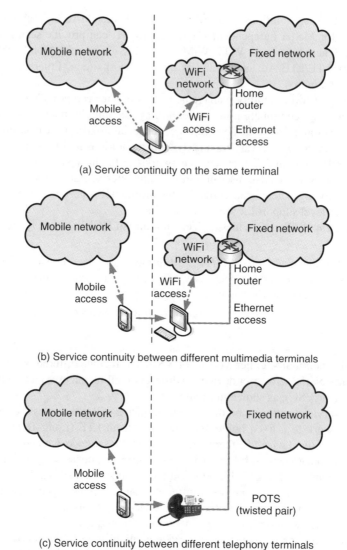

(a) Service continuity on the same terminal

(b) Service continuity between different multimedia terminals

(c) Service continuity between different telephony terminals

Figure 7.14 (a–c) Service continuity scenarios

7.8 Next Generation Service Overlay Networks

When a well standardized network infrastructure exists, such as NGN, then deployment of certain services may require an overlay network over the NGN, consisted of logically inter-connected nodes which perform certain functionalities to support that service. Such a network is referred to as a Service Overlay Network (SON). According to the ITU-T [30], the SON uses a logical service infrastructure to provide support for different applications. The SONs and their elements are connected logically regardless of the underlying networks.

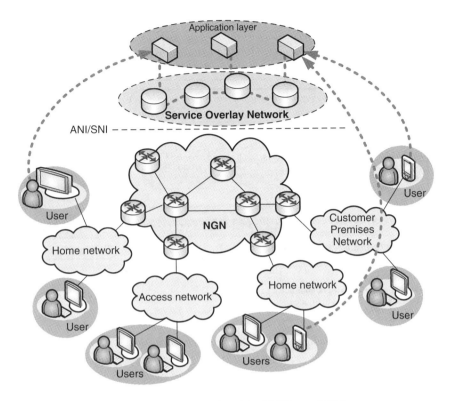

Figure 7.15   General model of SON over NGN

The general model of SON that is based on NGN is shown in Figure 7.15. In general, SON shall provide service-specific logical networking functions that can be used to establish services, further to modify them, as well as to provide new services. The SON is connected to NGN via ANI or SNI. The SNI is used in the case when service provider provides certain functions for the SON. All functions of the SON are positioned on the application layer regarding the protocol stack.

7.8.1   SON Framework

SON functional model is consistent with the NGN functional architecture. Different SON elements are distributed in a single-tier environment where they communicate between each other in a peer-to-peer manner.

Regarding the provided functions, SON includes the following main functional groups, as shown in Figure 7.16, and defined as follows:

- *ANI/SNI interworking function (between SON and NGN)*: This provides a channel for exchange of information between the application in SON and NGN by using ANI interface (to NGN). In the cases when service providers (outside the NGN) needs to exchange

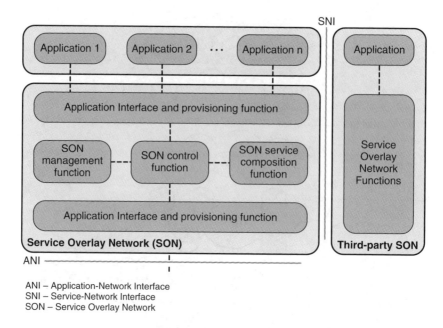

Figure 7.16   Functional model of service overlay network (SON) over NGN

information with NGN with aim SON to support control of service, then the SNI is used
by this interworking function.

- *Application interface provisioning function (between SON and application)*: This is an interface that is used by an application to provide to SON functional entities the required information for service creation and composition by the SON control function.
- *SON control function*: This is access control function within the SON that provides registration, authentication, and authorization of end users, service release, as well as personalization of applications according to the users' requests.
- *SON service composition function*: This is a complex function that provides service composition based on different possible requests, including: emergency conditions, user preferences and conditions, required QoS, required security level, as well as service adaptability.
- *SON management function*: This is used to provide SON service in NGN environments
  including session management, QoS management (either per flow, per session, or per CoS),
  accounting management (e.g., exchange of accounting information between NGN and
  SON), and security management.

SON are not novelty in telecommunications, because all signaling networks are in fact a kind
of SON. Therefore, usage of SON over the deployed NGN environments can be considered as
next generation SON (i.e., SON over NGN).

7.8.2   SON-Based Services

Several emerging services in NGN can be implemented and deployed by using SON architectures. Such examples include IPTV service, community-based services, virtual home network,

as well as cloud computing. IPTV can be provided with different scenarios (e.g., NGN or non-NGN, where NGN-based can be IMS or non-IMS), but in many cases IPTV contents and services are provided by third party IPTV providers. Customer-oriented third party IPTV service provisioning to NGN-attached customers can be provided by SON. The most important SON function in such scenario is SON session control function that should provide session mobility and security, as well as support to IPTV service community creation (e.g., IPTV provision to a closed-user group), user-centric SCFs (e.g., users interaction), and service personalization (e.g., the composition of media objects on the screen, including IPTV contents, Web-based information, messaging, etc.).

SON provides possibility for creation of so-called community services. Each community group has an owner which decides about the required level of QoS and security. So, in such scenario SON provides session configuration function according to the QoS parameters and the security setup for each created community group, where a community group can be one or more users in a home network, Customer Premises Network (CPN), public access network such as mobile network or WiFi hotspot, and so on.

Virtual home networks are logically interconnected physical networks in several physically different home locations. The virtual home network service is accomplished by interactions with SONs (over different home networks) to share services.

Cloud computing is approach that is used for different types of computing (e.g., file storage, office applications, etc.) which typically involves large number of end hosts (e.g., computers) connected through a network (e.g., IP-based network). Usually, it refers to services which provide client or server software to run on physically remote location, and to appear to the end user as it is running locally on the given host (e.g., computer, mobile device, etc.). Next generation SON (i.e., SON over NGN) are aimed to provide full transparency of provided services to the end users, without any concerns where, how, and by whom the computation is performed, or where the hardware that completes the computation is located (e.g., in some aspect cloud computing approach is similar to Web-based services, where the user does not have and does not need information about the physical location of the Web server providing the service). For cloud computing provisioning using SON (over NGN), the SON control function provides SLA negotiation, security configuration, provisioning, integration, and adaptation of cloud computing services to the end user and its terminal. Additionally, next generation SON can support different context and application aware features that are user-centric (i.e., targeted to requirements set by individual NGN users).

Overall, the next generation SON defines standardized approach for provisioning of many existing as well as new emerging services that require an overlay network for service provisioning over the NGN.

7.9   Discussion

The main purpose of a telecommunication network is to provide services to the end users. Contrary to the past century where telecommunication networks were service-centric (optimized for delivery of a certain service, such as telephony or television), the NGN is created to be a service-independent network with well-standardized interfaces to different applications and services.

The open service environments in NGN provides also new business opportunities in the relationships among end users, network providers, and third party service providers, which stipulate continuous development of new innovative services targeted to different types of users, including residential users and enterprise users.

However, certain real-time services with strict QoS requirements (e.g., voice, TV) need standardized signaling architecture in the network, which is standardized as IMS in NGN. It is using well standardized signaling and control protocols from the IETF, such as SIP and Diameter, which are an intrinsic part of the IMS and the NGN in general.

Finally, security mechanisms and IdM are increasingly important topics due to open service environments, heterogeneous services, and high intelligence in the end user equipment. Also, management of identities of end user, their devices, as well as network equipments, is crucial to provide service continuity in NGN among heterogeneous access and transport networks. That is further enhanced by the provisioning of SON over the NGN to provide even more possibilities for service creation, deployment, and management.

References

1. ITU-T (2008) Open Service Environment Capabilities for NGN. ITU-T Recommendation Y.2234, September 2008.
2. ETSI Standard (2007) ES 203915-3. *Open Service Access (OSA); Application Programming Interface (API); Part 3: Framework (Parlay 5)*, ETSI, January 2007.
3. Open Mobile Alliance (2011) OMA Global Service Architecture Overview, Version 1.1, March 2011.
4. OASIS Standard (2006) Reference Model for Service Oriented Architecture 1.0, October 2006.
5. ITU-T (2007) IMS-Based Real-Time Conversational Multimedia Services Over NGN. ITU-T Recommendation Y.2211, October 2007.
6. Rosenberg, J., Schulzrinne, H., Camarillo, G. *et al.* (2002) SIP: Session Initiation Protocol. RFC 3261, June 2002.
7. ITU-T (2010) NGN Capability Requirements to Support the Multimedia Communication Centre Service. ITU-T Recommendation Y.2216, March 2010.
8. ITU-T (2012) Functional Requirements and Architecture for the Next Generation Network Multimedia Communication Centre Service. ITU-T Recommendation Y.2023, April 2012.
9. Fajardo, V., Arkko, J., Loughney, J., and Zorn, G. (2012) Diameter Base Protocol. RFC 6733, October 2012.
10. Aboba, B., Blunk, L., Vollbrecht, J. *et al.* (2004) Extensible Authentication Protocol (EAP). RFC 3748, June 2004.
11. Aboba, B., Zorn, G., and Mitton, D. (2001) RADIUS and IPv6. RFC 3162, August 2001.
12. Dierks, T. and Rescorla, E. (2008) The Transport Layer Security (TLS) Protocol. RFC 5246, August 2008.
13. Tuexen, M., Seggelmann, R., and Rescorla, E. (2011) Datagram Transport Layer Security (DTLS) for Stream Control Transmission Protocol (SCTP). RFC 6083, January 2011.
14. Berners-Lee, T., Fielding, R., and Masinter, L. (2005) Uniform Resource Identifier (URI): Generic Syntax. RFC 3986, January 2005.
15. Schulzrinne, H., Rao, A., and Lanphier, R. (1998) Real Time Streaming Protocol (RTSP). RFC 2326, April 1998.
16. Handley, M. and Jacobson, V. (1998) SDP: Session Description Protocol. RFC 2327, April 1998.
17. Schulzrinne, H., Casner, S., Frederick, R., and Jacobson, V. (2003) RTP: A Transport Protocol for Real-Time Applications. RFC 3550, July 2003.
18. ITU-T (2010) Security Mechanisms and Procedures for NGN. ITU-T Recommendation Y.2704, January 2010.

19. ITU-T (2009) The Application of AAA Service in NGN. ITU-T Recommendation Y.2703, January 2009.
20. Dierks, T. and Rescorla, E. (2008) The Transport Layer Security (TLS) Protocol, Version 1.2. RFC 5246, August 2008.
21. Kent, S. and Seo, K. (2005) Security Architecture for the Internet Protocol. RFC 4301, December 2005.
22. Baugher, M., McGrew, D., Naslund, M. *et al.* (2004) The Secure Real-Time Transport Protocol (SRTP). RFC 3711, March 2004.
23. ITU-T (2009) NGN Identity Management Framework. ITU-T Recommendation Y.2720, January 2009.
24. ITU-T (2006) Security Assertion Markup Language (SAML 2.0). ITU-T Recommendation X.1141, June 2006.
25. W3C Specification (2008) Extensible Markup Language (XML) 1.0 (Fifth Edition), November 2008.
26. Liberty Alliance (2008) Web Services Framework: A Technical Overview.
27. OpenID (2007) OpenID Authentication 2.0, December 2007.
28. ITU-T (2011) Mobility Management Requirements for NGN. ITU-T Recommendation Y.2801, November 2011.
29. ITU-T (2009) Fixed Mobile Convergence with a Common IMS Session Control Domain. ITU-T Recommendation Y.2808, June 2009.
30. ITU-T (2012) Functional Model of a Service Overlay Network Framework Which Uses the Next Generation Network. ITU-T Y.2200 Series – Supplement 17, February 2012.

8

NGN Services

8.1 QoS-Enabled VoIP

Since the invention of telephony in 1876, it has become the most important service in telecommunications until today. The main reason for that lays in the fact that telephony provides the same type of communication between people as they have been accommodated to use verbally in real life from the beginning of civilization. So, telephony as a conversational two-way service, which provides voice communication between people on distance like they are talking facing each other, will remain as one of the most important services in the future as well. However, contrary to PSTN (Public Switched Telephone Network) and PLMN (Public Land Mobile Network) where telephony was a service around which the networks and technologies were developed and standardized, in the Internet-based telecommunication world it becomes an application. However, it is not any application, but one of the most important and most demanding. But, the Internet provides possibility for different types of multimedia services, either real-time (e.g., with strict delay requirements) and non-real-time (e.g., with more flexible delay requirements). To provide convergence of all different types of services, including the telephony, over IP-based networks, and Internet technologies, the QoS (Quality of Service) support with well standardized signaling protocols and mechanisms for end-to-end QoS guarantees is needed, and that is provided by NGN (Next Generation Network). In fact, the first release of NGN standards was primarily focused on telephony in all-IP environments (i.e., Voice over IP – VoIP), which provides the same or better quality (in some QoS aspects, such as available bandwidth) of telephony than in the Plain Old Telephony Service (POTS). However, in other QoS parameters, such as delay and jitter, the IP-based telephony "tries" to provide similar metric values as the ones found in circuit-switched POTS, where no significant additional delays are introduced after the end-to-end voice path is established (only the speed of the signals is a limitation, which is currently limited by the speed of the light as maximum possible with existing telecommunication technologies).

By definition, VoIP is the provision of voice communication in both directions (between the end users) over all-IP networks by using Internet technologies end-to-end. There are also several other terms that are commonly associated with VoIP in the literature, such as Internet telephony, IP telephony, voice over broadband, and so on. However, the term VoIP in this book is covering all forms of voice communication over all-IP networks.

For QoS provision to voice services in IP-environments the fundamental part is signaling, which needs to be standardized globally since the voice services are global. NGN

NGN Architectures, Protocols and Services, First Edition. Toni Janevski.
© 2014 John Wiley & Sons, Ltd. Published 2014 by John Wiley & Sons, Ltd.

provides exactly that, end-to-end QoS support for voice by using standardized architectures [by ITU-T (International Telecommunication Union-Telecommunications] and protocols [by IETF (Internet Engineering Task Force)]. However, different types of voice services is expected to continues to co-exist in the future as well, including NGN VoIP as well as best-effort voice services [primarily offered by third party service providers in peer-to-peer (p2p) manner, without QoS guarantees] which are using only the transport stratum in NGN.

8.1.1   Differences between VoIP and PSTN

VoIP (i.e., Internet-based telephony) and traditional circuit-switching telephony differ in several aspects. In all VoIP services voice is digitalized and converted into IP packets (in end user terminal or in a home gateway) which are transported over the IP-based networks to the recipient. VoIP user terminals are intelligent devices (e.g., computers, mobile terminals, IP phones, etc.) with an operating system and Internet protocol stack embedded in it. On the other side, PSTN telephony devices are simple analog handsets that have capabilities for line signaling (e.g., Dual-Tone Multi Frequency – DTMF) and analog voice transmission and reception. However, such simple telephone devices can be used in QoS-enabled VoIP in NGN [1], by performing analog to digital to IP packets conversion (and vice versa) in home routers or gateways [e.g., an xDSL (Digital Subscriber Line technologies) router, a PON (Passive Optical Network) home gateway, etc.].

Several important differences between PSTN and VoIP are the following:

- *Bit rates*: PSTN provides constant 64 kbit/s for digital telephony in each direction, obtained by using ITU-T G.711 codec (i.e., Pulse Code Modulation – PCM, with 8 bits per voice sample, and 8000 samples per seconds since voice is coded up to 4 kHz and sampling frequency must be at least two time higher than the maximum frequency in the band). On the other side, the VoIP supports G.711 codecs as well as other standardized codes [e.g., G.723, G.729, Adaptive Multi Rate – AMR from 3GPP (3G Partnership Project), etc.], and additionally provides better statistical multiplexing. However, protocol headers added to voice data increase the required bitrate for the same codec in PSTN [e.g., more than 64 kbit/s are needed for G.711 in the case of VoIP due to IP, UDP (User Datagram Protocol), and RTP (Real-time Transport Protocol) headers which are added to each VoIP packet].
- *Packet delay (i.e., latency)*: PSTN has controlled and constant packet delay end-to-end which is below 150 ms in each direction, while packet delay is higher in IP-based networks due to statistical multiplexing and higher processing time of IP packets in network nodes (e.g., buffering, shaping, scheduling, etc.). However, by using highest priority classes for VoIP its latency can be kept below the maximum allowed delay of 400 ms end-to-end in one direction.
- *Jitter (i.e., packet-delay variation)*: PSTN has very low jitter, while VoIP packets accumulate jitter due to statistical multiplexing and packet scheduling, which introduces variable delay to VoIP packets. Since the voice communication is very sensitive to jitter, such variations in the packet delay are treated by using the same path across Internet for all VoIP packets as well as by using playout buffers in receiving terminals (or gateways at user premises).
- *Network equipment*: PSTN uses intelligent exchanges and overlay signaling network, while VoIP in general can use intelligent terminals (e.g., computers) and simpler network nodes (e.g., routers). However, overlay signaling network is required with aim to provide

QoS-enabled VoIP in similar manner as voice service in PSTN, but with different protocols [e.g., SIP (Session Initiation Protocol) is used for VoIP, while SS7 is used in PSTN].

- *Convergence*: This is one of the main advantages of VoIP over PSTN. By convergence of different service-centric networks (e.g., telephone network, TV network, etc.) to Internet-based telecommunications, the overall capital and operational costs are reduced.
- *Service flexibility*: This is also very important advantage of VoIP over PSTN. While PSTN is not flexible or adaptable regarding the telephony, the VoIP provides possibilities to service provider to upgrade the service or extend the service by including additional multimedia contents (video, data, text, etc.) with the VoIP, without changing the deployed network infrastructure.
- *Service reliability*: The same requirements are set to VoIP as for PSTN, that is, 99.999% availability of the service during the year. While in PSTN that is achieved, in IP-based networks the reliability still lacks behind. For example, circuit-switched phone lines are powered from the operator's exchange which has 24 h uninterrupted power supply, while in the case of VoIP the power supply is left on the end users' side.

Overall, VoIP in NGN aims to implements architectures that will provide close values of QoS parameters to those obtained in its predecessor, the POTS, particularly regarding end-to-end delay and latency to which telephony is very sensitive due to its real-time conversational nature.

8.1.2 VoIP Protocols and QoS Aspects

The voice service requires protocols for signaling (in the control plane) and protocols for data transfer (in the user plane). Protocols used for signaling of the QoS-enabled VoIP in NGN are SIP and Diameter, which work on application layer regarding the OSI (Open System for Interconnection) protocol layering model (both protocols are covered in Chapter 7). For the voice data transfer the QoS-enabled VoIP uses RTP/UDP/IP protocol stack. Above the RTP [2], which belongs to transport layer together with the UDP, is the VoIP application that provides voice codec functionalities.

RTP is designed to provide end-to-end transport functions for provision of real-time services in Internet, such as audio (e.g., VoIP), video (e.g., IPTV – Internet Protocol Television), and multimedia data, over unicast (e.g., for VoIP) or multicast networks (e.g., for IPTV or multimedia streaming services). However, RTP is typically present at end hosts that have an established session, and hence it does provide QoS guarantees and resource reservation for the real-time services (they have to be implemented by the network, such as NGN, that is used to transfer real-time data). RTP is used together with RTP Control Protocol (RTCP) protocol, where RTP is used to carry media streams (e.g., audio, video, multimedia, etc.) while RTCP is used to provide control feedback between two end points of an RTP session regarding the transmission statistics, QoS, and synchronization of multiple streams. So, RTP standard [2], defines a pair of protocols, namely RTP and RTCP. Both protocols are created to be independent from underlying transport and network layers. Although RTP can run over TCP (Transmission Control Protocol) and UDP, it is typically used over UDP because it is targeted to be used by real-time applications which are sensitive to delay. The RTP also provides support for translators (between different types of media streams) and mixers (for creation content from several media streams coming from different sources, such as video conference).

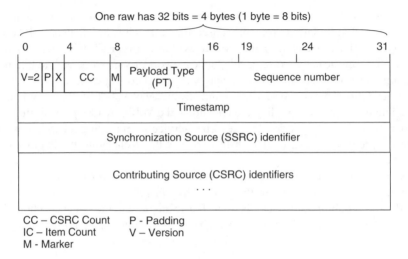

Figure 8.1   RTP packet format

The most important functionalities of the RTP are timestamping, sequencing, and mixing of data from different sources, which are implemented via relevant fields in RTP headers as shown in Figure 8.1. The RTP header is designed to be general enough to cover different real-time multimedia applications in the Internet.

The following fixed fields are defined in the RTP header:

- *Version ("V" bit)*: This identifies the current version of the RTP, which is currently 2 (version value of 1 was used for the first draft version of RTP).
- *Padding ("P" bit)*: If this bit is set it indicates whether there are padding bits at the end of the packet which are not part of the payload. The number of the padding octets (i.e., bytes) is written in the last padding octet of the packet.
- *Extension ("X" bit)*: When this bit is set it indicates a presence of exactly one extension header between the RTP fixed header and the payload.
- *Contributing sources count (four-bits field)*: This field denotes the number of contributors, where each contributor is identified with a contributor identifier. Maximum number of contributors is limited to 15, since the field has 4 bits ($2^4 = 16$, so the field can have integer values between 0 and 15).
- *Marker ("M" bit)*: This may be used for marking specific events such as packet which contains a frame boundary. However, more than one or none marker bits can be specified by changing the number of bits in the payload type field.
- *Payload Type (seven bits)*: This field indicates the format of the RTP payload and it is intended to be used by the application which generates or processes the payload of the packet.
- *Sequence number (16 bits)*: This is used to identify a sequence of RTP packets between sender and receivers. Sequence number starts with random initial value and increments by one for each RTP packet, hence it can be used to detect packet loss.
- *Timestamp (32 bits)*: This is used to provide information from the sender to the receiver to play back the received media samples at appropriate time intervals. The stamping rate must

be derived from a clock (which is dependent upon the application that uses RTP), which increments linearly and monotonically in time to allow synchronization as well as jitter calculations.

RTCP is used in parallel with RTP, and it is used for periodic transmission of control packets to all participants in a given session that uses RTP. RTCP is using the same mechanism as RTP, but on different ports. It performs four functions: (i) provides feedback on the quality of data transfer (from the receiver to the sender); (ii) carries a persistent transport-level identifier for a given RTP source called the canonical name (i.e., CNAME); (iii) it controls the rate at which packets are sent by the senders (which is individually calculated based on received RTCP packets); and (iv) optionally, RTCP can be used to carry participant's information that can be displayed in the user interface.

The format of the RTCP packet (which is based on UDP/IP) is shown in Figure 8.2. The Version number ("V") is currently 2 (the same version as RTP), while the Padding ("P") bit denotes whether padding octets are added at the end of the packet. The Item Count (IC) field is used to identify the number of items in RTCP packet (maximum number of items in a single packet is 31). The Packet Type (PT) field defines the type of the RTCP packet, and the Length field indicates the length of the packet after the standard header in number of 32-bits words. To perform the given control functions, RTCP uses five PTs which are defined by the PT field in the RTCP header, given as follows:

- *Sender Report (SR)*: This is sent periodically by all active senders to report transmission and reception statistics from active participants in an RTP session, to be reported from the sender to the recipients, including absolute timestamp, sender's packet count, cumulative number of packets lost, highest sequence number received, interarrival jitter, delay since last SR, as well as profile-specific extensions.
- *Receiver Report (RR)*: This is used for reception statistics from participants that are not active senders and in combination with SRs that are used by the active senders. The only

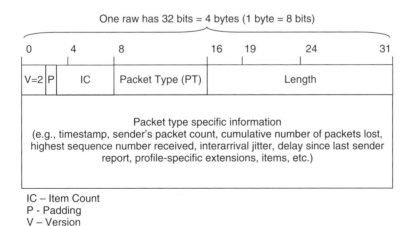

Figure 8.2   RTCP packet format

difference between RR and SR (i.e., both RTCP report types), besides the PT value, is so-called sender information section (that is used by all active senders).

- *Source Description items (SDES)*: This provides a description of the sender, including mandatory CNAME item, and optional items such as name (i.e., personal name) and e-mail address.
- *End of participation (BYE)*: This is used by a source to inform other participants that it is leaving the RTP session.
- *Application-specific functions (APP)*: This is used to design application-specific extensions (e.g., sending a document).

Generally, RTP can carry wide range of multimedia formats such as MPEG (Moving Pictures Experts Group) audio and video (e.g., MPEG-2, MPEG-4), H.261 encoded video, different voice codecs [e.g., G-711, G-723, GSM (Global System for Mobile communication) codec, etc.], and so on. However, RTP and RTCP can provide adaptation of the stream (e.g., sending rate) to the available bandwidth by using SR and RR, as well as mixing and synchronization of streams, but they cannot provide QoS guarantees since the full RTP functionalities are based on the end host (i.e., participants in the session). Therefore, RTP is used for the transport of real-time data over IP-based networks (e.g., VoIP, IPTV, etc.), while QoS-enabled session is provided by using control and signaling protocols (e.g., SIP, Diameter, etc.) prior to the data transfer.

8.1.3   QoS-Enabled VoIP in NGN

The QoS-enabled voice provision in NGN environments can be considered for mobile and fixed access networks. While in fixed access network the point of attachment does not change, in mobile access networks the terminal can change its point of attachment during an ongoing voice call, which requires session continuity support. QoS enabled VoIP in mobile environments is a mobile VoIP service that ensures QoS when a mobile terminal moves from one point of attachment to another one (e.g., to a neighboring cell in the mobile network).

When a mobile terminal changes the point of attachment in the mobile network (either in the same access network, which is referred to as horizontal handover, or in other mobile access network, which is referred to as vertical handover), the QoS parameters can be maintained or adapted (the latter case is typical for vertical handovers of VoIP calls). However, terminals can be single-mode or multi-mode. A single-mode terminal can perform only horizontal handovers, while multi-mode terminals can perform both horizontal and vertical handovers (e.g., between different mobile/wireless and fixed networks, dependent upon the availability of network interfaces).

Network QoS parameters include various information related to the quality support, such as available bandwidth (which is typically constant on average during the voice-only connections), maximum packet transfer delay, jitter (i.e., delay variation), packet loss rate, type of voice codec and voice packetization with output packets per second (ppt) rate, and so on.

Figure 8.3 shows a general architecture for QoS-enabled VoIP service in NGN heterogeneous mobile environment, consisted of different access networks such as LTE/LTE-Advanced, WiMAX, and WiFi. The QoS-enabled VoIP consists of two parts: (i) signaling for call establishment, management, and termination, and (ii) voice transfer path (for transfer of voice data in both directions). Signaling and user data traffic are using different protocols and

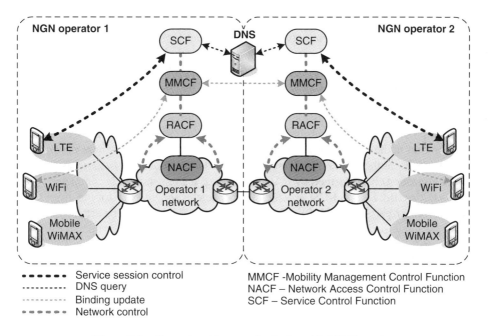

Figure 8.3 General architecture for QoS-enabled VoIP service

generally different paths across networks end-to-end. The signaling in NGN is out-of-band signaling, and NGN signaling networks are overlay networks with main entities located in the service stratum. It is standardized to be deployed by using common IP Multimedia Subsystem (IMS) functions and IETF protocols SIP and Diameter.

For QoS-enabled mobile VoIP in NGN are used Service Control Functions (SCFs), Mobility Management and Control Functions (MMCFs), Network Attachment Control Functions (NACFs), and Resource and Admission Control Functions (RACFs). SCF do terminal registration, service establishment, and service release, by using the Call Session Control Functions (CSCFs) of the IMS, which are: Proxy Call Session Control Function (P-CSCF), Serving Call Session Control Function (S-CSCF), and Interrogating Call Session Control Function (I-CSCF).

Typically, user and terminal have to be authenticated in each of the mobile and wireless networks during a given voice session. User terminals indicate possibility for handovers by measuring the signal strength in the wireless environments and signaling on Layer-2 to a given base station or access point. With the aim to have smooth transfer of voice session typical handover scenario is "make-before-break," which means establishing a new Layer-2/Layer-3 path (to the new location of the mobile terminal in the network) before handover, and then release of previous path for the given voice session.

Figure 8.3 shows three types of access networks: LTE/LTE-Advanced, Mobile WiMAX, and WiFi. When mobile terminal changes the type of access network (e.g., from WiFi to LTE/LTE-Advanced), then it applies changes in connecting capabilities such as available bandwidth, QoS, security, and so on. Regarding the QoS, that is, the quality for the given session after a handover, it may stay the same or can change (e.g., upgrade or downgrade). For example, different access network may have different bandwidth capacity (e.g., WiFi may have higher capacity than WiMAX), different network load, different QoS classes and policies for VoIP, and so on. Different QoS classes in different access networks require mapping of ongoing VoIP session from QoS class before a handover to corresponding QoS class after the handover [e.g., QoS class in LTE (Long Term Evolution) to QoS class in WiFi for a handover from LTE to WiFi].

In the NGN environment, registration of the mobile user and terminal includes network attachment procedures, including dynamic IP address allocation (either IPv4 or IPv6), and service stratum authentication and authorization procedures (based on IMS network registration). Network authentication and temporary IP address allocation (i.e., network location identifier allocation) is performed by NACF, while service registration is completed by using the IMS entities (P-CSCF to I-CSCF, which further selects the S-CSCF for the given session) and Home Subscriber Server (HSS). After registration, the service can be established.

Figure 8.4 shows service establishment from a user A connected to NGN operator A to a destination user B, connected to NGN operator B (which is different than operator A). The VoIP service is established by using signaling messages. Both NGN stratums are involved in the process. The originating user terminal (i.e., terminal A) sends service request for establishing

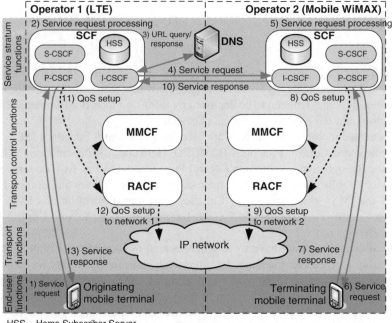

HSS – Home Subscriber Server
I-CSCF -Interrogating Call Session Control Function
MMCF -Mobility Management Control Function
NACF – Network Access Control Function
P-CSCF -Proxy Call Session Control Function
RACF – Resource and Admission Control Function
S-CSCF – Serving Call Session Control Function
SCF – Service Control Function

Figure 8.4   Establishing VoIP service between different NGN operators

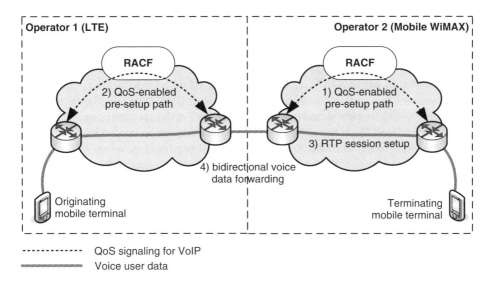

Figure 8.5   QoS-enabled VoIP data transfer in NGN environments

VoIP call (to the user B) to P-CSCF, which then forwards it to S-CSCF and finally reaches the I-CSCF of the NGN operator A (since the domain of the user B belongs to other NGN operator). I-CSCF performs DNS (Domain Name System) resolving (by using as an input the domain of the destination NGN operator) to obtain the IP address of the I-CSCF in the target NGN. The receiving I-CSCF (in NGN operator B) forwards it to S-CSCF entity in the destination NGN, with aim to obtain the P-CSCF to which the user is registered in the NGN network. Finally, service request from user A is forwarded to called user B (to terminating user terminal), which provides a response which is carried using the reverse path back to the originating user terminal. On that path, each P-CSCF (in both networks) communicates with the RACF for admission control (of the given service request) and QoS setup information in transport IP networks of each NGN operator. In the case of positive admission control decision and resource allocation by RACF (which communicates with MMCF for handling the mobility when it is necessary) the positive response is forwarded back to the VoIP call originator.

After the VoIP service establishment follows voice data transfer configuration which includes usage of RTP over UDP/IP for data transfer, as shown in Figure 8.5. Each RACF (of each of NGN operators) sets up the pre-established path (during service establishment) with aim to maintain the required QoS level at initial call establishment.

8.2   IPTV over NGN

According to the ITU-T [3], Internet Protocol Television is defined as multimedia services such as TV, video, audio, pictures/graphics, and data, delivered over IP-based networks that are designed to support the needed level of QoS and QoE (Quality of Experience), interactivity, reliability, and security.

In general, IPTV is a successor to digital TV broadcast networks (e.g., terrestrial TV broadcast, cable TV) in a convergent IP environment. Because digital TV requires higher bit rates for delivering the media (i.e., video synchronized with audio channels) the IPTV is becoming possible with deployment of broadband access networks (including fixed and mobile broadband). For example, with existing codecs, standard definition TV (SDTV) content requires 2–3 Mbit/s when it is delivered as IPTV to end users. Hence, unlike cable networks and terrestrial networks, in which all TV channels are delivered with broadcast (i.e., end users tune their TV receivers to a certain TV channel), in the case of IPTV typically is used multicast in the core and transport networks and unicast (or multicast) in the access networks. Unicast is a typical scenario for IPTV in the access part of the network due to limited bit rates (although the access is broadband) available to end users for all services, including VoIP, IPTV, and best-effort Internet services. With narrowband access to the Internet in the past [e.g., dial-up modems via PSTN or ISDN (Integrated Services Digital Network)], IPTV service was not possible. However, with broadband access to the Internet, IPTV has become possible and available, but the number of TV channels simultaneously delivered to the end user is limited by the available bandwidth that can be dedicated to IPTV service. That is accomplished by using admission control as a QoS mechanism for IPTV [e.g., admission of up to two SDTV channels over an ADSL (Asymmetric Digital Subscriber Line) bandwidth of 8 Mbit/s in downlink]. The downlink direction is more important due to asymmetrical character of IPTV. On the other side, the uplink direction is used for sending control information by the end user to the IPTV platform (e.g., TV channel change).

For the provision of an IPTV service (as shown in Figure 8.6) one may differentiate among four IPTV domains [4], which are: (i) content provider; (ii) service provider; (iii) network provider; and (iv) end user. However, a single provider can be involved in one or more IPTV domains.

8.2.1 IPTV Functional Architecture

IPTV functional architecture provides necessary network and service functions to deliver TV content via all-IP network to the end users. Generally, there are three different types of IPTV functional architectures:

- *Non-NGN IPTV functional architecture*: In this case IPTV services are provided by using existing protocols and network interfaces that are also used in the network for other IP-based

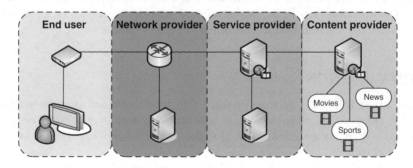

Figure 8.6 IPTV domains

services. This approach can be used as basis for further evolution of IPTV to NGN-based functional architectures.

- *NGN-based non-IMS IPTV functional architecture*: In this case IPTV services are provisioned by using NGN functional entities, but without deployment or usage of an IMS system in the network. The NGN-based IPTV uses RACF for resource reservation and admission control functions.
- *NGN IMS-based IPTV functional architecture*: In this architecture all components of the NGN including IMS functional entities are utilized for the purpose of IPTV service provisioning. This is also referred to as NGN-IMS IPTV, which uses core IMS functions as well as associated functions in the service stratum such as Service User Profile (SUP) functional entity.

IPTV functional architecture is shown in Figure 8.7, which is NGN IMS-based architecture. However, this is also a generalized architecture, because in NGN non-IMS IPTV architecture the core IMS functions are implemented by IPTV control functional block with similar functions as IMS. In non-NGN IPTV architecture the functions of NACF and RACF blocks (from NGN-based architectures) are provided via so-called authentication and IP allocation functional block and resource control functional block, respectively.

IPTV functional architecture is consisted of several groups of functions which are used to provide the service [4], given as follows:

- *End-user functions*: They are further grouped into two functional groups: (i) IPTV Terminal Functions (ITFs) and (ii) home network functions. ITFs collect control command from the end user, and interact with the application functions to obtain service information (e.g., Electronic Program Guide), licenses for the IPTV contents, as well as decryption keys. So, these functions interact with SCFs and Content Delivery Functions (CDFs) to provide IPTV service. On the end user's side they provide capabilities for service reception, decryption, decoding, and presentation. Home network functions provide connectivity between the home network (in which IPTV service terminates) and external network (e.g., NGN), and serve as a gateway between the ITFs and the network functions [including IP addresses allocation via DHCP (Dynamic Host Configuration Protocol), as well as configuration of the NAT (Network Address Translation) and IP forwarding/routing tables].
- *Application functions*: These functions provide possibility to select and where necessary to purchase content. IPTV application functions provide application authorization and execution by using user profile, metadata (e.g., for the content description), and so on. The application profile functional block provides end user settings, global settings (e.g., time zone, language, etc.), linear TV settings, list of subscribed linear TV packages, Video-on-Demand (VoD) settings, Personal Video Recorder (PVR) settings, as well as IPTV service action data (e.g., bookmark associated with pause action on a program, ordered VoD, list of contents recorded by the user, etc.). Application functions also provide content preparation as well as protection (e.g., authentication and authorization for service use, and optionally decryption of the content when it is encrypted for protection purposes).
- *Service control functions*: These functions handle initiation, modification, and termination of the IPTV service. For that purse IPTV control functions provide service access control as well as establishing and maintaining network and system resources for IPTV service provision. These functions use SUP functional block, which stores end user service profile

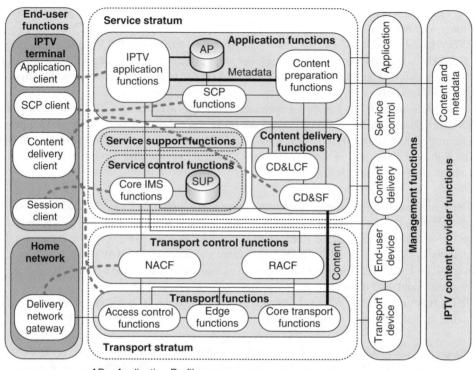

AP – Application Profiles
CD&LCF -Content Delivery and Location Control Functions
CD&SF -Content Delivery and Storage Functions
CDF – Content Delivery Functions
IMS -Internet Protocol Multimedia Subsystem
NACF – Network Attachment Control Function
RACF -Resource and Admission Control Function
SCF – Service Control Functions
SCP -Service and Content Protection
SUP – Service User Profiles

Figure 8.7 NGN IMS-based IPTV functional architecture

(e.g., subscribed IPTV services), subscriber data (e.g., ID of subscriber authorized for a given IPTV service), end user location data (e.g., network address), presence status (e.g., online, invisible, offline), as well as data management [e.g., response to queries for user profiles for AAA (Authentication, Authorization, and Accounting), location, presence, etc.].

- *Content delivery functions*: They provide cache and storage functionalities with aim to deliver the requested IPTV content. The CDFs consist of two groups of functions: Content Delivery and Location Control Functions (CD&LCFs) and Content Delivery and Storage Functions (CD&SFs). The CD&LCF handle interactions with IPTV service control functional block, gather information regarding the content distribution and storage (e.g., resource utilization, service status such as in-service or out-of-service, etc.), as well as perform selection of suitable content delivery based on gathered information about end user terminal capabilities. The CD&SF are responsible for content delivery to the client functions for content delivery by using network functions, such as unicast and/or multicast mechanisms.

- *Network functions*: These functions provide IP connectivity to provide network path for IPTV service provision, by including functions in access, core, and transport networks for allocation of network resources.
- *Management functions*: They handle monitoring and configuration of the whole system for IPTV service provision. Management functions are grouped into several blocks, including application management, content delivery management, service control management, end user device management, and transport management functions.
- *Content provider functions*: These functions are used for content preparation by using content and metadata sources for IPTV services, also including protection rights sources.

The given IPTV functional architecture can be used for different delivery mechanisms, including multicast delivery [5] and unicast delivery [6].

8.2.2 Multicast-Based IPTV Content Delivery

Multicast-based IPTV content delivery is using delivery of IPTV packets simultaneously to multiple destinations by using transmission from a single source. ITU-T defines four different functional models for multicast-based delivery of IPTV content [5], given as follows: (i) network multicast model; (ii) cluster model; (iii) p2p model; and (iv) hybrid model of cluster and p2p. An overview of the models is shown in Figure 8.8.

In network multicast model multicast capabilities are provided by the network provider domain (i.e., NGN provider) via so-called Multicast Transport Functions (MTFs). In this case each instance of the end user IPTV function registers to a specific multicast group by using multicast protocol messages for joining to the group. For example, standardized multicast protocol by IETF (for IPv4 multicast addressing scheme) is Internet Group Management Protocol (IGMP) [7], which establishes one-to-many multicast group membership among adjacent routers and hosts. For IPv6 multicast is used Multicast Listener Discovery (MLD) [8]. In this model network provider manages the multicast capabilities in its network, and the obtained performances of IPTV services are dependent upon the network conditions (e.g., traffic load).

Cluster model uses multicast capabilities provided by the service provider domain, and in particular, multicast is supported by CDF. The clusters for IPTV delivery (within the CDF) are placed in certain chosen locations. However, in this model CD&SF instances send IPTV content on a per user basis. So, the capacity of the CD&SF defines the maximum number of users that can be served by a given instance.

P2p model is based on multicast capabilities supported and controlled by the end user equipment and its functions. In this case an end user can receive IPTV content (from other users) and distribute the IPTV content (to other peer users) at the same time. Logically created multicast distribution paths between peers (in the p2p IPTV content delivery) are constructed by end user functions (e.g., within IPTV p2p application). However, end users distribute contents to other end users by using unicast transport since they cannot create multicast paths over the transport networks. Also, this model lacks efficient management mechanisms (e.g., for traffic management, presence, billing, etc.) since p2p delivery is not controlled by network providers nor by service providers.

Hybrid model of cluster and p2p uses the multicast capabilities from CDF in the service providers domain (provided in telecommunication equipment, and operated by a service provider) and at the same time assisting with p2p delivery in end user domain. This model

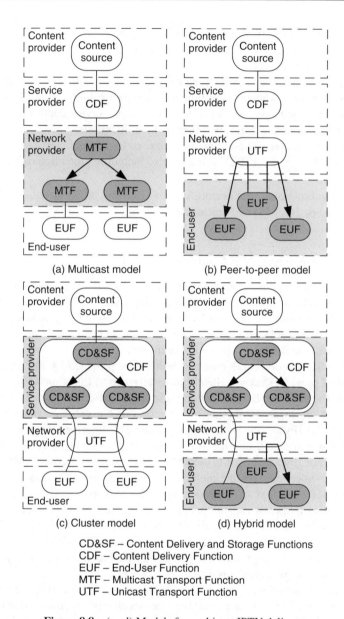

CD&SF – Content Delivery and Storage Functions
CDF – Content Delivery Function
EUF – End-User Function
MTF – Multicast Transport Function
UTF – Unicast Transport Function

Figure 8.8   (a–d) Models for multicast IPTV delivery

has the same characteristics as the cluster model and higher scalability due to IPTV content distribution in end users domain (i.e., with the p2p part of this hybrid model). On the other side, the hybrid model has less management issues than p2p model for IPTV content delivery.

8.2.3   Unicast-Based IPTV Content Delivery

Unicast-based delivery of IPTV is typical scenario in the cases where users are nomadic within the network, and when IPTV service provider is not located in the network provider domain of the end user. So, there are two scenarios for unicast-based IPTV content delivery [6]: non-roaming scenario and NGN roaming scenario.

In the non-roaming scenario it is assumed that ITF can use standard mechanisms (e.g., signaling and control protocols for IP networks) for communication with SCFs, including IMS-based (Figure 8.9) and non-IMS implementations. However, on long term IMS implementations are more likely to exist, while on short term non-IMS implementations for IPTV delivery prevail. In non-roaming scenario end user functions interact with the network functions located in the home NGN network, where NACF performs access authentication. In the case of successful authentication, the end user functions perform IPTV application and service discovery and selection by using application-based functions. After the service discovery and selection is completed, the end user functions setup an IPTV session via interaction with SCF in the home NGN. For QoS support, that is, admission control and resource reservation for the IPTV session, SCF exchanges information with the RACF of the home NGN. Finally, CDFs start sending IPTV content to end user functions (with unicast-based delivery), as shown

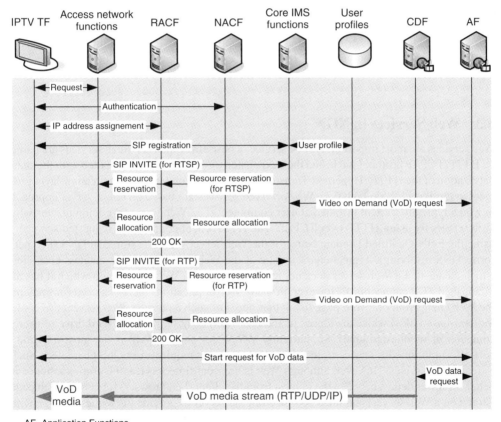

AF -Application Functions
CDF – Content Delivery Functions
IPTV TF – IPTV Terminal Functions
NACF – Network Attachment Control Function
RACF -Resource and Admission Control Function

Figure 8.9   IMS-based IPTV session setup

in Figure 8.9. The content is distributed by using video and audio codecs (e.g., MPEG-2, MPEG-4, etc.) over RTP/UDP/IP protocol stack on the end hosts (i.e., IPTV server for content distribution, and IPTV client in the end user equipment for IPTV receiving).

In the roaming NGN IPTV scenario both SCFs of both NGN operators should be capable to communicate with each other (e.g., both SCF to be IMS-based). In this scenario, the user request IPTV content while it is located in a visited NGN operator. In such case the end user functions interact with network functions of the visited NGN, and access authentication is performed by NACF in the home NGN through the visited NGN. After access authentication, the service discovery and selection is performed by using the home NGN application functions. Further, after completion of the IPTV service discovery and selection, the IPTV session is set up with SCF of the home NGN via SCF of the visited NGN. For the purposes of admission control and resource reservation SCF of each NGN (i.e., home and visited NGN) communicates with the RACF in the given network. Finally, CDF sends the content to the end user equipment (i.e., IPTV device, mobile terminal, etc.) by using network functions (i.e., network path) through home and visited networks for unicast-based delivery of IPTV content (by using codec/RTP/UDP/IP protocol stack at the end hosts for transfer of IPTV user data).

Overall, IPTV over NGN is a real-time service, which provides TV service in all-IP environments, with added QoS support with aim to have better quality than TV broadcast networks (e.g., satellite, terrestrial, or cable), and at the same time provides capabilities for different implementations and deployments of IPTV services including the possibility for innovations regarding combination of IPTV with other multimedia services in NGN.

8.3   Web Services in NGN

Web services are most important type of service in best-effort Internet. Such services are based on HTTP/TCP/IP protocol stack, and their expansion on a global scale has started with the standardization of the HTTP (Hypertext Transfer Protocol) as a fundamental application-layer protocol in the World Wide Web (i.e., WWW). Web services are based on client-server approach in which typically a client installed at user equipment (i.e., Web browser) communicates with Web servers by using HTTP over TCP/IP. The HTTP client contacts the requested server by using the Fully Qualified Domain Name of the Web server, which is stored in DNS (i.e., performs DNS resolving). Typical characteristic of the Web services is their stateless character, which means that a Web server does not record a state for a Web client (with which it has an established TCP connection). However, in many Web applications there is a need for tracking the Web client from one Web page to another one. For such a purpose Web servers use cookies (text-based files) which are stored in end user equipment (e.g., on the hard drive of user's computer or mobile terminal). So, currently WWW (or simply Web) is the most important best-effort multimedia service in Internet which does not require strict QoS guarantees, such as real-time services. However, although Web is non-real-time service, the user experience is better for lower delays between the request from the client (i.e., browser) for certain Web page (stored on a Web server) and the response from the server to provide the requested page with all embedded objects (e.g., text, audio objects, video objects, pictures, files, etc.).

According to the ITU-T definition, a Web service is a service provided by using Web services systems [i.e., Web clients and servers, including the DNS for resolving URLs (Uniform Resource Locators) of the Web servers] [9]. In NGN Web services enable different entities to communicate openly to each other (e.g., business applications, user applications).

Additionally, NGN provides possibility for a common open interface for service integration based on Web technologies.

There are several scenarios for Web services as extensions in NGN [9]: a loosely coupled connection between services [by using XML-based (Extensible Markup Language) messages], interoperability across open and proprietary platforms (e.g., in large enterprises with different standardized systems and platforms), application development [e.g., based on programming protocols such as SOAP (Simple Object Access Protocol) [10], which is typically used over HTTP], as well as open service delivery to users anywhere (e.g., access to different types of data provided by a given company).

Extension of the NGN with Web services is shown in Figure 8.10. For such an extension an appropriate NGN interface for Web service providers is the Application–Network Interface (ANI). The Web service providers register Web services in Web registry by using Web Services Description Language (WSDL) [11], which provides a machine-readable description of the services (including how the service can be called, service parameters, as well as data structures that the service returns). The Web service provider finds the interfaces of an NGN service by using the Web services registry. Further, it uses ANI with a standardized Web service access method, such as SOAP. With aim to provide such interaction between the Web service providers and NGNs, an interface of NGN (i.e., ANI) should be registered in Web services registry by using WSDL for description of service location, parameters (including their types), API (application programming interface) name, and so on. So, when Web services extension to NGN architecture is implemented, it provides possibilities for combining of best-effort Web services (e.g., Web-based photos, files, texts, maps, searches for Web pages, etc.) with NGN services to provide value-added service in NGN environments.

Regarding the implementation of Web services within the NGN, they can be associated with different service components. Initially NGN defines three service components: IPTV service component, PSTN/ISDN emulation service component, and IP multimedia service component. Additionally, a Web service component is newly added in the service stratum of NGN [12], and it enables NGN services to support legacy (i.e., best-effort) Web services, as well as convergent NGN services as a combination of other service components with the Web service component by using standardized protocols [e.g., HTTP, HTML (Hypertext Markup Language), XML, SOAP, WSDL, etc.].

Regarding the type of control, the Web services in NGN can be classified into two main groups:

- *Legacy Web services*: These are best effort services such as Web browsing, Web-based e-mail, Web portals, and so on.
- *Converged Web services*: These are services that are built up by interworking of Web services and NGN. They can be classified into: Web-based IMS services that are interworking with IP multimedia service component, that is, Multimedia over IP (MMoIP) using Web; Web-based streaming services (e.g., IPTV using Web services); and Web-based composition services (e.g., different services in a Web framework, such as IPTV or MMoIP using Web).

NGN functional architecture with Web service component is shown in Figure 8.11. It consists of two main parts, the Web Service Control Function (WSCF) and the Web Media Function (WMF), which are used for interworking of Web services component in NGN with service stratum and transport stratum, respectively. Both components (i.e., WSCF and WMF) are related

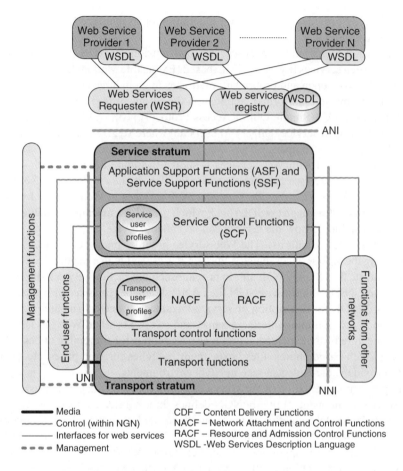

Figure 8.10 NGN architecture extension with Web services

to each other and contain shared functions. With the Web service component NGN can offer various Web services. Such services may include only the Web service component, or can be provided by a combination of the Web service component and other NGN service components, such as IPTV service component and/or IP multimedia service component.

The WSCFs include several functional entities in the NGN service stratum, as it is shown in Figure 8.11:

- *Web Media Resources Control Functional Entity (*WMRC-FE*)*: This is used for control of the resources to the end user of the Web services by interacting with the transport control functions.
- *Policy and Charging Control Functional Entity (*PCC-FE*)*: This interacts with corresponding entities in the transport stratum to obtain information regarding the resource usage of a given session flow (for the given Web service to the given end user).
- *Web Signaling Gateway Control Functional Entity (*WSGC-FE*)*: This acts as a session manager and signaling gateway.

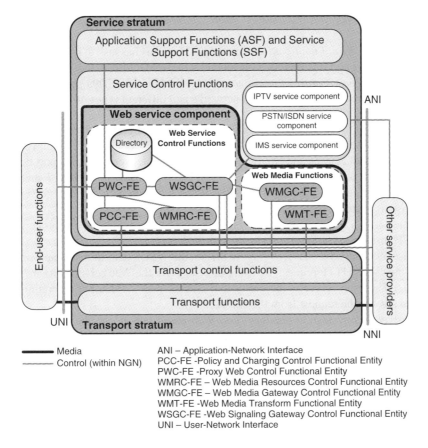

Figure 8.11 Functional architecture with Web service component in NGN

- *Proxy Web Control Functional Entity (PWC-FE)*: This acts as a proxy between the Web client and Web server [the Web server is located in the ASF/SSF (Application Support Function/Service Support Function), connected to NGN via the ANI as an interface] through interaction with the directory for the service (i.e., a database which includes profile information).

The WSGC-FE of the WSCF interacts with WMFs as well as other service components within the given NGN (i.e., IPTV, PSTN/ISDN emulation, IMS), or with other service providers. The service directory in the WSCF has a list of peer Web servers for the requested session. Summarized, the WSCF provides signaling mapping and translation (e.g., protocol translation between HTTP and SIP), session management, traffic measurement and monitoring, billing and charging, as well as use of Web interfaces (e.g., for Web service search and discovery through queries via the directory). Additionally, WSCF provides interactions with NGN SCFs (via the CSCF and profile management functions), WMF, Web components (i.e., interface with the Web server and Web service description, as well as capabilities for personalization of the Web services and social networking), and ASF/SSF (for service creation and composition).

The WMFs consists of several entities, given as follows:

- *Web Media Transform Functional Entity (WMT-FE)*: This has task to provide authorization of the media flow and it is directly controlled by the WSGC-FE in the WSCF.
- *Web Media Gateway Control Functional Entity (WMGC-FE)*: This provides functions of media adaptation to different transport networks and different types of user equipments.

Main functions of the WMF include conversion between different codecs (i.e., transcoding), packet filtering (based on IP addresses and port numbers), QoS control (i.e., resource allocation and policing the packet sizes and bandwidth limits), media delivery and transport between the NGN and the Web server. For that purpose, the WMF interacts with the RACF and NACF in NGN transport stratum. On the other side, the WMF interacts with the Web server by using media transport protocol (e.g., HTTP).

The framework for the Web services in NGN is general. The end users always communicate with WSCFs using the HTTP protocol.

In the case of legacy Web services (Figure 8.12a), the WSCF determine the appropriate Web server for the user's HTTP request, by using directory (for the Web servers list) and obtaining user profile. The Web server is the entity that creates Web service by generating response message to answer the request from the end user.

For Web-based conversation services (i.e., Web-based telephony) the end user also uses HTTP protocol to communicate with IP multimedia component via the WSCF (Figure 8.12b). Further, the IP multimedia service component creates the conversation service by using the relevant API to connect to an IMS call server, and afterwards sends response to the end user regarding the initiated HTTP request for a conversation service. If the requested codec from the user (e.g., voice codec) is not supported by the IMS call server, then WMFs provides transcoding for the given media flow.

Finally, the end user can request Web streaming services (e.g., IPTV) by using HTTP request/response approach to communicate with the IPTV service component in NGN (similar to other two cases, i.e., legacy and conversational Web services), as shown in Figure 8.12c. After the request, the IPTV streaming component is the one that creates and delivers video streaming to the end users. In the case when video format requested by the end user cannot be supported, the WMF provides transcoding functionality for the media data.

Overall, Web services in NGN are being standardized to provide all types of services, including legacy Web service (as a successor of the best-effort Web service in Internet), as well as conversational (e.g., VoIP) and streaming (e.g., IPTV) Web services in NGN. With the broadband access to Internet the packet latency is reduced, and there is a ground for Web-based real-time services in NGN.

8.4 Ubiquitous Sensor Network Services

One of the emerging services in NGN environments are Ubiquitous Sensor Network (USN) services. In the twenty-first century there are already many sensors and sensor networks deployed for various purposes (e.g., industry, public services, etc.). However, in NGN the USN is defined as a conceptual network built over the existing physical network which makes structured use of the sensed data to provide knowledge services to anyone, anywhere, and at anytime [13]. The term "ubiquitous" is derived from the Latin word "ubique" which means "everywhere."

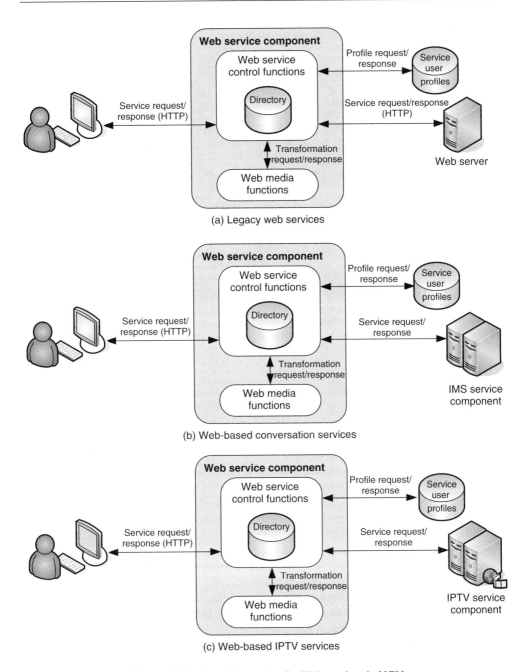

(a) Legacy web services

(b) Web-based conversation services

(c) Web-based IPTV services

Figure 8.12   (a–c) Interaction for Web services in NGN

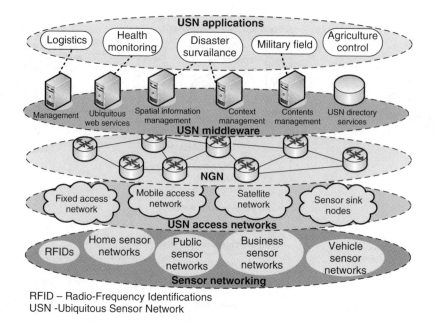

RFID – Radio-Frequency Identifications
USN -Ubiquitous Sensor Network

Figure 8.13   USN ecosystem with NGN infrastructure

The USN in NGN is consisted of wired or wireless sensor networks connected to the NGN functional entities. Wireless Sensor Networks (WSNs) are built of autonomous devices (i.e., sensors) that are connected to each other with aim cooperatively to monitor certain conditions and parameters in different locations and for different purposes (e.g., to monitor temperature, pressure, vibration, sound level, pollution, motion of objects, etc.). Before USN specification for NGN, such sensor networks (e.g., WSN) were implemented as isolated networks in a given environment (e.g., in a factory, shopping mall, forest, office, home, etc.). Generally, all such sensor networks are built by using simple devices with low battery consumption (in the case of WSN) connected to a centralized application system (with a database) for data collection, storage, and retrieval. During the past years many such isolated applications with sensor networks were developed and implemented on the ground. However, the technical developments (e.g., advanced hardware and software, integration of services over converged networks, context-aware approach, etc.) have provided possibilities for building a USN as the information infrastructure of sensor networks connected to the existing access networks. The USN opens possibility to provide usage of sensor networks and their data to different types of users globally, including individual users, enterprises, as well as governments and public organizations.

The USN has potential applications in civilian and military fields. For example, some civilian USN applications are environment monitoring, healthcare, intelligent transport systems, intelligent home, and so on. An overview of the USN ecosystem with NGN infrastructure is shown in Figure 8.13.

According to Figure 8.13, main components of the USN are the following:

- *Physical sensor network*: This consists of sensors with autonomous power source (e.g., battery, solar power), which are used for collecting information about the environment and transmitting it to a repository (i.e., a database).

- *USN access network*: This is a network of sink nodes that collects information from sensors in a given area on one side, and communicates with the external entities for provision of the USN.
- *NGN infrastructure*: This includes network architecture and NGN functional entities that provide necessary functions for USN application and services.
- *USN middleware*: This is software that collects and processes large volumes of data collected from many USN access networks (connected to an NGN).
- *USN applications and services*: This is a technology platform that has a target to provide effective use of the USN in a particular segment (e.g., industrial sector, end user application for weather forecasting, etc.).

In sensor networks the nodes may vary in sizes and capabilities. They can be connected with wired or wireless links. In the case of wireless transmission they usually use unlicensed spectrum. Hence, their coverage area is limited and therefore sensors transmit measured data to their sink node via other neighboring sensors. The USN as an information infrastructure uses heterogeneous sensor networks to deliver information (collected from the sensor networks) to anyone, anytime, and anywhere, but also it provides the ability to deliver information to "anything." To provide such approach USN utilizes the context-awareness approach, in which different parameters for location, time, user's preferences, as well as types of the sensed information (by sensor tags and/or devices), are used by the USN application. For example, the context-awareness may refer to the temperature of the sensed object (e.g., hot or cold), to time in the day, whether an object or a person is moving or being stationary, and so on.

The physical sensor network puts requirements on antenna technology and battery life, interfaces, sensor operating systems, as well as energy efficient network architecture. However, for USN it is important to have interworking of the sensor networks with NGNs as backbone infrastructure. Sensor networks can be IP-based or non-IP based. An example of IP-based sensor network standardization by IETF is 6LoWPAN (IPv6 based Low-power Wireless Personal Area Network) [14]. On the other side, some sensor networks are implemented with a non-IP platform, such as near-field communications (e.g., for mobile phone payments), Zigbee implementation of the standard IEEE 802.15.4 for Wireless Local Area Networks (WLAN), Bluetooth, and so on. For conversion of energy between different types of nodes in the USN, the IEEE has also standardized low cost transducers as IEEE 1451 standard.

8.4.1   USN Functional Architecture

The USN architecture is based on objects around us that are connected through the NGN via wired or wireless interfaces, either in fixed network environments (e.g., home, office) or mobile network environments (e.g., mobile networks). Physical objects (e.g., devices, terminals, etc.) are connected to logical objects through the NGN, by using relevant type of their identification and further their virtual presentation. In such case, there can be a gateway placed between the objects and the NGN (e.g., gateway in a home network, in a building, etc.).

From the architectural point of view for USN services, there are needed certain capabilities in NGN including the following: end-to-end connectivity among different interconnected networks (e.g., sensor networks and NGN), multi-networking capabilities, Web-based open service environments, context-awareness, as well as connecting-to-anything capability [15]. Connecting-to-anything capability refers mainly to object-to-object communication, which focuses on ubiquitous networking functions at the end user equipments. The architectural

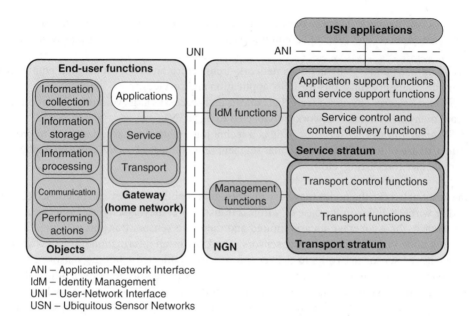

ANI – Application-Network Interface
IdM – Identity Management
UNI – User-Network Interface
USN – Ubiquitous Sensor Networks

Figure 8.14   Architectural model for NGN object-to-object communication

model for object-to-object communication is shown in Figure 8.14, used to interconnect various types of objects.

In the given architectural model the end user functions provide capabilities for connecting objects on the end user side, with or without a gateway node between the objects and the NGN. The objects may provide functionalities individually or in collaboration with other objects (e.g., information collection, processing and storage, and performing actions such as sensing given parameters). On the other side objects need identifiers for their discovery and management.

NGN transport stratum provides additionally a bridge between heterogeneous sensor networks. Also, transport stratum provides address mapping for the objects as well as transport resources for exchange of messages among different types of objects and the corresponding application. When non-IP based sensor networks are connected to the NGN, then IP-capable gateway is used and NACF (in transport stratum) provides connectivity between the IP-based NGN and non-IP based sensor networks through the gateway. The transport stratum also provides location management (at IP layer) with the NACF. Also, NACF is used for authentication of the USN end users.

NGN service stratum provides functions to support highly scalable, available, and secure information system, which aggregates the information from different objects, and further filters it according to the accuracy and density that is required by the given USN application. Also, the service stratum provides monitoring and management of the objects (via the transport stratum) since they change during their lifespan due to mobility, change of ownership, as well as change of business and regulation environments for such services.

On top of the USN ecosystem are positioned ubiquitous networking applications. They use the data that has been collected from the objects and stored in a repository (i.e., a database) to support various applications that provision the stored data (from the object) in a given context required by the end USN user.

The objects in the USN include different types of physical devices, such as: Radio-Frequency Identifications (RFIDs), sensors, information devices, personal devices (e.g., for measuring the pulse or body temperature), contents, appliances (e.g., in the home), transportation objects (e.g., vehicles, trains, airplanes, etc.). The objects that are associated with individuals or groups (i.e., with humans) are connected to the NGN via the User-Network Interface (UNI). For provisioning of the USN in NGN different types of gateway nodes can be used between the objects on one side and the NGN on the other side [16]. In general, the objects can have different capabilities and belong to different types, such as: physical or logical object (i.e., resource or content), fixed or mobile object, with constant power supply or limited power supply (e.g., a battery), managed by human or other device (i.e., machine), IP or non-IP (regarding the networking capabilities), different access interfaces (to the objects themselves), types of protocols (e.g., for power efficiency of the object), as well as regarding the amount of data (e.g., number of kilobytes) exchanged via transaction messages with sink nodes or NGN entities.

8.4.2   USN Applications

The potential of sensor networks and associated applications is increasing. They are also main contributor to the Internet of Things (IoT) paradigm. There are different applications based on the usage of data supplied by the sensor networks. In general, all such applications can be classified into three main groups:

- *Monitoring*: This covers a wide range of monitoring goals, such as the monitoring of human body parameters (e.g., blood pressure), presence (e.g., in the classroom, in the office), monitoring the environment (e.g., volcanoes, hurricanes, earthquakes, etc.), monitoring the structural parameters of a building, monitoring the behavior of animals, and so on.
- *Detection*: This includes detection of any change for a certain parameter and certain objects, such as a change of temperature in a room, field or human body, intruder detection, mine detection in war zones, fire detection in a building, and so on.
- *Tracking*: This is used for tracking humans and different objects. Many examples exist, such as vehicle tracking in intelligent transportation systems, tracking animals in the wild or cattle in the food chain, and so on.

There are endless varieties of applications and services that can be provided with USN in NGN, which can be personal (e.g., health monitoring, smart home) or public services (e.g., bushfire monitoring, earthquake monitoring and alarming, smart cities, forest monitoring, etc.). With availability of low cost end devices (e.g., sensors, tags) and NGN information infrastructure, the USN services emerge to different types of end users (e.g., residential users, enterprises, governments, military, police, hospitals, different public services, etc.).

8.5   VPN Services in NGN

Virtual Private Networks (VPN) solutions are crucial in carrier grade all-IP transport networks. In fact, the solution that is used practically for IP-based transport networks since the beginning of the twenty-first century is MPLS/BGP IP VPN, standardized by the IETF [17]. Multi Protocol Label Switching (MPLS) is used as a tool for QoS provision in IP-based transport networks (as discussed in Chapter 6), while Border Gateway Protocol (BGP) [18], is a routing protocol among Autonomous Systems (AS), that is, inter-AS routing protocol. Each carrier has its own Autonomous System, which communicates with other Autonomous Systems by using BGP (its current version which is deployed globally is BGP-4). The BGP is in fact a speaking system, which exchanges reachability information with other BGP routers. Such reachability information includes list of Autonomous Systems that BGP packets traverse on their path from one to another BGP system. BGP-4 supports Classless Inter-Domain Routing (CIDR) which is based on use of IP prefixes (i.e., the network part of the IP address) for announcing set of destination (that have that IP prefix in their IP addresses). In best-effort Internet architecture BGP-4 is essential for connecting the Autonomous Systems to each other and existence of the flat Internet architecture on a global scale. In MPLS/BGP scenario, the BGP is used by the service providers for exchange of the routes for a particular VPN among the provider's edge routers. In such scenarios each route within a given VPN is assigned an MPLS label, and BGP distributes such label with the route.

The NGN provides possibility for VPN services to all types of end users (not only to enterprises). There are several possible types of VPN services in NGN [19]:

- *Site-to-site VPN*: This is a solution that links branch offices of a given company, or individual users in communities of mutual interest (e.g., gaming, collaboration, information exchange, etc.).
- *Access VPN*: This is used by mobile workers (e.g., work from home) to access their corporate network from other networks.
- *Multiservice VPN*: This is a type of multimedia service with QoS provision on the path among end users.

VPN framework architecture in NGN is shown in Figure 8.15, which can be used for provision of VPN services in NGN environments, including fixed and mobile ones.

VPN functions in NGN service stratum include application support and service functions to provide the requested capabilities by the end users. They work together with the VPN service control, which provides registration, authentication, and authorization of VPN members, security solutions, QoS provision, as well as management functions such as session and mobility management, multicast service control, as well as membership management (i.e., creation, management, and release of a VPN, joining and leaving VPNs, and partitioning of a VPN in multiple groups).

On the other side, VPN functions in NGN transport stratum are targeted toward VPN admission and resource control (e.g., admission of a VPN, resource reservation for the VPN, mapping of VPN logical network to transport network, policy management, etc.), VPN network attachment control based on VPN user profiles (e.g., QoS profile information), as well as VPN connection management and traffic forwarding within the NGN.

Over the top are VPN end user functions which provide several important functionalities for the VPN services, including VPN traffic engineering mechanisms (e.g., service priority, traffic shaping, rate limiting, admission control, and resource reservation in the transport stratum of the network), VPN join and leave functions (including multicast), as well as security mechanisms (e.g., end-to-end VPN authentication, user data encryption and decryption, etc.).

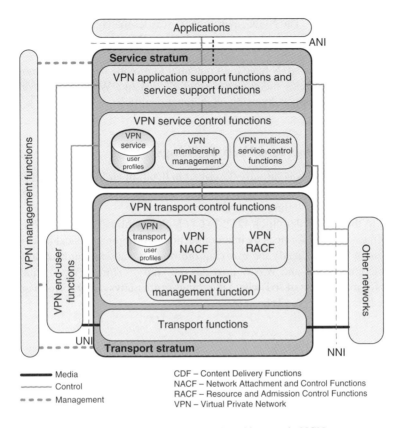

Figure 8.15 VPN framework architecture in NGN

There are different scenarios for VPN services in NGN, which include the following ones [19]:

- *Virtual corporate office*: This is a VPN service that is practical to be used in scenarios where all corporate resources are centralized in so-called hub, while mobile workers from different branch offices connect to the central office by using VPN tunnels through the NGN. However, this VPN service is also suitable for multiservice VPNs which are used to carry real-time voice and video traffic as well as data traffic with appropriate QoS and security support end-to-end.
- *Personal end-to-end VPN*: In this scenario VPN tunnels are established between the end users. This approach is available in best-effort Internet (different proprietary applications exist), but it is not standardized and has no possibility for QoS support. Personal end-to-end service in NGN may be optionally accompanied with certain QoS level for the traffic in the VPN tunnels among the end users.
- *Community-based VPN*: This scenario consists of so-called community groups where end users are connected via community VPNs, similar to the Web-based communities (e.g., social network groups). End user may join or leave certain community group. In each community group users have access to the same set of applications and services which define the

given community group. The NGN environment provides management of the QoS, security mechanisms and applications that are provided within the group.

Overall, VPN services in NGN have high potential to provide networking among individuals based on different interests and contexts, similar to existing Web-based services, but with added QoS and security support by using NGN functionalities.

8.6 Internet of Things and Web of Things

The vision behind the IoT was initially described in detail in an ITU report [20], which covered the potential technologies, market potentials, challenges, and implications, as well as the benefits to all countries, including the developing ones.

According to the ITU-T definition [21], IoT is a global infrastructure for the information society enabling advanced services by interconnection of different physical and virtual things, based on existing (standardized) and evolving information and communication technologies (ICT). The merging of IoT and WWW provides the Web-of-Things (WoT) which is defined as a concept for making use of IoT where the things (either physical or virtual) are connected and controlled via the WWW [21].

8.6.1 Internet of Things

The IoT paradigm is expected to have a long term influence on the technologies as well as society. In that manner, the IoT can be considered as standardization of information architecture (based on the NGN concept) for enabling information society in practice. The IoT adds another dimension, referred to as "any-thing communication" to the ICT, besides the other two dimensions, "any-time communication" and "any-where communication," as shown in Figure 8.16.

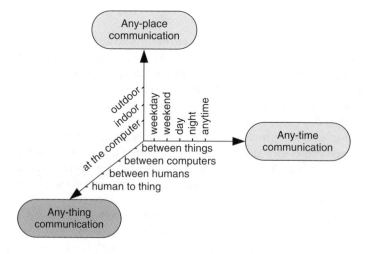

Figure 8.16   Internet of Things (IoT) dimensions

The "things" can have associated information, which can be either static or dynamic. Also, the "things" can be physical or virtual objects. The virtual things (e.g., multimedia content, application software, etc.) are present in the information world, and such things (i.e., objects) can be stored, processed, and accessed. In the IoT system physical objects can be represented via certain virtual objects, but also virtual objects can exists in the NGN without any association with physical objects. Physical objects are all types of devices that can provide certain information to other devices or to humans (via applications or services). The devices (i.e., physical objects) can communicate through a gateway toward the network (e.g., the NGN), or without a gateway (which is possible if they have IP connectivity capabilities), or directly to communicate with other devices via local network (e.g., Ethernet, sensor network, ad-hoc network, etc.). In IoT framework devices are categorized into several categories, including data-carrying devices (used to connect physical things with the network), data-capturing devices (it refers to a reader/writer device), sensing and actuating devices (used to detect or measure certain type of information and convert it into digital signals, while in the opposite direction to convert digital signals from the network into certain operations), as well as general device (e.g., industrial machines, smart phones, electric appliances in the home, and all such devices that have capabilities to communicate with wired or wireless access networks).

The IoT applications are very similar to USN applications, or one may note that USN services are a subset of IoT services and applications. Emerging examples of IoT applications are e-health (i.e., electronic health systems for online health monitoring and diagnostics), smart grid, smart home, and intelligent transportation system. In the beginning such applications are typically based on proprietary application platforms. However, NGN provides common functionalities for AAA, devices and resource management, and common application and service support systems.

The IoT are characterized with huge scale of applications and devices that are connected to the global information infrastructure (e.g., via the NGN). With "things" becoming Internet hosts, the number of Internet hosts (as unique devices connected to the Internet) will go much beyond the number of individual human users and their devices.

The reference model for IoT includes four layers, given as follows:

- *Application layer*: This contains the IoT applications (e.g., e-health, smart home, smart city, etc.).
- *Service support and application support layer*: This may include general support capabilities (e.g., data processing and storage in databases) as well as specific support capabilities.
- *Network layer*: This provides access and transport resource control functions, AAA, mobility management (in mobile environments), as well as connectivity for the transport of the IoT services.
- *Device layer*: This layer consists of device capabilities and gateway capabilities.
    - The device capabilities are used for direct or indirect (e.g., via a gateway) interaction with the network, as well as ad-hoc networking.
    - The gateway capabilities include multiple interfaces support, which can be provided on device layer (e.g., WiFi, Bluetooth, Zigbee, etc.), or on network layer (e.g., PSTN, 2G/3G/4G mobile networks, Ethernet, xDSL, optical access networks, etc.). The gateway may provide protocol conversion functionality, which is needed in cases when device uses access on device layer (i.e., OSI layers 1 and 2), while the access network (fixed or mobile) has different interfaces and protocols (e.g., connection of Zigbee through 4G mobile network).

The IoT builds a huge ecosystem around itself, which is composed of different business players. There are device providers, network providers, software platform providers, application providers, and finally there are IoT application users. The central role in the IoT ecosystem belongs to the network providers (e.g., NGN providers), which connect all other providers, that is, device providers on one side with service and application providers on the other side. Also, end users are connected to network providers to access IoT applications. Single player in the IoT ecosystem may play one or more provider roles (e.g., single telecom operator can be owner of devices, network, service platforms, and applications). Hence, there are possible different business models for the IoT, depending on the ownership of different types of providers in the IoT ecosystem.

In general, IoT is an ongoing evolution of the ICT that is targeted to connect all everyday objects to Internet that can provide any form of benefit to the individuals or groups and the society.

8.6.2   Web of Things

The IoT is targeted to solutions for interconnecting things by using interoperable ICT. But, creation of application that run on the top of heterogeneous devices is a problem due to many heterogeneous networks and technologies behind them, lack of interoperability across many proprietary platforms (e.g., hardware platforms, operating system, databases, middleware, and applications), as well as different types of data formats. The practical solution is to use for IoT a technology that is applicable and already world wide deployed in different types of devices.

The WWW is used as a global platform to deliver services to the end users, which is the "core" of the best-effort Internet. It is based on HTTP/TCP/IP protocol stack which uses request/response HTTP application protocol for communication between the end hosts (HTTP clients and servers) over reliable TCP on the transport layer and IP on the network layer. The discovery of the web sites is performed by using URLs, and DNS for resolving the IP address of a given URL. On one side WWW uses standardized protocol stack for communication end-to-end (which is independent of the type of access and transport networks, but IP connectivity is required), while on the other side Web has program language independent properties and enables various business entities (e.g., enterprises) and applications to intercommunicate openly with each other and with end users. The Web technology provides possibility for users to interact with devices (i.e., the things) in the IoT by using Web interfaces. Such approach is referred to as Web-of-Things [22].

A conceptual model for WoT is shown in Figure 8.17. According to the model end user applications (which will be Web clients, such as Web browsers) will access the physical devices directly or through a WoT broker which has agents which provide an adaptation between interfaces of physical objects and Web interfaces. Each agent within the WoT broker is dedicated to a specific interface (e.g., WiFi, Bluetooth, etc.).

Considering device capabilities regarding the Web interfaces, in WoT one may distinguish among two general types of devices:

- *Constrained device*: This is a device that cannot connect directly to Internet by using Web technology, so it requires a WoT broker.
- *Fully-fledged device*: This is a device that has all necessary Web functionalities to connect with services on the Web without a need for the WoT broker, but it can also interact with it.

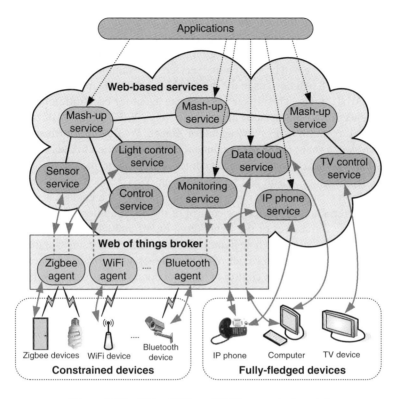

Figure 8.17   Web of Things (WoT) conceptual model

The WoT broker has functionality for communication between the WoT end user (e.g., Web client) on one side, and devices on the other side (either constrained or fully-fledged ones). So, WoT broker is exposing devices on the Web and thus provides its seamless integration onto the WWW. Each WoT broker contains agents, which communicate with the physical devices and provide control functions for them. So, when certain Web application sends a request to access certain physical devices, the WoT broker perform adaptation of such request to the interface toward the device (i.e., the thing).

Generally three types of services can be distinguished within the WoT:

- *Web service*: This is a service that can be directly accessed on the Web.
- *WoT service*: This is a service provided via an adaptor that provides 1 : 1 mapping with the services of the physical device.
- *Mash-up service*: This is a service that integrates WoT with other services (e.g., IPTV, VoIP, cloud services, etc.) via the WoT broker.

In WoT model all interfaces between each pair of WoT broker, mash-up service, and Web application, are based on HTTP protocol. Fully-pledged devices also use the HTTP protocol for communication with all other entities in the WoT model (i.e., WoT broker, mash-up service, and Web application). However, there is also possibility for use of proprietary protocol between the fully-pledged device and the WoT broker. On the other side, constraint device communicate only through the WoT broker, and such communication is based on device-dependent proprietary protocol. In this case all other communication between the WoT broker and mash-up service or Web application is also performed with HTTP.

One typical WoT service will be smart home service. An example of smart home WoT service is shown in Figure 8.18. All devices in the home which need to be monitored or controlled are connected to a WoT broker. The broker has separate agent for each different technology that is used for connection to the devices in the home (e.g., WiFi, Zigbee, Bluetooth, etc.). So, there will be WoT agent for WiFi, WoT agent for Zigbee, WoT agent for Bluetooth, and so on. The WoT devices in a smart home may include (but are not limited to) heater, air conditioner, refrigerator, TV set, cooker, lights, security cameras, front door, picture display on the wall, robot cleaner, temperature sensor, and so on. Different WoT services are created for different "things" in the home, such as cooking service for control of the cooker, heating service for control of the heater, cooling service for control of the air-conditioning system in the home, light control service for control of the lights (e.g., on/off and adjusting the light density for each light in each room, or in the whole home), cleaning service for control of the robot cleaner, temperature monitoring service for control of the temperature sensor, picture display on the wall service, front door service for control of the front door, security camera service for control of the security camera (e.g., on/off, online home surveillance) and so on.

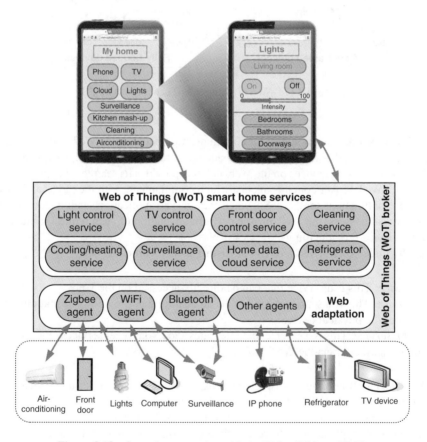

Figure 8.18   Smart home service with the Web of Things (WoT)

8.7   Business and Regulation of Converged Services and Contents

NGN provides basis for converged service and contents based on Internet technologies. However, the regulation and business aspects for many services have influence on their development. For example, legacy services such as telephony and television are being transferred from PSTN/PLMN and broadcast TV networks to IP-based access and transport networks in form of VoIP and IPTV, respectively. Other services such as WWW, which are Internet-native and deployed in best-effort manner in the Internet, are being used for development of broad range of services, such as Web services, IoT, and WoT. The IoT/WoT has also impact on the regulation and the business aspects of network and service providers, as well as end users and the society in a given country. All services are converging onto IP-based networks infrastructure (such as NGN) and broadband access (including fixed and mobile/wireless access networks). That requires certain adaptation of existing business and regulative environments (e.g., for legacy services) as well as provision of new business models and regulation (where needed) in the converged telecommunication world.

For implementation of a certain service there are three layers from vertical perspective: services and applications (the top layer), production (the middle layer), and research and development (the first layer). Regarding the provision of the services and applications one may distinguish among four main stages, given as follows:

- *End user equipment*: This includes customer interfaces such as user devices and home or enterprise networks.
- *Transport networks*: This includes all type of operator networks such as access, core, and transit networks.
- *Platforms*: This includes software platforms for provision of services and applications, as well as content to end users.
- *Content*: This is the provided content (e.g., Web, video, multimedia), including its creation and packaging.

Typically, telecom operators (as successors of POTS providers in the past) are primarily interested in provision of legacy services (e.g., via PSTN/ISDN emulation in IP-based networks) and best-effort services (e.g., WWW, e-mail, p2p, etc.) through broadband access to Internet. However, with broadband access to Internet, the TV provision as IPTV is added to the list of services offered by telecom operators.

8.7.1   Business Models for NGN Services

The current business models as well as regulation are targeted to three main types of services: digital telephony (e.g., VoIP), digital TV [e.g., DVB-T (Digital Video Broadcasting–Terrestrial), IPTV], and broadband Internet (e.g., all other services besides QoS-enabled VoIP and QoS-enabled IPTV). They can be divided into services over fixed and mobile networks, but with Fixed Mobile Convergence (FMC) the difference between services in the two network environments disappears. Emerging services that request innovations are based on IoT and WoT services (which include as a subset so-called Machine-to-Machine communication, i.e., M2M; or in other words, physical object-to-object communication).

Regarding the VoIP there are several business models that can exist or coexist, including the following ones [23]:

- *Fully-fledged QoS-enabled VoIP*: In this case VoIP services are provided with QoS support and can be bundled with other services (e.g., IPTV, best-effort Internet) in the network owned by the operator. This is a typical scenario of PSTN/PLMN replacement by QoS-enabled VoIP provision, which is based on NGN mechanisms such as IMS in control plane (i.e., service stratum) and IP transport mechanisms in user plane (i.e., transport stratum).
- *Bundled provision of VoIP service*: In this case VoIP services are bundled with broadband access to Internet as a resale product of the service provider.
- *Separate provision of VoIP service*: This is business model in which VoIP is partly bundled with broadband access to Internet, where the VoIP service can be offered to customers connected via other broadband access networks to Internet (i.e., VoIP provider and broadband access providers can be different business entities).
- *Best-effort VoIP service*: This is a scenario in which there is no business relation between the network provider (e.g., an Internet Service Provider-ISP) and the VoIP service provider (e.g., p2p VoIP service providers, such as Skype, Viber, etc.). In this business model typically there is no QoS support for VoIP service, so it is provided in best-effort manner.
- *Mobile VoIP service*: This is a scenario for QoS-enabled VoIP provision in mobile environments, which provides possibilities for roaming (i.e., use of VoIP through visited mobile or fixed networks).

VoIP can be offered as audio only, or combined with video (e.g., video conference) or data (e.g., text, data transfer). Additionally, there are different codecs for VoIP with different subjective quality levels (e.g., Mean Opinion Score – MOS). NGN provides the basis for all different VoIP business models since it has capabilities for call and session control and signaling, as well as QoS support end-to-end, which are necessary for the VoIP services with highest demands (e.g., QoS-enabled VoIP).

Unlike voice the video has much higher bandwidth demands. On the other side, video streaming and IPTV are mainly unidirectional types of traffic (in the downlink direction) and can use multicast in the transport networks for higher efficiency (i.e., less bandwidth utilization). Regarding the TV, it is not likely that all TV will fully migrate to IPTV in the second decade of the twenty-first century. For example, analog to digital TV (i.e., DVB-T) transition is scheduled to be completed by 2020 in the whole world. So, it is expected that terrestrial TV broadcast, satellite TV, as well as cable TV will continue to exist in parallel with emerging IPTV (as TV service over all-IP networks, such as NGN). However, broadband access to Internet and convergence of all service to Internet provides basis for increasing market share of IPTV in future, including fixed and mobile networks. NGN provides all needed functionalities for IPTV and different business models for its provision. Regarding the TV provision in NGN there are several business models that can be used in NGN (but are not limited to them), given as follows:

- *Fully-fledged architecture for IPTV services*: In this model the network provider (e.g., telecom operator) is also an IPTV service provider. The content is converted to IP packets within the operator's network (i.e., in house), and it is provided to customers with end-to-end QoS support (e.g., with admission control, priority over best-effort Internet services, traffic shaping to limit used bandwidth per stream, and playout buffering at the receiving side to eliminate delay variations due to statistical multiplexing and scheduling mechanisms in the network). Typically, this business model restricts the service offering to the customers of the telecom operator.

- *IPTV service provider*: In this case IPTV service provider provides IPTV streams to network providers (e.g., NGN providers). In such case there can exist roaming and non-roaming scenarios for IPTV users. NGN defines interfaces that can be used for IPTV provision to end users by third party IPTV service providers. This is a scenario that requires standardized information infrastructure which is provided with the NGN.
- *Internet TV (or Web-based best-effort IPTV)*: This is the provision of the same IPTV services, but in best-effort manner (i.e., without QoS support). They can be provided free of charge by network providers that also provide and charge for QoS-enabled IPTV provision, but also they can be provided by using Web business models (e.g., adding advertisements onto the Web page with embedded IPTV player).

In all business models above, IPTV can be provided with different media quality (e.g., SDTV, High Definition TV – HDTV), which can be offered to end users for certain TV channels. IPTV is usually referred to a linear TV provision over IP networks, while non-linear video provision is regarded as VoD. Regarding the IPTV content distribution there are also different business models, including the following:

- *Content from traditional broadcast stations*: This is the case when traditional linear TV content is provided via all-IP network to the end users.
- *Content Delivery Networks (CDNs)*: These can be used to provide IPTV content from all over the world.
- *Un-managed content*: This is a business model which extends the IPTV ecosystem with end user generated contents.

Different business models for IPTV services create market segmentation, but on the other side provide more options for end users. One may expect that a given user will use different models of IPTV service (e.g., Internet TV while sitting on the computer, and IPTV on the TV screen while resting on the sofa in the living room).

Overall, Internet TV (i.e., best-effort IPTV) as less expensive approach can be a competitor to IPTV (provided with QoS support by network operators, which is more user-friendly and more reliable service provision), but also both types can be included in a bundle of IPTV services offered to the end users.

The third fundamental component of the NGN services portfolio is WWW and Web-based services. Typical business model for WWW is best-effort model, without direct business relations between the Web service providers and network operators. NGN offers possibility to add such relation where it is necessary. However, Web is also a framework for provision of other emerging services in NGN such as USN and IoT. In fact, combination of IoT and Web is called Web of Things which provides basis for innovative services (e.g., smart cities, smart homes, intelligent transportation systems, e-health, etc.). The WoT services can generate new business models by using a private–public partnership (e.g., smart city, e-health) or can be offered as services to end users (e.g., smart-home) by network providers or/and device (or home network solution) providers.

Table 8.1 Business models in the Internet of Things ecosystem

Business model for IoT	Device provider	Network provider	Platform provider	Application provider
Model 1	Player 1	Player 1	Player 1	Player 1
Model 2	Player 1	Player 1	Player 1	Player 2
Model 3	Player 2	Player 1	Player 1	Player 2
Model 4	Player 2	Player 1	Player 2	Player 2
Model 5	Player 3	Player 1	Player 2	Player 3

In general, regarding different providers in IoT ecosystem there are possible several business models as shown in Table 8.1 [24], which are outlined as follows:

- *IoT business model 1*: Single business player (e.g., telecom operator) operates devices, network, platform and applications, and serves the end user (i.e., the customer).
- *IoT business model 2*: One business player operates devices, network, and the platform for the IoT (e.g., a telecom operator), while other business player is application provider (i.e., a third party IoT service provider).
- *IoT business model 3*: One business player operates the network and the platform for IoT services, while other business player operates the devices on the end users' side and applications on the other side.
- *IoT business model 4*: In this model one business player operates the network, while other business player acts as device provider on one side and platform and application provider on the other side (i.e., has vertically integrated businesses).
- *IoT business model 5*: In this model one business player operates the network, second business player provides the platform, and third player operates devices and provides IoT applications to the customers.

Regarding the business approaches in all-IP telecommunication world, one must consider the replacement of leased lines [e.g., SDH/PDH (Synchronous Digital Hierarchy/Plesiochronous Digital Hierarchy) based leased lines] with VPN in the NGN. For example, VPN services are used for connecting branch offices of enterprises on different locations. They are offered in wholesale and retail markets. However, NGN provides possibility to offer VPN services to individual users or community groups.

8.7.2 Regulation of NGN Services

NGN includes broad range of services, including traditional telecommunication service which transit to all-IP environments (e.g., VoIP, IPTV), traditional Internet services provided in best-effort manner (e.g., WWW, e-mail, p2p services, etc.), VPN services, and emerging new services such as IoT and WoT.

Provision of NGN services requires broadband access and its regulation, which is covered in Chapters 4 and 5 for fixed and mobile broadband, respectively.

Regulation of traditional services telephony and television is well established. Since telecom operators are the ones that become NGN providers, it is straightforward to continue to use the existing regulation for PSTN/PLMN telephony in the case of QoS-enabled VoIP as its replacement in all-IP telecommunication world. However, best-effort VoIP (e.g., Skype, Viber) which is offered by third parties by using proprietary platforms (i.e., not standardized and not open ones) is used in deregulated manner. There are several issues that may evolve or change regarding the VoIP regulation [25]. For example, the QoS-enabled VoIP (with SIP signaling) should support all additional service features available in PSTN/ISDN such as call line identification, call forwarding, call blocking, and so on. But, utility independent power supply becomes obsolete, which can be justified with battery support (e.g., in mobile devices). Emergency calls are mandatory, but they can be supported only by certain VoIP business models in NGN (e.g., models based on carrier grade VoIP provision). Unlike PSTN which were designed to meet certain QoS constraints such as end-to-end delay, there are QoS-enabled as well as best-effort VoIP business models. However QoS provision to VoIP can be mandated to telecom operators (e.g., NGN providers) in a given country, while third party VoIP service providers cannot provide such guarantees (e.g., the performance of best-effort VoIP service depends on the traffic load on the links in the voice data path). Also, legal interception of VoIP is not equally applicable in all business models. The regulation framework in most of the countries includes telephony in the universal service, which is a baseline level of telecommunication services to every resident in a country. The QoS-enabled VoIP service provided by telecom operator is suitable option for PSTN replacement in the universal service, while on longer term universal service should be based on broadband access to Internet.

Regulation of IPTV services in NGN will continue in similar manner as regulation of TV provided via cable network or terrestrial broadcast networks (e.g., DVB-T), since they will coexist as concurrent options for TV services within each country and globally. The content regulation of TV programs is well established in each country and it is dependent upon the population, its language, culture, and political environment.

Content in Internet is provided also by third parties, typically in best-effort manner, based mainly on Web services (e.g., web sites with certain contents, including text, pictures, audio, video, multimedia, and data). IPTV can also be provided via third parties, either in client–server or p2p manner. However, Web contents, including best-effort IPTV and VoD (provided as Web services), are unregulated since the WWW is global. However, each country can apply its own regulation for access to certain contents and sites that provide such contents (e.g., prohibit certain type of contents for given user groups).

In the past there was no need for a distinction as to whether a particular obligation in the regulation is applied to the network provider or to the service provider since both were inseparable (e.g., PSTN, PLMN, broadcast TV networks). However, in the NGN such approach cannot be used in all cases (e.g., in different business models for various services). In NGN there are many business models extending from telecom operators which are network and service providers at the same time, to the pure service provider (without its own network facilities). In such cases, economies of scope between network operation and service provision (as NGN offers as a possibility) might become relevant for regulation. On the other side, unlike today where it is usual that there is single dominance in given market for traditional telecommunication services (e.g., telephony, for which typically the incumbent operator is dominant market power for a certain time period), a dominance of global third party service providers (e.g., for Web services) or joint dominance of several companies might become subject to certain regulation [25].

Generally, NGN does not necessarily eliminate significant market power, but to some extent it changes its influence due to variety of business models available to different business players in the value chain (e.g., device providers, network providers, platform providers, application providers, and content providers). So, NGN may rearrange some kinds of market power, but it may also provide new forms of market power. The separation of transport and services in NGN opens up competition for services to third parties, thus reducing the market power associated with those services provided by the network operators and provides basis for deregulation of the services over time. However, regulation of the fixed and mobile broadband access, as basis for all services including existing ones as well as emerging ones, is crucial to ensure the availability of all types of NGN services and applications.

8.8   Discussion

NGN is designed to have capabilities for all telecommunication services, either provided within the NGN or via third party service providers. For continuity of telephony and television in IP environment NGN includes call session control functionalities through the standardized IMS. For QoS support which is needed for traditional telecommunication services that transit to all-IP networks, NGN provides common mechanisms for AAA for access to and use of services, as well as admission control and resource reservations for QoS support in the access, core, and transit networks (i.e., end to end). Also, leased lines for business users (e.g., enterprises) are replaced with VPN services in NGN, which are further extended to individual end users and community groups.

On the other side, NGN incorporates the best from the Internet world, that is, the open service environment for innovation of new emerging services and modification of the existing services and applications without a need for change of the underlying transport infrastructure. That is accomplished by the separation of transport and service functions in the two stratums of NGN. Further, the most important best-effort Internet service, the WWW, is important part of the NGN service portfolio where the Web is incorporated as basis for many mash-up services in NGN (i.e., combination of two or more existing services to create another service, such as Web and IPTV, or Web and IoT).

Finally, NGN provides information infrastructure for the IoT, that is, connection, control, and management of different objects, either physical or virtual, with aim to provide certain information or to accomplish given tasks. Combination of the IoT and the WWW provides even more practical deployment of such services, known as the Web of Things, which introduces practical "any-thing connectivity" as a new dimension in the telecommunication world.

References

1. ITU-T (2010) Functional Model and Service Scenarios for QoS-Enabled Mobile VoIP Service. ITU-T Recommendation Y.2237, January 2010.
2. Schulzrinne, H., Casner, S., Frederick, R., and Jacobson, V. (2003) RTP: A Transport Protocol for Real-Time Applications. RFC 3550, July 2003.
3. ITU-T (2010) Terms and Definitions for IPTV. ITU-T Recommendation Y.1901, March 2010.
4. ITU-T (2008) IPTV Functional Architecture. ITU-T Recommendation Y.1910, September 2008.
5. ITU-T (2011) Framework for Multicast-based IPTV Content Delivery. ITU-T Recommendation Y.1902, April 2011.

6. ITU-T (2010) IPTV Services and Nomadism: Scenarios and Functional Architecture for Unicast Delivery. ITU-T Recommendation Y.1911, April 2010.
7. Cain, B., Deering, S., Kouvelas, I. *et al.* (2002) Internet Group Management Protocol, Version 3. RFC 3376, October 2002.
8. Fenner, B., He, H., Haberman, B., and Sandick, H. (2006) Internet Group Management Protocol (IGMP)/Multicast Listener Discovery (MLD)-Based Multicast Forwarding ("IGMP/MLD Proxying"). RFC 4605, August 2006.
9. ITU-T (2008) NGN Convergence Service Model and Scenario Using Web Services. ITU-T Recommendation Y.2232, January 2008.
10. W3C (2007) SOAP Version 1.2 Part 1: Messaging Framework (Second Edition). W3C Recommendation, April 2007.
11. W3C (2007) Web Services Description Language (WSDL) Version 2.0 Part 1: Core Language. W3C Recommendation, June 2007.
12. ITU-T (2012) Functional Requirements and Architecture of the Web Service Component in Next Generation Networks. ITU-T Recommendation Y.2024, July 2012.
13. ITU-T (2010) Requirements for Support of Ubiquitous Sensor Network (USN) Applications and Services in the NGN Environment. ITU-T Recommendation Y.2221, January 2010.
14. Montenegro, G., Kushalnagar, N., Hui, J., and Culler, D. (2007) Transmission of IPv6 Packets Over IEEE 802.15.4 Networks. RFC 4944, September 2007.
15. ITU-T (2012) Framework of Object-to-Object Communication for Ubiquitous Networking in Next Generation Networks. ITU-T Recommendation Y.2062, March 2012.
16. ITU-T (2012) Functional Requirements and Architecture of the Next Generation Network for Support of Ubiquitous Sensor Network Applications and Services. ITU-T Recommendation Y.2026, July 2012.
17. Rosen, E. and Rekhter, Y. (2006) BGP/MPLS IP Virtual Private Networks (VPNs). RFC 4364, February 2006.
18. Rekhter, Y. and Li, T. (2006) A Border Gateway Protocol 4 (BGP-4). RFC 4271, January 2006.
19. ITU-T (2009) Requirements and Framework for the Support of VPN Services in NGN, Including the Mobile Environment. ITU-T Recommendation Y.2215, June 2009.
20. ITU Report (2005) The Internet of Things, November 2005.
21. ITU-T (2012) Terms and Definitions for the Internet of Things. ITU-T Recommendation Y.2069, July 2012.
22. ITU-T (2012) Framework of the Web of Things. ITU-T Recommendation Y.2063, July 2012.
23. Marcus, J.S. and Elixmann, D. (2008) The Future of IP Interconnection: Technical, Economic, and Public Policy Aspects, WIK-Consult, Final Report, Study for the European Commission.
24. ITU-T (2012) Overview of the Internet of Things. ITU-T Recommendation Y.2060, June 2012.
25. Elixmann, D., Figueras, A. P., Hackbarth, K. *et al.* (2007) The Regulation of Next Generation Networks (NGN): Final Report, Wik-Consult, Study, May 2007.

9

Transition to NGN and Future Evolution

9.1 Migration of PSTN Networks to NGN

The Public Switched Telecommunication Networks (PSTN) including its enhancement ISDN (Integrated Services Digital Network) are migrating to all-IP networks with QoS (Quality of Service) support end-to-end for traditional telecommunication services (i.e., telephony) and signaling capabilities required for QoS-enabled real-time services. With aim to provide inter-networking and communication between any pair or group of users globally (something that is possible in all types of POTS-Plain Old Telephone Service), such transition of circuit-switching telephone networks (PSTN/ISDN) to all-IP (i.e., packet-switching networks) must be implemented via globally standardized and accepted approach. Capacity for global harmonization of all-IP networks for all telecommunication services has the ITU (International Telecommunication Union), which has standardized the NGN (Next Generation Network) for such purpose. So, initially NGN is targeted for replacement of circuit-switching telephony with IP-based telephony end-to-end with all capabilities provided previously in PSTN/ISDN. Migration from PSTN/ISDN to NGN is also referred to evolution to PSTN/ISDN evolution to NGN [1].

Evolution of PSTN to NGN is a process in which parts or segments of the existing POTS networks are upgraded or fully replaced to corresponding NGN network elements providing similar or better functionalities than its predecessors, and at the same time attempting to maintain the services as provided by the original network (e.g., PSTN/ISDN). However, NGN provides additional capabilities and functionalities to existing services in PSTN/ISDN. The NGN maintains the traditional services in their original form by using PSTN/ISDN emulation and simulation [2].

9.1.1 Evolution of PSTN/ISDN to NGN

To consider the evolution toward the NGN one needs to analyze the migration of different entities in PSTN/ISDN. Generally, it consists of the following (as outlined in Chapter 1):

- *Transport network*: This include access and core networks with User Access Module (UAM), Remote User Access Module (RUAM), Access Network (AN), as well as switches (i.e., exchanges) in the core network and their interfaces.

NGN Architectures, Protocols and Services, First Edition. Toni Janevski.
© 2014 John Wiley & Sons, Ltd. Published 2014 by John Wiley & Sons, Ltd.

- *Signaling network*: This is overlay signaling network with functionalities typically integrated with exchanges in the network.
- *Management*: This refers to the management of the PSTN exchanges.
- *Services*: Different services are implemented within exchange hosts [e.g., roaming in PLMN (Public Land Mobile Network), value-added services, etc.].

So, all functionalities in PSTN/ISDN are located in an exchange, although there are exchanges on different layers such as Local Exchanges (LEs), Transit Exchanges (TEs), and international exchanges. On the other side, in NGN different functionalities may be spread across many network elements. For migration from PSTN/ISDN to NGN there are several aspects that need to be considered, such as transport, leased-lines, services, and supplementary services (e.g., call forwarding, call barring, etc.), E.164 numbering scheme for telephony services, interworking, and so on. The main characteristics of PSTN/ISDN are covered in Chapter 1 of this book.

The migration to NGN includes migration of PSTN/ISDN access networks and core networks. While there can be different access networks (including fixed and mobile ones), there is convergence regarding the core network of the NGN. In general, there are several possible migration scenarios of PSTN/ISDN core network to NGN [1], given as follows:

- Call Server (i.e., Soft Switch) based scenario for transition of PSTN core network with three different possible scenarios:
  - *Scenario 1*: Migration starts from LEs.
  - *Scenario 2*: Migration starts from TEs, with immediate use of packet switched network to which are connected LEs.
  - *One-step approach*: This is the evolution of both LEs and TEs to packet-based (i.e., IP-based) network nodes.
- One step evolution to IP Multimedia Subsystem (IMS): This is based on implementation of Core IMS entities in the network.

In scenario 1 for migration of PSTN/ISDN to NGN, as a first step, some of the LEs are replaced by Access Gateways (AGs) which provide access network connection to NGN as a packet-based network. Some of the access elements, such as UAM, RUAM, and Private Branch eXchange (PBX), which were originally connected to the removed LEs, are connected directly connected to AGs. Trunking Media Gateways (TMGs) and Signaling Gateways (SGs) are deployed for interconnection between the NGN and the TEs of the legacy network (as well as other PSTN/ISDN operators). Call server controls all AGs and Trunking Media Gateways. In PSTN/ISDN emulation scenarios for NGN a call server is a network entity that is responsible for call control, call routing, as well as authentication, authorization, and accounting (AAA). As a second step in this scenario the remaining LEs are replaced with AGs.

In scenario 2, as a first step, all TE functions are performed by the Trunking Media Gateways (TMGs) and the SGs, to which are connected LEs of the legacy network, under the control of the Call Server. Further, as a second step, all LEs are replaced by AGs (controlled by the Call Server).

Scenario 3 is one step approach, in which LEs are replaced with AGs and LE functions are transferred to the AGs and the Call Server. In this case all access elements, such as UAMs, RUAMs, and PABXs, are connected to AGs. Also, access networks are connected through

AGs, or they are being replaced by AGs. The TE functions are replaced by Trunking Media Gateways (TMGs) for handling user data traffic and SGs for control traffic. The TMGs and SGs provide also interconnection to other PSTN/ISDN operators.

Unlike Call Server evolution to NGN, in IMS-based evolution to NGN the PSTN/ISDN directly evolves to packet-switching network based on the IMS core architecture, as shown in Figure 9.1. In this approach end-users access the network and the services by using NGN-capable user equipment or more often using legacy terminal equipment connected via an AG to the NGN. Such access includes access to all services, including IMS-based services [e.g., VoIP (voice over IP), IMS-based IPTV (Internet Protocol Television), etc.] as well as non-IMS services (e.g., best-effort Internet services such as WWW-World Wide Web). As in the previous scenarios, Trunking Media Gateways (TMGs) and SGs are deployed for interconnection between the NGN and other PSTN/ISDN-based networks. However, there is also possibility to have concurrent CS-based approach and IMS-based approach in a same NGN (e.g., Call Server solution is implemented initially, while IMS is used for new services

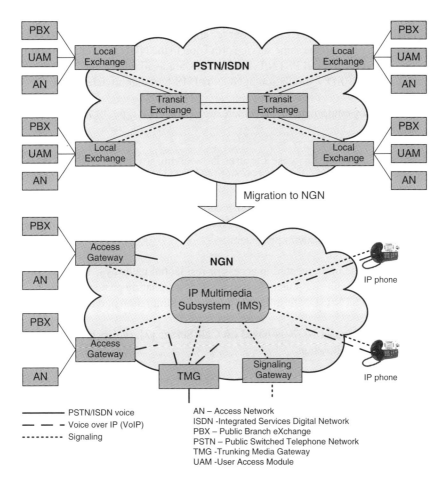

Figure 9.1   IMS-based evolution of PSTN/ISDN to NGN

and for takeover of the services provide through the Call Server), where interoperation can be implemented by using Session Initiation Protocol (SIP) as main signaling protocol in both approaches. However, on long term IMS evolution to NGN is the preferred approach, while Call Server solutions can be an intermediate approach for transition of PSTN telephony to NGN-based VoIP services (with final target set to IMS-based network).

9.1.2 PSTN/ISDN Emulation and Simulation

PSTN/ISDN provides capabilities and features (e.g., ISDN supplementary services such as call line identification presentation/restriction, call waiting, call barring, call forwarding, and many more), to which telephony users are accommodated. Hence, NGN need backward compatibility to PSTN/ISDN, and primarily to voice services (i.e., telephony). In general, there are two implementations in NGN for provision of PSTN/ISDN-alike voice services [2]:

- *Emulation*: This refers to provision of most of the existing PSTN/ISDN service capabilities and interfaces by using their adaptation to IP environment in NGN. Typical approach for PSTN/ISDN emulation in NGN is Call Server approach.
- *Simulation*: This refers to the same service provision in the NGN as emulation approach, but there is no guarantee that all PSTN/ISDN features are available to end customers. Typical approach for PSTN/ISDN simulation in NGN is IMS-based approach.

For access of legacy equipment to NGN there are defined two types of Adaptation Functions (ADFs), which are:

- *ADF1 for simulation*: In this case the user is receiving a regular NGN services that are the same as services received with NGN end-user equipment. This ADF is implemented in residential gateways (e.g., in user homes).
- *ADF2 for emulation*: In this case the user receives standard (i.e., legacy) PSTN/ISDN services, although the provider's network is NGN. The ADF2 is typically implemented via AGs (located at NGN provider's premises).

So, the ADFs may be implemented in access or residential gateways. The ADF on one side interfaces to IP multimedia component by using SIP as a signaling protocol, while on the other side it interfaces to PSTN/ISDN terminals and provides simulation and emulation services, as shown in Figure 9.2.

PSTN/ISDN emulation in NGN is based on Call Server [3], which has several functional entities, including the following:

- *Access Call Server (ACS)*: This is used for control of the AGs, call control, and subscriber registration.
- *Breakout Call Server (BCS)*: This is used for control of the Trunking Media Gateways (TMGs) to interwork with PSTN/ISDN.
- *IMS Call Server (ICS)*: This controls the border gateways that are used to interwork with other NGN and IP-based networks.
- *Gateway Call Server (GCS)*: This provides interoperability between different Call Server-based PSTN/ISDN Emulation Service component (as a NGN service component) to provide end-to-end service (e.g., voice connection).
- *Routing Call Server (RCS)*: This provides routing functions between Call Servers.

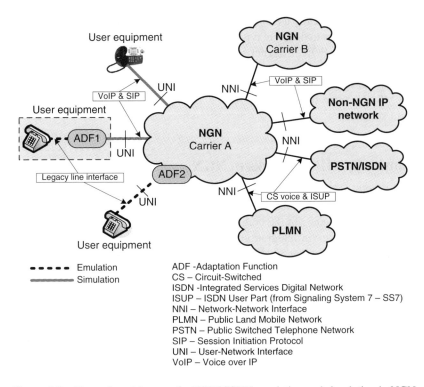

Figure 9.2   Network architecture for PSTN/ISDN emulation and simulation in NGN

So, PSTN/ISDN emulation is more complex and requires several different network nodes for its implementation, which are dedicated for that purpose. Also, emulation approach requires usage of different signaling protocols, such as H.248 between ACS and AGs, and SIP between ICS and IMS-based networks as well as between GCS and other PSTN/ISDN emulation subsystems. However, the communication between the SG and external PSTN/ISDN networks is implemented by using ISUP (ISDN User Part) signaling (as part of the SS7 signaling), while SIGTRAN (Signaling Transmission, i.e., ISUP/SS7 over IP-based networks) is used between SG and BCS.

The PSTN/ISDN simulation in NGN is based on IMS, which is used for different service components in NGN (e.g., IPTV, VoIP, Web-based services, etc.), so it is not dedicated to PSTN/ISDN simulation purposes only. Also, IMS uses unified signaling between all signaling entities in the operator's signaling network, including SIP communication between IMS entities and IP phones (at the end-user premises) and AGs (in cases when legacy user equipment is used).

Additionally, for PSTN/ISDN to NGN migration (or evolution) the practical approach is to install so-called Multi Service Access Nodes (MSANs) in the LEs at Main Distribution Frame (MDF) sites (the MDF sites are locations where begins the "last mile" toward the end users) [4]. However, the MSANs can also be installed in remote cabinets, which depend upon the network design approach. The MSAN (or Multi Service Access Gateway – MSAG) is typically a device that connects customer lines (e.g., twisted pairs, optical fiber, etc.) to the core network to provide telephone, ISDN, and broadband access to Internet from a single platform. So, instead of deploying DSLAMs (Digital Subscriber Line Access Multiplexers) for xDSL (Digital Subscriber Line technologies) access networks, the operator can use MSANs to provide more flexibility to different types of access networks (e.g., PSTN/ISDN, xDSL, Passive Optical Networks – PON, Ethernet, wireless access networks, etc.) by using TDM (Time Division Multiplexing) and Ethernet aggregation over fiber, copper, and wireless access links. In the final stage of migration to NGN, the traditional PSTN/ISDN services will have a minor role, hence MSANs (also referred to as Access Hubs) are better, faster to deploy, and more cost-efficient solution.

9.2   Transition of IP Networks to NGN

The IP-based NGN brings full network transformation that includes not only the service provider's entire network, but also it has influence on its business. The main approach in NGN is network convergence in the core and transit parts. Also, NGN provides applications and services convergence via integration of voice, video (including TV), and data services over any access networks by using the principles for communication anytime, anywhere, and to anything (as discussed in Chapter 8). One of the goals of the NGN is also provision of personalized rich multimedia services which can be implemented using the broadband access networks (including fixed and mobile broadband access). However, the only part of the NGN that is not converging is the access part, which consists of heterogeneous environments.

So, the transition of IP networks to NGN is based on changes in network structure, including core and transit IP networks. While in typical Internet networks in the past two decades the IP routers were interconnected by using the underlying SDH (Synchronous Digital Hierarchy) technology over copper, fiber, or wireless point-to-point links, the broadband access to Internet (also referred to as Next Generation Access – NGA) uses core and transit networks based on IP routers interconnected through Gigabit Ethernet over fiber [e.g., using WDM (Wavelength Division Multiplexing)]. However, change of SDH network with Ethernet-based technologies is not necessarily related to NGN.

Traditional IP networks consists of an Autonomous System (AS) connected with other ASs via border gateway which use BGP-4 for inter-AS communication regarding the routing of the traffic. Within each AS, the operator is free to implement any routing protocol as well as to choose whether to implement certain QoS solution in the network [e.g., MPLS (Multi-Protocol Label Switching), Differentiated Services, Integrated Services, policy-based routing, etc.]. Transition to NGN requires implementation of certain network entities which are related to resource and admission control function (i.e., RACF) and network attachment control function (i.e., NACF), as well as standardized storage and retrieval of end-user profiles [e.g., via HSS (Home Subscriber Server)], which provide interrelation between transport network, offered services, and end-users. Also, an IMS is important part of IP network transition to NGN (although it is also important part in PSTN/ISDN evolution to NGN), since user AAA, as well as resource requests by end-users (for certain services, such as VoIP and IPTV) are provided by core IMS functionalities with use of SIP for signaling, and Diameter for communication between its entities and databases which store user and service profiles. So, transition of traditional IP networks to NGN requires implementation of transport control functionalities (in NGN transport stratum) and service control functionalities (in NGN service stratum) between transport IP networks (consisted of IP switches and routers) and applications (e.g., WWW, e-mail, streaming, VoIP, etc.) which are provided in traditional Internet in best-effort manner without signaling between end-user terminals and network entities [except DNS (Domain

Name System) use for URIs (Uniform Resource Identifiers) resolving, and DHCP (Dynamic Host Configuration Protocol) use for dynamic assignments of IP addresses to attached user terminals].

Besides changes in the network architecture within the operator's network, with its transition from traditional best-effort Internet to NGN there are required changes in IP interconnection (e.g., between different Internet Service Providers-ISPs). NGN implementation has implication on IP interconnection [4], since for full interoperability between NGNs it has to be assured at the service level, control level (e.g., signaling protocols and procedures), and transport level.

There are several requirements for NGN interconnection related to different functional elements in both stratums. Regarding the transport stratum there is requirements for end-to-end QoS support (which includes interconnection) by using resource reservation and admission control for such tasks [i.e., NACF and RACF in the NGN transport stratum, as well as MMCF (Mobility Management and Control Function) for mobile environments]. The QoS (regarding packet delay, jitter, and packet losses) are particularly needed for real-time conversational services, such as voice (i.e., QoS-enabled VoIP) and video conferences/telephony (video is accompanied and synchronized to voice in both directions between end-users).

QoS is most challenging across all-IP core and transit networks since BGP (Border Gateway Protocol), as a protocol used at the gateway edge nodes, does not specify treatment of the QoS, such as prioritization of certain types of traffic over others. However, BGP-4 is the most important routing protocol for inter-AS routing, hence one may expect that it will remain on longer time scale. In such case typical solution for interconnection between ASs is to use aggregation of different types of traffic in MPLS/IP networks in separate VPN (Virtual Private Network) tunnels, which are used for interconnection of the traffic between the operators' edge routers. There are several standards generated by the IETF (Internet Engineering Task Force) regarding the BGP-4 and MPLS/VPN, including unicast [5] and multicast traffic [6]. Also, certain traffic engineering attributes are added to BGP, such as maximum and minimum bandwidth for a given Label Switched Paths (LSP) as well as priority level [7].

Overview architecture of the main NGN interconnection architecture is shown in Figure 9.3. Unlike traditional IP-based networks of the ISPs (targeted mainly to best-effort broadband access to Internet), NGN introduces standardized interfaces between services and applications and NGN, as well as between NGN and end-user equipment (e.g., Network-Network Interface – NNI, User-Network Interface – UNI, and Application-Network Interface – ANI).

However, in NGN environments, either migrated from PSTN/ISDN or from best-effort Internet, there is a need for use of different components that are commercially available for creation of network and services solutions. That is accomplished by Carrier Grade Open Environments (CGOE).

9.3 Carrier Grade Open Environment

The traditional telecommunication networks (e.g., PSTN, PLMN) lack openness in telecommunication solutions. But, many times there are Commercial-Off-The-Shelf (COTS) components that enable creation of different networks, platforms, and applications. For such purpose there is needed a common reference model. CGOE is intended to be such common reference model, which is based on COTS components and standards accepted by the industry.

COTS components that are used by telecommunication operators must comply with carrier-grade requirements for their networks and services from several different aspects

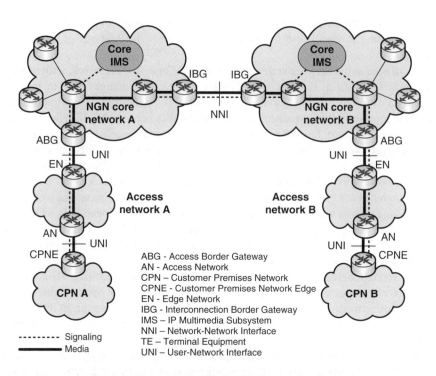

Figure 9.3   Architecture for NGN interconnection

such as [8]: very high performance (i.e., large number of simultaneous sessions), very high availability (e.g., five or six nines, i.e., 99.999% or 99.9999% or greater uptime for services), high scalability (from smaller to larger systems and capacities), hardware changes and software updates without interruption of the services, efficient and uniform management systems, high security level, easy adaptation of protocols across different systems, controlled lifecycle of the resources, rapid development and testing of the COTS components, as well as high cost-efficiency (which is very important from the business point of view). For the purpose of establishing common reference model for CGOE three types of providers are abstracted, which include the following (going from services and applications to hardware) [9]:

- *Service providers*: These are responsible for delivering services to the end-users.
- *Solution providers*: These deliver solution building blocks to service providers for composition of the offered services.
- *Technology providers*: These deliver functional components to the service providers for creation of their building blocks.

The COTS technologies provide basis for rapid development and fast delivery of new IP-based applications and services. Also, the evolution of open industry standards provides possibilities for creation of "plug-and-play" solutions by using components from different vendors. Although open systems and environments are being implemented by certain providers, the global utilization of the COTS within CGOE concept is possible with transition of service providers to NGN environments.

Each COTS component is part of its ecosystem. In general, COTS ecosystem consists of the following main areas:

- Networking platforms, software and hardware platforms;
- Sessions and events management support functions;
- Management infrastructure and interfaces;
- Security and carrier grade functions.

Each COTS component is built from one or several CGOE components. Further, each CGOE component [10], is defined by a category regarding the CGOE reference model (as shown in Figure 9.4), programmatic interfaces (e.g., based on existing standards), internal functional properties which describe what the component does as its function (e.g., based on standards), and non-functional properties which define features that a building block must provide with aim to provide certain behavior within the service architecture.

The CGOE reference model is presented in Figure 9.4. The reference model identifies CGOE categories. On the top of the CGOE reference model are industry solution-specific applications

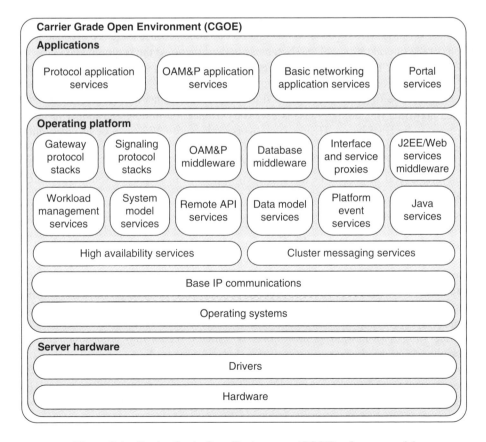

Figure 9.4 Carrier Grade Open Environment (CGOE) reference model

(i.e., NGN applications), which are divided into three main groups: control plane applications (e.g., call processing servers, nodes for providing network reliability of five to six nines, etc.), management plane applications (e.g., network development, optimization, administration, etc.), and service plane applications [e.g., SMS (Short Message Service) centers, content delivery servers, download servers, Web portals, etc.].

On the bottom of the reference model is the hardware, which includes transmission media (e.g., copper cables, fiber, wireless links) and physical devices used for service and application provision (e.g., network storage, servers, etc.). Also, drivers for different hardware components (provided by the vendors) are essential for enabling plug-and-play carrier grade hardware.

So, going from top to bottom of the CGOE reference model one moves from industry-specific CGOE components toward industry-agnostic components. Between the two ends (the industry application on the top, and server hardware on the bottom) is located the technology usage, that is, the operating platform with many CGOE categories ranging from services on the top of the operating platforms to standard and embedded Operating Systems (OSs) at the bottom. An overview of CGOE categories with existing standardized CGOE component for each category is given in Table 9.1.

9.4   IPv6-Based NGN

NGN is packet-based network which uses Internet protocols. Since there are two standardized versions of the Internet protocol, NGN can use both of them. At the beginning NGN networks are based on IPv4. However, since IPv6 is well established as a standard and IPv4 address space has drained out (although DHCP together with Network Address Translation – NAT have extended use of IPv4 over longer time period), the transition of IPv4 to IPv6 in all IP-based networks (including NGN) is necessary. Hence, on middle and longer terms NGN is expected to be IPv6-based. Hence, IP networks relevant to NGN can be grouped into several groups regarding the type of the Internet protocol [11], such as IPv4-based NGN and IPv6-NGN, as well as non-NGN IPv4 and IPv6 networks.

There are several important features in IPv6 important in NGN environments, which are absent in IPv4. Such features include:

- *Simplified header format*: IPv6 has fewer fields than IPv4 in the packet header.
- *Expanded addressing scheme*: IPv6 has much larger addressing space, enough to cover all devices including the ones used by humans as well as different objects. Also, IPv6 introduces autoconfiguration (i.e., automatic IP address assignment to a network interface without DHCP or manual configuration), as well as introduces anycast addresses (which are important for sending IP packet to any node in a given group of nodes).
- *Quality of Service (QoS)*: Besides traffic class field (which is present in IPv4 as Type of Service-ToS), IPv6 introduces flow label, for which a sender may request certain type of special handling by the network nodes (e.g., for real-time traffic).
- *Security*: IPv6 provides support for built-in IPsec services including authentication, integrity, and confidentiality of data by using IPsec options such as Authentication Header (AH) and Encapsulated Security Payload (ESP). Such approach enables end-to-end security services via global IPv6 addresses.

Table 9.1 Carrier Grade Open Environment (CGOE) categories

Category group	Category	Examples/description
Industry applications	Control plane applications	Call processing servers, radio access network servers, and so on
	Management plane applications	Tools for network development, optimization, monitoring, reporting, and administration
	Service plane applications	SMS center, MMS center, WAP gateway, content delivery servers, mobile location center, presence servers, Web servers, and so on
Operating platform	Protocol application services	Accessing the protocols, combining primitives, providing an abstracted interface to protocols, enabling differentiation according to the specific NGN layer and application layer
	OAM&P (Operations, Administration, Maintenance and Provisioning) application services	Common Information Model (CIM), Web-Based Enterprise Management (WBEM), Web Services Distributed Management (WSDM)
	Basic networking application services	Accessing the database, defining the network addressing models (e.g., E.164/ENUM), address and route analysis
	Portal services	Developing applets and portlets, and customizing Web pages
	Gateway protocol stacks	SCTP, RTP
	Signaling protocol stacks	SIGTRAN, SIP, Diameter, COPS (Common Open Policy Service), Megaco/H.248, BGP
	OAM&P middleware	Tools to create user, management, and provisioning interfaces for installation, configuration, update, monitoring, lifecycle management, billing, accounting, and charging
	Database middleware	Object Database Connectivity (ODBC), Java DataBase Connectivity (JDBC)
	Interface and service proxies	Brokering and mediating the APIs between clients and servers
	J2EE/Web services middleware	Simple Object Access Protocol (SOAP), Universal Description, Discovery, and Integration (UDDI), Service Oriented Architecture (SOA), HTTP
	High availability services	Starter service, graceful shutdown service, fault detection mechanism, IP address recovery units and check pointing, system state management service, failure detection, isolation, recovery, and repair services
	Cluster messaging services	TCP/IP messaging
	Workload management services	Overload control functions, load balancing functions

(continued overleaf)

Table 9.1 *(continued)*

Category group	Category	Examples/description
	Remote API services	Distributed objects communication and execution
	System model services	Installation, configuration, updating, monitoring, and lifecycle management
	Data model services	XML, Relational Database Descriptor (RDD), Lightweight Directory Access Protocol (LDAP)
	Platform event services	Managing events and logs
	Java services	
		Java 2 Platform Enterprise Edition (J2EE) Web services
	Operating system	Carrier grade operating system
	Base IP communications	TCP, UDP, IPv4, IPv6, IPSec, PPP, ARP, FTP, DNS, NFS (Network File System), RTP, RTCP, SSL, HTTP, SMTP, SNMP
Server hardware	Drivers	Drivers for carrier grade hardware
	Hardware	Variety of carrier grade hardware platforms

MMS, Multimedia Messaging Service; HTTP, Hypertext Transfer Protocol; XML, Extensible Markup Language; PPP, Point-to-Point Protocol; ARP, Address Resolution Protocol; FTP, File Transfer Protocol; RTP, Real-time Transport Protocol; RTCP, RTP Control Protocol; SMTP, Simple Mail Transfer Protocol; and SNMP, Simple Network Management Protocol.

- *Mobility support*: IPv6 has Next Header option which provides possibility of concatenated IPv6 headers which can be used for mobility support in all-IP environment by using destination option, mobility, and routing extension headers.

In general, IPv6-based NGN is defined as a NGN-based network which supports addressing, routing, protocols, and services related to IPv6. Overview of IPv6-based NGN is shown in Figure 9.5, where three relationships can be identified. The first relationship is between the end-user equipment and NGN transport stratum, which provides possibility IPv6-based NGN to handle multiple heterogeneous access interfaces (and multiple IPv6 addresses) by using multihoming (with IPv6) for mobility management of the users. The second relationship introduced with IPv6 in NGN is between end-user functions and applications, such as key management for IP security solutions. The third relationship is the relation between applications and transport stratum in NGN regarding the request for certain resource reservation and QoS support to a given application.

Regarding the signaling, IPv6-based NGN uses IPv6 enhanced features and at the same time it continues to use the same signaling protocols (which also used in IPv4-based networks) [12], without a need for their modification, which is defined as strategy by IETF regarding the two versions of the Internet protocol [13].

9.4.1 Multihoming in IPv6-Based NGN

Multihoming refers to multiple connections from a given node or host in the network. In core and transport networks multihoming is convenient for increasing reliability (e.g., in a case of a

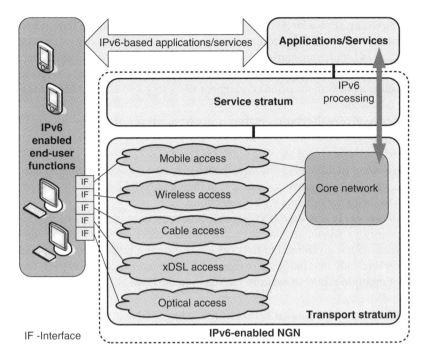

Figure 9.5 IPv6-based NGN

link or node failures in a given path the traffic continues via the other established connections), load balancing, higher utilization of network resources, and so on. In IPv4 environments multihoming is provided by Stream Control Transmission Protocol (SCTP) on the transport layer as well as BGP between network domains. Although the goals for multihoming in IPv4 and IPv6 are the same, IPv6 provides many more possibilities for multihoming due to its enhanced features such as the Next Header option in the packet header as well as enhanced addressing schemes.

9.4.1.1 Horizontal Multihoming

In NGN one may expect that a network node has multiple network interfaces and multiple IPv6 addresses (e.g., for renumbering), as well as multiple network prefixes for routing scaling. In general, in IPv6-based NGN the transport stratum consists of heterogeneous IPv6-based access networks (including fixed and mobile ones) and heterogeneous IPv6-based core networks. In such an NGN environment, end-user terminals may use multiple network interfaces and multiple IPv6 addresses to be multiply connected to the NGN (i.e., with several parallel connection via the same or via different access networks from multi-interface terminals).

With aim to provide multihoming in IPv6-based NGN there are several requirements that shall be met by the network, including multiple access capabilities in transport stratum offered to the end-users (from heterogeneous access networks), multiple network connections with the NGN and other networks by using a single user terminal or single user site (e.g., a home network gateway), as well as possibility for additional IPv6 addresses acquire by the host (or relinquish of existing IPv6 address). On the network's side multihoming requires capabilities of network nodes to dynamically acquire or relinquish an IPv6 address by using router renumbering protocol [14], which are carried in ICMPv6 (Internet Control Message Protocol) packets and provide mechanisms for address prefixes on routers to be configured and reconfigured in the same manner as the combination of neighbor discovery and address autoconfiguration is used for IPv6-based hosts.

From the perspective of the end-users there are two main types of multihoming in IPv6-based NGN, given as follows:

- *Site multihoming*: In this case a site (which also may be a network, including user network, access network, or core network) connects to the same NGN provider or to separate NGN providers by using multiple network connections.
- *Host multihoming*: In this case IPv6-capable host with multiple network interfaces might have capability to simultaneously access multiple access networks at the same time to communicate with a given remote host (which is different than approach in traditional IP networks where only one interface is selected for communication at a given time even in the case of multiple available network interfaces in the host).

Host multihoming in IPv6-based networks (including NGN) is shown in Figure 9.6, from which one may conclude that multihoming refers to capabilities located in different protocol layers (e.g., network layer, transport layer, etc.). Hence, according to the protocol layer in which it is implemented, the multihoming can be further classified into two models [15]:

- *Network-layer multihoming*: This is based on IPv6 protocol, and it provides possibility for both types of multihoming, that is, site-based and host-based.

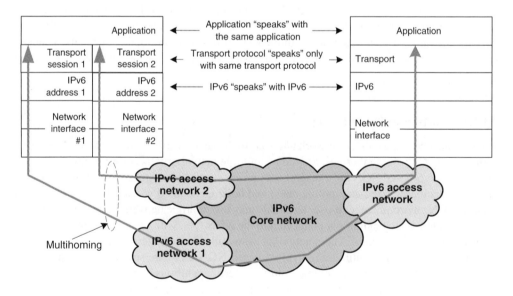

Figure 9.6   Host multihoming in IPv6-based networks

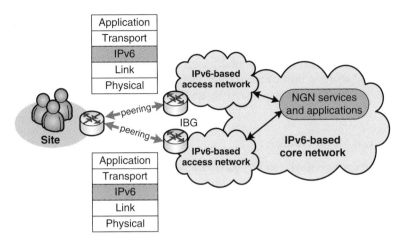

Figure 9.7 Site-based network-layer multihoming in IPv6-based NGN

- *Transport-layer multihoming*: This is based on transport-layer protocols SCTP [16], and Datagram Congestion Control Protocol (DCCP) [17], and it is targeted to host-based multihoming.

Typical applications of the network-layer multihoming model in site-based multihoming scenario (Figure 9.7) are: traffic load sharing (for higher network utilization), and traffic engineering (for network and routing optimization). In the case of host-based network-layer (i.e., IPv6) multihoming, typical applications are traffic load sharing (for higher utilization of the bandwidth assigned to the host), and higher reliability (to recover the host connection from failures in the network).

Transport-layer multihoming is based on two protocols, SCTP and DCCP, which have evolved from traditional TCP (Transmission Control Protocol) and UDP (User Datagram Protocol), respectively. Unlike SCTP which uses the TCP SACK (Selective Acknowledgment) mechanism for congestion control, the DCCP is based on UDP with added optional congestion control mechanisms but without support for reliable in-order delivery (to decrease end-to-end delay), and hence it is targeted to real-time applications which are sensitive to delays (e.g., online games, streaming media, etc.). In IPv6-based NGN the SCTP/DCCP multihoming operates on the IPv6 multihoming, so a given host in transport layer is multihomed to the access network and further to corresponding IPv6 host over multiple IPv6 addresses. This is also referred to as end-to-end multihoming since the multiple connections are established between the end hosts for the communication between their application peers. Illustration of SCTP usage for multihoming is given in Chapter 2. Typical application that can benefit from transport layer multihoming include fault tolerance with redundancy (i.e., use of redundant path in a case of network failure), and seamless vertical handovers (i.e., session continuity) in heterogeneous wireless and mobile environments. However, for use of transport-layer multihoming both end hosts must support certain transport protocols (i.e., SCTP, DCCP).

Multihoming in IPv6-based NGN in implemented in transport stratum since the transport functions provides connectivity for all connections. The NACF in transport stratum is responsible for dynamic allocation of IPv6 addresses to multiple network connection (in IPv6-based

NGN) [18]. Multihoming has to be supported also by end-user functions on one side, as well as access network functions, edge functions (between access and core networks) and core network functions in IPv6-based NGN transport stratum.

Overall, multihoming has high potential for several important application on the network side, such as load sharing, traffic engineering, and fault tolerance (i.e., reliability), as well as end-user side, such as always-on connectivity and session continuity.

9.4.1.2 Vertical Multihoming

Although IPv6-based network can provide multiple network connections (i.e., multihoming) there is single connection used at a given time, while other connections are used as backup connections in special cases such as fault tolerance, load balancing, as well as mobility support (e.g., at handovers). NGN environments include heterogeneous access networks and most of the user terminals have different network interfaces (for different access networks). In such case multi-interface node or terminal may have ability to connect simultaneously to different access networks. So, multiple physical/MAC (Medium Access Control) interfaces [e.g., WiFi, WiMAX, UMTS (Universal Mobile Telecommunications System), LTE (Long Term Evolution), Bluetooth, Ethernet, etc.] can be simultaneously used by a single terminal or node. From network point of view in IPv6-based NGN there can be multiple IPv6 addresses (with same or different IPv6 prefixes) assigned to different physical/MAC interfaces (i.e., network interfaces), which can simultaneously establish multiple network connections to other nodes for a given service. Regarding the transport layer, there is multiple transport session that can be simultaneously established. So, each protocol layer (physical/MAC layer, network layer, transport layer, and application layer) has a specific role for multihoming. However, to provide simultaneous use of multiple end-to-end connections between two (or more) nodes or hosts, there is required interaction between different layers, which is referred to as "vertical multihoming" [19].

In vertical multihoming concept each of the layers has specific role for support of the multihoming, and at the same time many network resources (from single or different networks) are used to establish simultaneous multiple connections. Regarding the number of IPv6 addresses used in vertical multihoming there are two main methods:

- *Unique IPv6 address per interface*: In this case there are multiple network connections that have unique IPv6 address assigned to each interface. So, multiple connections simultaneously use multiple network interfaces (with unique IPv6 addresses).
- *Shared IPv6 address*: In this case a host is assigned a shared IPv6 address across multiple interfaces. However, in the case with single IPv6 address the application (on the top protocol layer) "sees" only one address and hence cannot utilize multiple network connections by using only one socket interface (between the application and underlying protocol stack which is typically embedded in the operating system of the host). Hence, for this case is needed particular function for the establishment of multiple network connections (via single IPv6 address), which will have task to manage through which network interface a given application flow should be serviced.

Besides the network layer in IPv6-based NGN, other layers should also be considered for the vertical multihoming. Concept for vertical multihoming and relationship between different protocol layers in a given host is shown in Figure 9.8. In vertical multihoming each layer

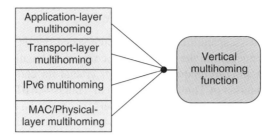

Figure 9.8   Vertical multihoming concept in IPv6-based NGN

can have (or not have) multihoming capabilities. Regarding the multihoming features in different protocols there are four cases (since protocols are grouped into four layers in vertical multihoming approach), which are the following:

- Multihoming features in physical/MAC layer only (e.g., carrier aggregation feature in LTE-Advanced and Mobile WiMAX 2.0 mobile networks).
- Multihoming features in physical/MAC and network layers.
- Multihoming features in physical/MAC, network and transport layers (e.g., SCTP/DCCP are used since traditional TCP and UDP protocols does not provide features necessary for multihoming).
- Multihoming features in physical/MAC, network, transport, and application layers (this is ultimate multihoming since in this case multihoming features are present in all protocol layers).

Finally, let compare multihoming, in particular vertical multihoming, and traditional TCP/IP communication over Internet. In traditional TCP/IP approach, single IPv6 address and single port are bound in the socket (in the Application Programming Interface – API in the host). In such case selection of an IPv6 address from multiple IPv6 addresses assigned to multiple interfaces is completely independent of selection of a port number from multiple port numbers (i.e., ports), but combination of IPv6 address and port should be used during the whole duration of an established connection without possibility for modification (i.e., does not provide support for multihoming). This is referred to one-to-one mapping (between IPv6 address and port number). In such case (the traditional IPv6 approach) for mobility support are used additional protocols such as Mobile IPv6 and Proxy Mobile IPv6 (the reader may refer to Chapter 5 for more details). So, a traditional TCP/IP layering mechanism cannot simultaneously utilize multiple IPv6 addresses and/or different session/port numbers. In vertical multihoming approach the vertical multihoming function (which can be embedded in the operating system of the host) enhances one-to-one mapping from traditional TCP/IP to many-to-many mapping among IPv6 addresses (assigned to the given host) and session/port numbers, and dynamically update the relation between IPv6 addresses and session/port numbers upon changes in network environments (e.g., available access networks). Regarding the application layer, vertical multihoming functions (on one side located in the user terminal, while on the other side implemented in the transport stratum of the NGN, including the NACF as well as access, edge, and core transport functions) dynamically control the required QoS for each simultaneous connection.

9.4.2   Object Mapping Using IPv6 in NGN

IPv6-based NGN provides capabilities and addressing space that is necessary for different types of object to be connected to the network and contacted globally. The identification of all objects and their mapping to unique object IDs is crucial for end-to-end communication between objects (e.g., Internet of Things – IoT, Web of Things – WoT).

In general, there are many types of possible identifiers for objects on different protocol layers [e.g., URIs/URLs (Uniform Resource Locator), E.164 numbers, IP addresses, MAC addresses, etc.]. By using certain Identifiers (IDs) for objects they can be reachable to users and other devices (i.e., other objects). Different types of object identifiers are shown in Figure 9.9, which are grouped into three groups: transport stratum identifiers (e.g., session/port number, IPv6 address, MAC address, etc.), service stratum identifiers (e.g., content ID, device ID, E.164 number, URL, etc.), and user and object identifiers on application layer. The DNS translates identifiers of objects to IP addresses. However, the most important aspect in IPv6-based NGN is mapping of different object IDs to IPv6 address (which can be many-to-one mapping) [20].

One host may have several objects or devices associated with it, where each such object needs to be reachable in the network [e.g., for a certain application, such as USN (Ubiquitous Sensor Network) application]. Regarding the mapping between the host and objects there are two types:

- *Host = object (one-to-one mapping)*: In this case the host is the object (e.g., a computer). Telephone devices also belong to this group since telephony uses E.164 numbers as service IDs, and the device can be treated in the same manner as an object.
- *Host ≠ object (one-to-many mapping)*: In this case there are many objects included with the host. However, such objects can be virtual objects (e.g., content objects) or physical objects (e.g., non-IP remote objects).

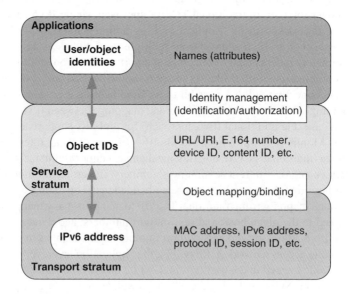

Figure 9.9   Object mapping in IPv6-based NGN

Mappings between the objects and their IDs refer to both stratums in NGN, since there are object identifiers in each stratum. IPv6-based NGN uses layered architecture for objects mapping, which consists of three layers:

- *Application layer*: The objects in this layer are identified by Object IDs. Such IDs are used by services and applications to address an object.
- *Network layer*: The objects in this layer are mapped to globally unique network identifier (i.e., IPv6 address), which are IP termination points (i.e., locators in the IP-based NGN). In this layer can be provided separation of locator role (LOC role) (toward the network) and ID role identifier role (toward the applications) of the IPv6 address (i.e., ID/LOC separation) [21].
- *Connectivity layer*: This layer covers the Points of Attachment (PoA) of objects to the network, and provides routed path between them.

The network layer in IPv6-based NGN can provide separation of node ID and LOC roles of IP address. In traditional IP-based networks IP address has a dual role. From the network point of view, the IP address identifies current topological location of a given host, and that role is called the "locator role". So, when a host is mobile and moves from one IP network to another (with different address prefix) the IP address changes. On the other side, from the application point of view the IP address identifies the host to the peer host in the communication (hence, it should be unchanged during the duration of the connection between the end hosts), and such role is called identifier role. For separation of the two roles of the IP address, the solution is introduction of an additional Node ID between the transport layer and IP layer in the protocol stack, so the end-to-end communication session is not bound anymore to IPv6 address, but to Node ID. This introduces a mapping function between the network layer and transport layer, which may be considered also as a new layer (e.g., IETF has experimented with Host Identity Protocol – HIP, which does ID/LOC separation of IP addresses by using Host Identity tags [22]). However, introduction of Node IDs will introduce a new name space (a third name space in IP-based network, besides IP address name space and domain name space). But, many researches share the opinion that it is hard to predict how such new name space, needed for separation of locator and identifier roles of IP addresses, will affect the global network infrastructure (e.g., IETF has standardized HIP and its mechanisms in the experimental track of the Request For Comments).

The objects mapping can be further classified into direct mapping and indirect mapping. In direct mapping approach the object is directly located in the end host, that is, it is directly reachable via the PoA to which the IP flow toward the object is terminated. Example of objects with direct mapping is contents stored on a given server (e.g., a Web server). Indirect mapping refers to cases where the object is not necessarily located at the PoA to which the IPv6 address refers. In such cases an intermediate element is forwarding the communication to the actual location of the object. Example of indirect object mapping are mobility schemes such as Mobile IPv4 where Care-of-Address is associated with an intermediate node (Foreign Agent) which has that IP address (assigned to one of its interfaces) and forwards locally the packets addressed to the mobile object by using the Home Address of the mobile object.

The central role in objects mapping play DNS, which provides the mapping between two existing name spaces in Internet (and accordingly in NGN), IP addresses and domain names. So, object mapping uses DNS mapping information between the domain name and IPv6 address of a given gateway (a node or a host to which the objects are connected). Further, gateway node maintains an object–gateway table for mappings between object IDs and the IPv6 address of the gateway, and an object–port table for mappings between object IDs and port numbers. The DNS as a standardized solution provides accessibility to the gateway (in the way toward the end objects), which maintains the object-gateway and object-port tables to provide end-to-end communication to objects over the NGN.

9.4.3   Migration to IPv6-Based NGN

IPv4-based and IPv6-based network will coexist for a given time period, but the final target on longer term is all-IPv6 network. However, migration to IPv6 means at the same time continuation of IPv4 networks and services (i.e., backward compatibility).

Generally, there are three basic approaches for IPv4/IPv6 interworking and integrations, which include: dual IP layer (i.e., dual stack), network address and protocol translation, and tunneling. These mechanisms or their combinations can be used for IPv6 migration of NGN operators [23]. In dual-stack approach, IPv4 and IPv6 coexist on one device (all newer operating systems in hosts and network nodes provide the dual-stack). Translators from IPv4 to IPv6 (and vice versa) can be network level translators [e.g., Stateless IP/ICMP Translation Algorithm [24], Network Address Translation–Protocol Translation (NAT-PT) [25], Bump in the Stack [26], etc.), transport level translators (e.g., IPv6-to-IPv4 Transport Relay Translator [27], etc.), and application level translators (e.g., Bump in the API [28], SOCKS64 [29], etc.). Finally, tunneling is performed by adding additional IP headers for transfer of IP packets from one network to another. In the beginning there is tunneling of IPv6 packets across IPv4 networks (i.e., IPv4 header is added to IPv6 packet at the start point of the tunnel, and added IPv4 header is dropped at the end point of the tunnel). Later, there will be more IPv4-to-IPv4 tunneling over IPv6 networks, when IPv6 addressing scheme will prevail over IPv4. There are two approaches in tunneling, which include configured tunnels (i.e., router to router), and automatic tunnels (e.g., Tunnel Brokers [30], 6 to 4 [31], 6 over 4 [32], etc.). Also, NGN operators may use Carrier-Grade NAT (CGN) to provide sharing of IPv4 addresses where it is necessary (e.g., for IPv4-only hosts).

Migration from IPv4-based NGN to IPv6-based NGN from operators' perspective can be divided into several phases, given as follows [23]:

- *Phase 0*: NGN with IPv4 only, no IPv6 networks deployed.
- *Phase 1*: IPv6 is introduced and IPv6 NGNs are connected across the IPv4-based NGNs (e.g., with tunneling across IPv4-based MPLS backbone).
- *Phase 2*: Connecting IPv4-based and IPv6-based NGNs via dual-stack NGN, where IPv4 and IPv6 logical planes are isolated between each other, but NGN devices may use both logical planes (i.e., IPv4 and IPv6).
- *Phase 3*: Connecting IPv4-based NGNs across IPv6-based NGNs, which will happen when IPv6 will become major addressing scheme globally due to exhausted IPv4 addresses.
- *Phase 4*: Fully migrated IPv6-based NGN, which is completely IPv6-based, where IPv4 may still exist locally as isolated islands in some enterprise or home networks. This is final phase of migration, and after it the NGN is fully IPv6-based.

Overall, IPv6 will completely replace IPv4 in all networks, because it has much bigger IP addressing space and many enhanced capabilities such as QoS, security and mobility support. It is certain that IPv4 will coexist with IPv6 at least in the following decade and IPv6 will completely takeover the telecommunication world after certain time period.

9.5 Network Virtualization

There are many network infrastructures deployed, consisted of interconnected network nodes (e.g., switches, routers) and network hosts (e.g., servers, databases). All networks converge to Internet technologies (e.g., NGN) which provide possibility for innovation due to their openness for development of new applications without changing the underlying network architecture (in traditional best-effort Internet and NGN) as well as to add new services by using existing functions in the network for QoS, mobility, and security support (in NGN, but not in best-effort Internet). However, such network architectures provide possibility of changes only at the edges (e.g., application and service providers). So, recently researches started to ask questions such as: What are the types of nodes in the network? What type of information they exchange? What are the nodes power and bandwidth limitations? What are the specific requirements from different services from the network (e.g., online banking requires security, IPTV requires bandwidth, voice and gaming require low delay and loss, etc.)?

The answers to the above question lead to the answer that there is no single solution in the network that can fit all different requirements from heterogeneous services, users and network environments. That leads further to virtualization of the network resources [e.g., isolation of resources in a given physical node, such as Central Processor Unit (CPU) and bandwidth]. Then, there can be designed multiple networks using different virtual components that exists in physical network elements (e.g., routers, servers, etc.) for different goals (e.g., different target services, different customers, etc.). Such an approach in network design is referred to as a network virtualization.

By definition [33], network virtualization is a method that provides possibility to multiple virtual networks, referred to as Logically Isolated Network Partitions (LINPs), to coexist in a single physical network. Virtual resources are created on physical objects in the network, such as switches, routers, and hosts. An LINP uses virtual resources by using their programmability features (which are mandatory for design of such virtual networks). Regarding the end users (e.g., residential users, enterprises) LINPs can provide the same services to them as networks without virtualization.

Conceptual architecture for network virtualization is shown in Figure 9.10. It consists of multiple interconnected LINPs, where each LINP is built from multiple virtual resources. In this approach each physical resource may have multiple virtual instances, so it can be shared among multiple LINPs. To provide such sharing of physical resources among multiple virtual networks (i.e., LINPs) there are needed Virtual Resource Managers (VRMs), which will interact with physical resource managers and will provide tools for management and control of virtual resources (e.g., virtual routers). Each LINP can have various characteristics regarding the partitioning (of the resources and networks), isolation [e.g., VPN and VLAN (Virtual Local Area Network) are examples of isolated networks], programmability (e.g., standardized interface through the VRM will have access to virtual resources), AAA, and so on. Overall,

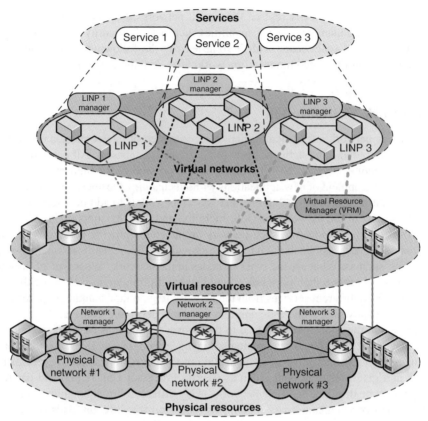

LINP -Logically Isolated Network Partitions

Figure 9.10   Architecture for network virtualization

network virtualization aims to provide higher utilization of the physical resources (by their sharing using the virtual resources approach) and better adaptation of the networks to the requirements from services and end-users.

What has brought attention to network virtualization? Certainly that are the problems that current IP-based networks are facing. First one is coexistence of multiple networks. In IP transport networks VPNs are used; and in IP access networks VLANs are used, which provide isolated logical networks over shared physical networks (e.g., offices of a given company can be interconnected by using VPNs over the operator's transport network, which is a typical example for enterprises). VPNs security is based on IPsec protocol which is well established, but it is less flexible for dynamic changes of the VPNs upon request. In the access networks existing VLANs require manual configuration settings by network administrator. Mobility support in all-IP networks is based on standardized protocols such as Mobile IPv4 and Mobile IPv6 which require home agents (both protocols) and foreign agent (in Mobile IPv4). The introduction of LINPs, besides existing VPNs, VLANs, and agents, should provide higher flexibility and dynamics of the networks (including access and core ones).

Another issue is complexity in management of network resources, such as switches and routers, due to heterogeneous equipments. Network virtualization should provide simplified access to resources. However, for that purpose are needed standardized interfaces for access to virtual resources in future networks, and that requires changes in the existing equipment that is already installed in the network. Lifecycle of the equipment in telecommunication networks is several years (e.g., four years), so longer usage of the network components is less important issue due to development of the processing power (e.g., according to Moore's Law, in average, processing power of devices doubles every 18 months) as well as increase of the capacities in access and core networks (with deployments of broadband access to Internet). But, flexibility in provisioning, that is, network adaptation to the existing and newly deployed services by internal or external changes, is important advantage of network virtualization. In fact, network virtualization allows reuse of network resources, as well as adding (or extracting) and aggregating logical resources to provide increased capabilities or capacities at significantly lower costs than approach with network change with new physical resources. Finally, network virtualization provides possibility to researches to use existing network resources (which are used to provide commercial services to end users) to create testbed LINPs for experimental purposes, thus eliminating the need for creation of isolated testbeds for testing new technologies, as well as provides real-world environment for the experiments.

There are several design goals for the LINPs, given as follows:

- *Isolation*: Network virtualization should provide isolation among different LINPs regarding power consumption and capacities of each LINP.
- *Network abstraction*: Underlying network technologies should be hidden from the virtual resource used by LINPs.
- *Reconfigurability*: This is referred to topology awareness of virtual resources (and associated physical resources) used by the LINP, which is needed for interaction among virtual resources and their reconfiguration when needed.
- *Performance*: Network virtualization introduces so-called virtualization layer between hardware and software in each physical resource for creation of isolated partitions in it (i.e., virtual resources). Such virtualization consumes CPU time in the physical device, thus leading to higher CPU utilization and lower bandwidth available when it is compared with non-virtualized network with the same physical resources.
- *Programmability*: LINPs require new control schemes and mechanisms for control of the virtual resources (e.g., via programmable control plane). But, network virtualization must also support data plane programmability for control of routing and forwarding of data traffic within given LINP.
- *Management*: Each LINP is isolated from others, so it has to be managed separately. Also, in network virtualization there are many complex mappings between physical resources and virtual resource (defined within the physical ones), so there is a need for development of integrated management system in a virtualization plane that can support monitoring, dynamic reconfiguration, topology awareness, as well as resources discovery, allocation, and scheduling.
- *Mobility*: In network virtualization mobility refers to movement of virtual resources, including end-users (with their mobile terminals) and services (used by the end users). To provide service continuity to users, the services must be moved together with users when they change the LINP or attachment point within the LINP (e.g., single LINP can be a mobile network

operator or a virtual mobile network operator). For this purpose resources required by given LINP for a given mobile user have to be identified in real-time and moved to the requesting LINP within limited time period.

- *Wireless*: Network virtualization also targets the wireless channels as virtual resources. However, wireless environments suffer from signal interference (which is absent in fixed networks). Hence, within a single wireless device with several active wireless connections to different LINPs should be used scheduling that will prevent transmitting while receiving in a given frequency range (and vice versa).

There are several types of network virtualization that are emerging, which include cloud computing on one side and software defined networks on the other side, as part of future network concepts.

9.6   Future Packet Based Network

With the evolution of telecommunications the requirements to networks change over time. Some requirements, such as fair competition among different business players (e.g., network providers, service providers) remain important, but there are also new requirements that arise regarding the network architectures. So far, IP protocol and related Internet technologies (e.g., DNS, DHCP, routing protocols, transport protocols, etc.) have succeeded to provide flexibility in telecommunication networks by hiding underlying technologies from the network and upper layer protocols, and at the same time providing QoS support, identity management, and security. All these elements are accomplished in the NGN architectures with many functional elements in transport stratum and service stratum. There is a lot of investment for building such networks, so someone may ask whether new emerging application can generate revenues which will cover necessary investment to change the networks for them.

The main objectives and design goals for future networks are given in Chapter 3, which summarized means evolution of NGN toward maximum flexibility and awareness to services, data, environment, and society. There are several possible enabling technologies for future networks, such as cloud computing, Software-Defined Networking (SDN), as well as energy saving of networks. Since energy saving is beyond the scope of this book, in the following sections the focus is given to cloud computing and SDN as the main concepts for future packet-based networks.

9.6.1   Cloud Computing

Cloud computing, as one accepted definition in the telecommunication world, is a model for enabling convenient, on-demand network access to a shared pool of configurable computing resources (e.g., networks, storage, servers, applications, and services) which is provisioned in real-time with minimum service provider's interaction. In such approach hardware to end users is provided in a form of virtual hardware components (with all features and applications as real hardware, such as computers) simulated by software running on real (i.e., physical) machines. Cloud computing is already present in Internet, and it is provided by telecom operators (usually offered to enterprise users, with Service Level Agreement) as well as global service and application providers (e.g., Web-based cloud services).

Why cloud computing? There are several possible answers to this question, such as reduced costs to enterprises (less costs for hardware and software and easier migration to new services), increased storage (i.e., less dependence upon real storage capacity, no planning in advance), highly automated system (i.e., no need for operation and maintenance people, while software is up to date), flexibility (e.g., easy change of the platform), higher mobility (e.g., possibility to work on distance, from different locations), and more possibilities for innovation.

There are several different service models for cloud computing, which include the following:

- *Infrastructure as a Service (IaaS)*: In this model cloud providers offer to the end users virtual machines, virtual storage, virtual operating systems, and applications software, by using large data centers (with physical computing resources, i.e., large computers). Connectivity to such clouds is performed either through the Internet (for residential users) or carrier clouds (via VPNs). When cloud providers provide network and computing resources in it as a single product offered to the end users, it is referred to as Network as a Service (NaaS) or cloud computing networking [34]. Examples of NaaS are flexible VPN allocations and bandwidth on demand, as well as Virtual Network Operators (VNOs) and Mobile Virtual Network Operators (MVNOs).
- *Platform as a Service (PaaS)*: In this model the cloud provider provides to end user a platform which includes operating systems, execution environment for a given set of programming languages, databases, and Web servers. In this model application developers can develop, test, and provide applications without a need for buying expensive underlying equipment for their provisioning.
- *Software as a Service (SaaS)*: In this model cloud providers install and manage application software in the cloud thus reducing the need to run the software on cloud users' own computers or devices (the user access cloud applications by using cloud clients).

Emerging cloud computing services are media clouds and context-aware computing. In media clouds (shown in Figure 9.11) multimedia processing is performed in the cloud (e.g., audio, video, images, etc.) in the CPU and GPU (Graphic Processor Unit) clouds, content is

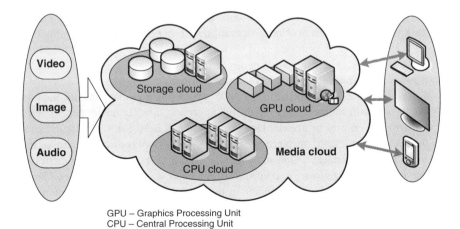

GPU – Graphics Processing Unit
CPU – Central Processing Unit

Figure 9.11   Media cloud

stored in the storage cloud and delivered in different forms to heterogeneous user devices (e.g., computers, TV sets, lap-tops, mobile terminals) across heterogeneous networks (with different capacities and QoS capabilities). In this approach the network intelligence for media delivery moves away from network core (e.g., NGN core network) to network edges (i.e., to cloud providers). Also, media storage is moving from local storage on devices at home or office to storage in cloud, which provides service mobility and access to media content anytime, from anywhere, and from any device (with certain capabilities regarding the media contents).

Another emerging trend in cloud computing services is context-aware computing. In the first half of 2010s context-aware cloud computing is presented via social networking services, location-based services (e.g., tracking objects or humans), as well as simple augmented reality. In the period until 2020 one may expect context-enriched contents, complex context brokers, as well as context delivery architectures.

Besides many advantages, cloud computing has also certain disadvantages, which include requirements for reliable Internet connection, broadband infrastructure, and threats for unauthorized access to customer's data. However, NGN provides high network reliability and broadband access, and incorporates a well established management system, which overcome the possible disadvantages of cloud computing. Hence, cloud computing is emerging trend in telecommunications, which provides lower hardware and software costs, instant software updates, almost unlimited storage capacity, user device independence, and universal access to applications and services.

9.6.2   Software Defined Networking

Another emerging implementation of network virtualization is SDN. The idea behind the SDN is the existing situation in all networks in which different networks nodes (e.g., switches, routers) have their own operating systems which perform autonomous tasks within the node. Each node, such as switch or router, has hardware with specialized packet forwarding for many protocols [e.g., OSPF (Open Shortest Path First), RIP (Routing Information Protocol), BGP, multicast protocols, differentiated services, MPLS, NAT, firewalls, etc.] and operating system with many features (and large source code that is typically specific to the vendor of the equipment). The SDN concept is creation of Network Operating System (NOS) that will perform the same task to the network as host/node operating system performs to the host/node.

In Figure 9.12 is shown the evolution of networks with autonomous operating systems in network nodes to SDN, where network nodes (e.g., switches and routers) are simplified to perform forwarding of packet and protocol-based communication between each other, while control of their processes is performed by a NOS which has a global network view (since it controls many nodes in the network). With this approach there is no longer a need for the design of distributed control protocols. However, there is a need for a standardized interface between the network nodes and the Network OS for the control of traffic (i.e., similar to drivers between host OS and interfaces). One such standardized interface in SDN is OpenFlow [35].

OpenFlow is a standardized way to control flow tables in switches and routers. Commercial equipment can include OpenFlow by update of the firmware. So, if a node (switch or router) is planned to become part of an SDN, then it needs to have OpenFlow.

Figure 9.13 shows OpenFlow use in SDN. Each network node in SDN needs to have Open-Flow client which communicates with OpenFlow controller by using SSL (Secure Sockets Layer)/TCP. Regarding the number of OpenFlow controller nodes, there are two main types of control in SDN: centralized control (with single OpenFlow controller in the SDN), and

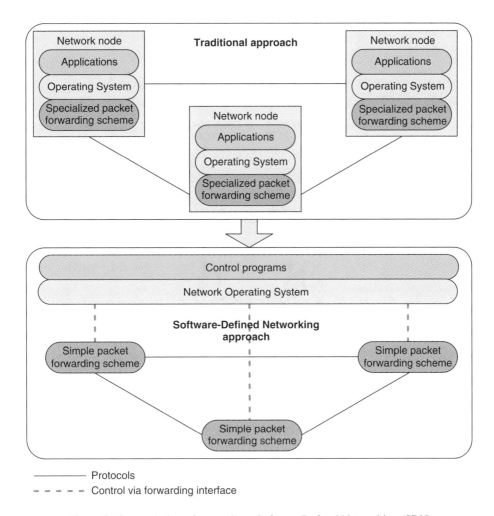

Figure 9.12 Evolution of networks to Software Defined Networking (SDN)

distributed control (with two or more OpenFlow controllers in the SDN, which provide load balancing and/or higher reliability).

Summarized, SDN is a system-layered abstraction of the network (which is programmable and flexible) while OpenFlow is the interface between network nodes (i.e., switches and routers) and controllers that enables the SDN. With such concept, the SDN provides possibilities for innovation in deployed network architectures, including also optical transport networks [36], as well as software-defined mobile networks [37].

9.7   Business Challenges and Opportunities

The deployments and transition to NGN and future network developments initiate certain business challenges and provide new opportunities in continuously changing environment of telecommunication networks and services.

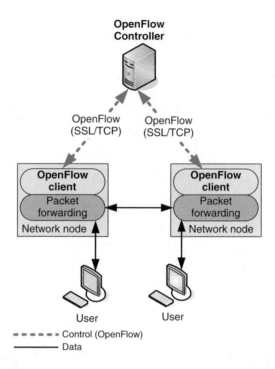

Figure 9.13   OpenFlow for Software Defined Networking (SDN)

The first challenge to the business environment in NGN is the change of network cost. Access networks remain with a high participation in overall costs, similar to traditional telecommunication networks in the past, but the costs for core networks are decreasing. While in PSTN/PLMN/ISDN environments the telecom operators were selling services (e.g., telephony) and owning the core networks at the same time, in NGN the telecom operators become network providers (i.e., NGN providers) which sell connectivity to end users through access and core networks. The services can be offered by network providers (similarly as in traditional approaches by telecom operators) or by third party service providers. While the costs for core network operation and maintenance in NGN decrease, the costs for their management and billing increase. Deployment of broadband access networks (e.g., optical access networks, 4G mobile networks, etc.) also increases the costs. The costs of electronic network equipment (e.g., switches, routers) are decreasing due to convergence of networks, services, and devices.

The second business challenge for NGN is investment in broadband access (i.e., NGA) which is required for NGN [38]. That means deployment of optical access networks on long term, as well as using existing copper networks (e.g., twisted pairs via xDSL) in short- and mid-time intervals. In developing countries the operating costs are lower, but also the potential revenues are lower. Also, there is no point of investments in new copper-based access networks, since the difference between copper and fiber is becoming insignificant, while fiber-based access networks have much higher capacities than copper-based access solutions.

The third business challenge is investment in NGN interconnection. Broadband access network require higher capacities in the interconnection links as well. However, in developing countries and regions as well as isolated locations (e.g., islands) international interconnection can be a significant issue. In such cases there may be needed public funding for NGN transit interconnections (e.g., based on national broadband plans).

The fourth challenge refers to NGN competition and its regulation. The migration to NGN does not change the role of the regulators as players in telecommunication business environment. However, with the introduction of NGN and network virtualization in the future, differentiation of the costs is becoming harder. But, optical access networks can be regulated, because it requires large investments and it cannot be substituted with the mobile broadband networks (and copper is not considered as one of the options for the future networks). Mobile broadband access and fixed broadband access are two different markets (although NGN provide Fixed Mobile Convergence with it functional entities, such as IMS), which have different capabilities (e.g., optical access networks have higher capacities than mobile networks) and requirements (e.g., mobile networks require a certain amount of frequency spectrum, which is subject to licensing, as discussed in Chapter 5).

Demographics can be considered as another business challenge for NGN migration. Developed countries have higher penetration of fixed-line infrastructure from the past, so they can use existing access ducts for laying fiber (e.g., fiber to the cabinet, fiber to the curb, etc.) which is needed for fixed broadband access (on longer time periods). On the other side, developing countries with lack of fixed network infrastructures have higher growth in mobile networks. Additionally, for growth in broadband access in developing countries, one may also consider government control funding via "broadband fund" which is supported by other players in the telecommunication markets (e.g., similar to universal services funds). Such funding approach can be further extended (if needed) to centralized infrastructure, platforms, and contents.

Finally, profits are important business challenge for network providers (e.g., telecom operators). Although core network costs in NGN are lower, migration to NGN requires large investments in the core networks and in broadband access networks. Incumbent operators shall maintain PSTN/ISDN users in parallel with NGN (e.g., with MSANs in the access networks), while new alternative network providers may become "pure NGN" providers from the beginning. However, in a competitive market alternative operators go with lower prices to attract customers, sometimes with smaller product ranges and focus on smaller target groups (e.g., business environments). That requires an operator specific approach for investments in NGN and their dynamics.

On the other side, in general, NGN brings almost unlimited business opportunities. There are possible all different relationships between end-users, network providers, and service providers, as well as new service possibilities, including the following:

- *Fixed-mobile convergence*: This provides new business possibilities for offering the same services over fixed and mobile access networks, that is, access independent services.
- *Ubiquitous networking*: This opens completely new business possibilities by enabling person to object and object to object communication, using any service, any time, anywhere, and any device communication capabilities.
- *Communication to anything*: This provides business opportunities for implementation of IoT and WoT to different providers, such as device, network, platform, and application providers.
- *Open service environment*: This provides new business opportunities by integrating third-party applications/services and telecom infrastructure.

- *IPTV services*: NGN provides different possible business relationships between network providers, services providers, and content providers.
- *QoS-enabled VoIP*: This provides possibility for different business models of voice offerings, including voice only, video telephony, video conferences and presence, to all types of end users.
- *IMS-based service creation environment*: The IMS is used for support of VoIP and IPTV, and at the same it provides environments for creation of innovative services by using its control plane and identity management functions in NGN, which may boost different business opportunities.
- *Enhanced Web-services*: NGN provides converged Web services which may provide new business offerings by combining a Web service component with other NGN components, such as IP multimedia and IPTV components.
- *Seamless migration from PSTN/ISDN*: NGN provides seamless migration of voice business models from traditional networks, where voice remains as one of the most important revenue generator to telecom operators, although such revenues are decreasing while revenues from broadband access to Internet are increasing.
- *Future networks*: Novel concepts in future development of the NGN, such as cloud computing and SDN, provide new business possibilities by offering cloud services such as virtual infrastructures, platforms, and software as services to end users, as well as network resource sharing and different business relationships that become possible with virtualization of network resources in different types of networks (e.g., optical transport networks, mobile networks, etc.).

Overall, NGN provides a whole new world in telecommunications. On one side, there are business challenges that should be considered during migration to NGN and later during its operation. On the other side, there are limitless business opportunities based on existing and new innovative applications and services, thus proving the business logic behind NGN.

9.8   Discussion

The transition to NGN from traditional PSTN/ISDN (including PLMN) is an inevitable process. For ensuring transparent migration to NGN for voice service (as main service in traditional telecommunication networks) NGN provides functions for PSTN/ISDN emulation and simulation, thus providing the possibility for end users to continue to use telephony as they are accustomed to do, either with their legacy telephone devices or with NGN-capable devices.

On the other side, best-effort Internet also migrates to NGN, since NGN is an all-IP based network that supports heterogeneous services, including best-effort access to Internet as well as QoS-enabled services and new innovative services, by using common signaling platform (e.g., IMS), and common functional entities for management of networks, services, and users (e.g., identity management). Additionally, NGN provides a standardized approach for building CGOE by using existing standardized COTS components.

However, to provide full functionality the NGN shall evolve to an all-IPv6 based network, due to the many advantages of IPv6, including bigger and new addressing schemes (e.g., needed for IoT) as well as enhanced features regarding QoS, security, reliability (e.g., multihoming), and mobility support at the network protocol layer.

The evolution of telecommunication networks and services continues with further developments of NGN toward future networks. In that direction network virtualization is an approach which further segments the deployed physical network into virtual network components which may be used for building LINPs to suit different new services. In that manner, future packet-based networks are targeted to cloud computing (e.g., virtual networks, platforms, and applications offered to the end users) and SDN as system-layered abstraction of the network which introduces NOS connected to switches and routers via standardized interface (i.e., OpenFlow).

Migration to NGN and future networks brings many challenges to telecom operators, service providers, regulators, vendors, and other related business segments, but at the same it provides endless possibilities for innovation, with possibilities for development, testing, and deployment of new networks and services faster than ever before.

References

1. ITU-T (2006) PSTN/ISDN Evolution to NGN. ITU-T Recommendation Y.2261, September 2006.
2. ITU-T (2006) PSTN/ISDN Emulation and Simulation. ITU-T Recommendation Y.2262, December 2006.
3. ITU-T (2006) Call Server-Based PSTN/ISDN Emulation. ITU-T Recommendation Y.2271, September 2006.
4. Marcus, J. S. and Elixmann, D. (2008) The Future of IP Interconnection: Technical, Economic, and Public Policy Aspects, WIK-Consult, Final Report, Study for the European Commission, January 2008.
5. Rosen, E. and Rekhter, Y. (2006) BGP/MPLS IP Virtual Private Networks (VPNs). RFC 4364, February 2006.
6. Rosen, E. and Aggarwal, R. (2012) Multicast in MPLS/BGP IP VPNs. RFC 6513, February 2012.
7. Ould-Brahim, H., Fedyk, D., and Rekhter, Y. (2009) BGP Traffic Engineering Attribute. RFC 5543, May 2009.
8. ITU and OCAF Focus Group (2005) The Carrier Grade Open Environment Reference Model, Basis Version 1.0, May 2005.
9. ITU-T (2006) The Carrier Grade Open Environment Reference Model. ITU-T Recommendation Y.2901, December 2006.
10. ITU-T (2008) Carrier Grade Open Environment Components. ITU-T Recommendation Y.2902, November 2008.
11. ITU-T (2008) General Overview of IPv6-Based NGN. ITU-T Recommendation Y.2051, February 2008.
12. ITU-T (2008) Framework to Support Signaling for IPv6-Based NGN. ITU-T Recommendation Y.2054, February 2008.
13. Brunner, M. (2004) Requirements for Signaling Protocols. RFC 3726, April 2004.
14. Crawford, M. (2000) Router Renumbering for IPv6. RFC 2894, August 2000.
15. ITU-T (2008) Framework of Multi-Homing in IPv6-Based NGN. ITU-T Recommendation Y.2052, February 2008.
16. Stewart, R. (2007) Stream Control Transmission Protocol. RFC 4960, September 2007.
17. Kohler, E., Handley, M., and Floyd, S. (2006) Datagram Congestion Control Protocol (DCCP). RFC 4340, March 2006.
18. ITU-T (2008) Functional Requirements for IPv6 Migration in NGN. ITU-T Recommendation Y.2053, February 2008.
19. ITU-T (2011) Framework of Vertical Multihoming in IPv6 Next Generation Networks. ITU-T Recommendation Y.2056, August 2011.

20. ITU-T (2011) Framework of Object Mapping using IPv6 in Next Generation Networks. ITU-T Recommendation Y.2055, March 2011.
21. ITU-T (2011) Framework of Node Identifier and Locator Separation in IPv6-Based Next Generation Networks. ITU-T Recommendation Y.2057, November 2011.
22. Moskowitz, R., Nikander, P., Jokela, P., and Henderson, T. (2008) Host Identity Protocol. RFC 5201, April 2008.
23. ITU-T (2011) Roadmap for IPv6 Migration from the Perspective of the Operators of Next Generation Networks. ITU-T Recommendation Y.2058, November 2011.
24. Nordmark, E. (2000) Stateless IP/ICMP Translation Algorithm (SIIT). RFC 2765, February 2000.
25. Tsirtsis, G. and Srisuresh, P. (2000) Network Address Translation - Protocol Translation (NAT-PT). RFC 2766, February 2000.
26. Tsuchiya, K., Higuchi, H., and Atarashi, Y. (2000) Dual Stack Hosts using the "Bump-In-the-Stack" Technique (BIS). RFC 2767, February 2000.
27. Hagino, J. and Yamamoto, K. (2001) An IPv6-to-IPv4 Transport Relay Translator. RFC 3142, June 2001.
28. Lee, S., Shin, M-K., Kim, Y-J. *et al.* (2002) Dual Stack Hosts Using "Bump-in-the-API" (BIA). RFC 3338, October 2002.
29. Kitamura, H. (2011) A SOCKS-Based IPv6/IPv4 Gateway Mechanism. RFC 3089, April 2011.
30. Durand, A., Fasano, P., Guardini, I., and Lento, D. (2001) IPv6 Tunnel Broker. RFC 3053, January 2001.
31. Carpenter, B. and Moore, K. (2001) Connection of IPv6 Domains via IPv4 Clouds. RFC 3056, February 2001.
32. Carpenter, B. and Jung, C. (1999) Transmission of IPv6 over IPv4 Domains without Explicit Tunnels. RFC 2529, March 1999.
33. ITU-T (2012) Framework of Network Virtualization for Future Networks. ITU-T Recommendation Y.3011, January 2012.
34. Azodolmolky, S., Wieder, P. and Yahyapour, R. (2013) Cloud computing networking: challenges and opportunities for innovations. *IEEE Communications Magazine*, **51** (7), 54–62.
35. Open Networking Foundation (2012) OpenFlow Switch Specification, September 2012.
36. Gringeri, S., Bitar, N. and Xia, T.J. (2013) Extending software defined network principles to include optical transport. *IEEE Communications Magazine*, **51** (3), 32–40.
37. Pentikousis, K., Wang, Y. and Hu, W. (2013) Mobileflow: toward software-defined mobile networks. *IEEE Communications Magazine*, **51** (7), 44–53.
38. ITU (2012) Strategies for the Deployment of NGN and NGA in a Broadband Environment – Regulatory and Economic Aspects, December 2012.

10

Conclusions

When all media types are transferred with digital signals, as series of ones and zeros, there is no need to have different networks for different types of services, because the same digital signals are used for transferring different media (e.g., text, images, audio, video, data, multimedia, etc.), voice, and television. Concurrent with these processes of digitalization which happened separately for the telephony and the television/radio in the past two or three decades, the Internet has evolved as a network for the transmission of multimedia data contents, based on simplicity and low cost of the network elements (e.g., switches, routers) and without explicit support for Quality of Service (QoS), unlike the one provided in traditional telephone networks and TV broadcast networks. Such best-effort Internet principle, which means that each IP packet is admitted to the network and any IP connection is established, is the main reason for the success of the Internet, but also it has become the main obstacle for full integration of traditional telecommunication services (e.g., telephony and TV) to the Internet.

The integration and convergence of traditional telecommunication networks and the Internet is realized with the standardization of the Next Generation Network (NGN) by ITU-T (International Telecommunication Union-Telecommunications). However, as stated in the introductory chapter, the NGN is not something that happens in a given moment of time, but it is an ongoing process (umbrella of principles and recommendations for synergy of different existing and future technologies), a "living thing" which grows and adapts to the continuously changing environment regarding the technologies as well as business and regulation environments. For that purpose, NGN uses architectures, protocols, and services, as standardized by the ITU (International Telecommunication Union), by using the existing Internet technologies as standardized by the IETF (Internet Engineering Task Force), as well as other standardization organizations [e.g., 3GPP (3G Partnership Project), IEEE, etc.].

In summary, the main drivers for NGN include development of broadband Internet access including fixed broadband and mobile broadband, convergence of telecommunication markets to Internet, technological convergence to IP-based networks and services, requirements for end-to-end QoS provisioning in Internet, as well as transition of traditional telecommunication networks [e.g., PSTN/ISDN (Public Switched Telephone Network/Integrated Services Digital Network), PLMN (Public Land Mobile Network), TV broadcast networks] to the Internet environment. In that manner, the ITU has identified "broadband" as a key target initiative in the current decade.

NGN Architectures, Protocols and Services, First Edition. Toni Janevski.
© 2014 John Wiley & Sons, Ltd. Published 2014 by John Wiley & Sons, Ltd.

In general, broadband access means access to Internet with high bit rates by using ubiquitous networks, devices, and applications, in fixed and mobile environments. The fixed broadband has started with ADSL (Asymmetric Digital Subscriber Line) standards over legacy access networks, while the Next Generation Access (NGA) is based on passive and active optical access networks, as well as Metro/Carrier Ethernet. The mobile broadband is becoming reality with 4G mobile networks, that is, LTE-Advanced and Mobile WiMAX 2.0 are providing higher bitrates to end-users as well as many novelties in radio access networks such as carrier aggregation. The 4G is based on NGN principles regarding the separation of the transport networks and services [e.g., IMS (IP Multimedia Subsystem) is used for signaling], so with its deployments starts the implementation of NGN principles in mobile broadband environments. However, mobile broadband needs more spectrum bands to provide higher bitrates, which requires more dynamic spectrum management approaches in the future.

NGN is completely heterogeneous network consisted of different access and core networks, which are based on Internet technologies with IP as the default network-layer protocol. In such environment fundamental part of the NGN is QoS provision for certain real-time services [e.g., VoIP (Voice over IP), IPTV (Internet Protocol Television), etc.]. That is accomplished in NGN with RACF-based (Resource and Admission Control Function) QoS control and signaling, and applied traffic engineering based on legacy Internet technologies for QoS support (e.g., Differentiated Services, Integrated Services, and Multi-Protocol Label Switching-MPLS). In that regard, NGN can be defined as an all-IP network with end-to-end QoS provision to requesting services.

Unlike traditional telecommunication networks which are service-centric (optimized for a certain service, such as telephony or television), the NGN is service-independent network with standardized interfaces to third-party applications and services. Such open service environment in NGN creates new business opportunities in the ecosystem consisted of end users, network providers, and third-party service providers, thus leading to rapid and cost-effective development of new innovative services. On the other side, NGN uses IMS signaling architecture for real-time services with QoS requirements (e.g., voice, TV), which is based on well standardized signaling and control protocols from the IETF such as SIP (Session Initiation Protocol) and Diameter. Additionally, NGN provides standardized management of identities of end user, their devices, as well as network equipments, which is crucial for development of new services by using Service Overlay Networks (SON).

The focus of NGN is in creation of standardized network environments in user plane (for data traffic) and control plane (for signaling traffic) which are created to be common for different services provided within the NGN or via third-party service providers. For example, NGN includes call session control functionalities through the core IMS (e.g., for VoIP, IPTV, etc.) and VPN (Virtual Private Network) services which are extended to individual end users and community groups. On the other side, NGN incorporates the most important best-effort Internet service, the WWW (World Wide Web), as basis for many mash-up services in NGN [i.e., combination of Web and other services, such as Web and IPTV, or Web and Internet of Things (IoT), etc.].

As a completely new dimension in telecommunications NGN introduces practical "any-thing connectivity" by providing information infrastructure for the IoT and Web of Things (WoT). Such evolution goes toward the Internet-of-Everything concept, which provides possibility for connection of countless devices and sensors for various purposes (e.g., smart cities, smart buildings, smart transportation, smart industry, smart environment, connected gaming and entertainment, etc.). With networking of peoples, data, business processes, and things, the IoT/WoT provides large scale communication and exponential increase of the digital data, leading to many new innovations and opportunities. However, such development is possible with transition to IPv6-based NGN and Internet in general.

Transition to NGN from traditional PSTN/ISDN is an ongoing process. However, for provision of transparent migration for legacy voice service, the NGN provides functions for PSTN/ISDN emulation and simulation, thus providing possibility to users to continue to use traditional telephony service with both, legacy telephone devices and NGN-capable devices. On the other side, best-effort Internet also migrates to NGN, including best-effort access to Internet as well as QoS-enabled services. For carriers, the NGN also provides Carrier Grade Open Environments (CGOE) for building industry applications by using existing standardized Commercial-Off-The-Shelf (COTS) components.

With aim to provide the full functionality to different types of services (e.g., IoT) the NGN have to migrate further to all-IPv6 based network, due to IPv6 bigger and new addressing schemes as well as enhanced features regarding QoS, security, reliability, multihoming, and mobility support.

Evolution of NGN continues further with future networks, which are targeted to service awareness including virtualization of resources, data awareness including identification, environmental awareness including energy consumption, as well as social and economic awareness including service universalization. Network virtualization is an emerging approach which provides segmentation of the deployed physical networks into virtual network components which may be used for building Logically Isolated Network Partitions (LINPs) to suit different new services. Hence, future packet-based networks are targeted to cloud computing (e.g., virtual networks, platforms and applications offered to the end users) and Software Defined Networking (SDN) as system-layered abstraction of the networks.

Currently telecommunications and media regulators inhabit a dynamic world which is dominated by end-to-end IP-based communication, where global over the top service providers are delivering applications and services in which voice is just one of the applications, while every user is becoming content generator. The dominant business players in the telecommunication world are no longer the national operators, since largely free reigned international companies as service providers dictate to the infrastructure players. NGN deployments and its future evolution provide many opportunities for development of new relationships among different business players nationally and globally.

Finally, migration to NGN and future networks brings many challenges to network and service providers, telecommunications and media regulators, equipment vendors, and other related business segments, but at the same time it provides endless possibilities for rapid innovation of new networks, protocols, and services.

Index

NGN Architectures, Protocols and Services, First Edition. Toni Janevski.
© 2014 John Wiley & Sons, Ltd. Published 2014 by John Wiley & Sons, Ltd.